Jesuits and the Book of Nature

Jesuit Studies

MODERNITY THROUGH THE PRISM OF JESUIT HISTORY

VOLUME 25

The titles published in this series are listed at *brill.com/js*

Jesuits and the Book of Nature

Science and Education in Modern Portugal

By

Francisco Malta Romeiras

BRILL

LEIDEN | BOSTON

Cover illustration: Jesuit Scholastics in Setúbal (1892–3). © Arquivo da revista Brotéria, Lisbon.

Library of Congress Cataloging-in-Publication Data

Names: Romeiras, Francisco Malta, author.
Title: Jesuits and the book of nature : science and education in modern Portugal / by Francisco Malta Romeiras.
Description: Leiden ; Boston : Brill, [2019] | Series: Jesuit Studies - Modernity through the prism of Jesuit history, 2214-3289 ; volume 25 | Includes bibliographical references and index. | Summary: "Jesuits and the Book of Nature: Science and Education in Modern Portugal offers an account of the Jesuits' contributions to science and education after the restoration of the Society of Jesus in Portugal in 1858. As well as promoting an education grounded on an "alliance between religion and science," the Portuguese Jesuits founded a scientific journal that played a significant role in the consolidation of taxonomy, plant breeding, biochemistry, and molecular genetics. In this book, Francisco Malta Romeiras argues that the priority the Jesuits placed on the teaching and practice of science was not only a way of continuing a centennial tradition but should also be seen as response to the adverse anticlerical milieu in which the restoration of the Society of Jesus took place"-- Provided by publisher.
Identifiers: LCCN 2019026742 (print) | LCCN 2019026743 (ebook) | ISBN 9789004382350 (hardback) | ISBN 9789004382367 (ebook)
Subjects: LCSH: Jesuits. | Life science--Study and teaching--Portugal--History--19th century. | Nature--Religious aspects--Portugal--History--19th century.
Classification: LCC QH320.P67 R66 2019 (print) | LCC QH320.P67 (ebook) | DDC 508.469--dc23
LC record available at https://lccn.loc.gov/2019026742
LC ebook record available at https://lccn.loc.gov/2019026743

Typeface for the Latin, Greek, and Cyrillic scripts: "Brill". See and download: brill.com/brill-typeface.

ISSN 2214-3289
ISBN 978-90-04-38235-0 (hardback)
ISBN 978-90-04-38236-7 (e-book)

This book is printed on acid-free paper and produced in a sustainable manner.

Uxori filiisque meis carissimis

Contents

Acknowledgments

The idea of revising and expanding my doctoral research on Jesuit science and education in modern Portugal into a monograph dates back to 2015, when I published an abridged version of my thesis in Portuguese: *Ciência, prestígio e devoção: Jesuítas e ciência em Portugal* (*séculos XIX e XX*) (Cascais: Lucerna, 2015). Since then, I had discovered new sources; worked in different archives, libraries, and universities; discussed my book project with colleagues in Portugal and abroad; and published some of my findings in edited volumes and peer-reviewed journals. This book has greatly benefited from the comments and suggestions I received throughout this period. As academics know only too well, it takes a village to write a book.

Henrique Leitão introduced me to the fascinating field of the history and philosophy of science in 2011 and has been a great mentor and source of inspiration ever since. His support and friendship have been invaluable. He read earlier versions of these chapters with patience, rigor, and acumen, and I am deeply grateful for his guidance. Ana Simões welcomed me into the Centro Interuniversitário de História das Ciências e da Tecnologia (CIUHCT), University of Lisbon, and introduced me to the historiography of science and scientific popularization in 2011. At CIUHCT, I have had the opportunity to work closely with many talented colleagues and friends. For suggesting readings in natural history, the history of education, Jesuit history, and the philosophy of science, and for commenting on some of the chapters of this book, I am especially indebted to Catarina Madruga, Inês Gomes, Luís Tirapicos, Ana Duarte Rodrigues, and Antonio Sánchez.

Fundação para a Ciência e Tecnologia funded my doctoral research between 2009 and 2013 (Grant: SFRH/BD/61883/2009), and the Institute for Advanced Jesuit Studies, Boston College, granted me a senior research fellowship in 2017 to write the book. Casey Beaumier, S.J., Seth Meehan, Claude Pavur, S.J., Cristiano Casalini, Eugenio Menegon, Bart Geger, S.J., Elisa Frei, and Sean McLaughlin welcomed me into Boston College and were very helpful during our family's stay in Boston. Claude Pavur and Seth Meehan were tireless readers, even after my departure, and I am most indebted for their insightful comments, meticulous proofreading, and continuous support.

Since the very beginning, I benefited from the knowledge of many librarians and archivists at Jesuit institutions, including Carlos Maria Vasconcelos, S.J., and Francisco Correia, S.J. (Arquivo Português da Companhia de Jesus), António Júlio Trigueiros, S.J., and António Vaz Pinto, S.J. (*Brotéria*), Mauro Brunello, and Robert Danieluk, S.J. (Archivum Romanum Societatis Iesu), and

Christian Dupont (John J. Burns Library, Boston College). I am especially grateful to my dear cousin Jorge Brito e Abreu for opening the doors of his private archive and for sending me copies of a series of unknown letters of Carlos Zimmermann, one of the leading figures of this book.

Finally, I would like to express my deep gratitude to Robert A. Maryks for agreeing to publish my book in Brill's Jesuit Studies series and for his valuable help and advice from the very early stages of writing up until the production of the book. I would also like to thank Tim Page for copyediting the manuscript, Raymond DePew-Paas for assembling the index, and the anonymous reviewers for their helpful suggestions.

Abbreviations

AB	Arquivo da revista *Brotéria*, Lisbon
AHPL	Arquivo Histórico do Patriarcado de Lisboa, Lisbon
ANTT	Arquivo Nacional da Torre do Tombo, Lisbon
APSI	Arquivo da Província Portuguesa da Companhia de Jesus, Lisbon
ARSI	Archivum Romanum Societatis Iesu, Rome
ASV	Archivio Segreto Vaticano, Vatican
BNP	Biblioteca Nacional de Portugal, Lisbon
BNR	Biblioteca Nazionale Centrale di Roma, Rome
DCHJ	Charles O'Neill and Joaquín M. Domínguez, eds., *Diccionario histórico de la Compañía de Jésus: Histórico-biográfico*, 4 vols. (Rome: Institutum Historicum Societatis Iesu, 2001)
FMR	Arquivo Francisco Malta Romeiras, Lisbon
HML	Hemeroteca Municipal de Lisboa
JBA	Arquivo Jorge Brito e Abreu, Lisbon
Mendizábal	Rufo Mendizábal, *Catalogus defunctorum in renata Societate Iesu ab a. 1814 ad a. 1970* (Rome: Apud Curiam Praepositi Generalis, 1972)
OAL	Arquivo do Observatório Astronómico de Lisboa, Lisbon
Sommervogel	Carlos Sommervogel, ed., *Bibliothèque de la Compagnie de Jésus: Nouvelle édition*, 10 vols. (Brussels: Oscar Scherpens, 1890–1900; Paris: Alphonse Picard, 1890–1909)
Verástegui	Nicolàs Verástegui, *Catalogus defunctorum in renata Societate Iesu (II), 1970–1985* (Rome: Institutum Historicum Societatis Iesu, 1986)

Figures and Tables

Figures

Tables

Introduction

At the college, along with the exaggerated religious life that took our time and hijacked our energies, there was a good humanistic and scientific education, which was not more perfect merely by virtue of being restricted to state programs, some of them hardly recommendable. To that orientation, I owe much of my achievement in my university career. The mental discipline they imposed upon the students in the exact sciences and in others was well designed. In those distant times, the college already used to give some attention to physical exercises. The equilibrium of prayer, physical exercise, and study ought to have been better organized, but I benefited from the education they gave me in mathematics, physics, chemistry, and the biological sciences. They brought about a certain development on the experimental side, which contrasted with most high schools at the time. The chemistry laboratory and the physics cabinet were sufficiently equipped, and the teaching [of the natural sciences] was based on experiments whenever possible. I am happy to leave here my impartial testimony.[1]

•••

A document has recently been circulated, bearing the names of twelve Portuguese naturalists who have been exiled from their country by the new government "on the pretext that they are Jesuits." Five have gone to Brazil, three to Belgium, two to Spain, and two to Holland. These men were professors in the colleges of S. Fiel and Campolide, in Lisbon, and were known for their work in different branches of biology, and especially for the journal *Brotéria*, which they published.[2]

∴

1 António Egas Moniz, *A nossa casa* (Lisbon: Paulino Ferreira Filhos Lda, 1950), 254. I would like to thank Claude Pavur for his precious help with the English translation.

2 Theodore Dru Alison Cockerell, "The Exiled Naturalists of Portugal," *Science* 34 (1911): 714–15, here 714–15.

After being awarded the Nobel Prize in Physiology or Medicine in 1949 for his discovery of the value of leucotomy in the treatment of certain psychoses, the Portuguese physician António Egas Moniz (1874–1955) published his long-awaited memoirs. In the passages describing his childhood, Egas Moniz praises the education he received at the Colégio de São Fiel, a Jesuit college in a sleepy provincial town. The book recalls how the Jesuits had offered him an education in the humanities and, most importantly, an excellent scientific education, one "based on experiments whenever possible" in contrast to what was by then the usual practice in public state schools. By stating his belief that the Jesuits' experimental approach to science teaching had played a particularly valuable role in his scientific career, Egas Moniz's memoirs reignited a debate about Jesuit education. As a result, his book not only reached a vast readership eager to know more about the life of the celebrated physician and Nobel laureate but also helped to awaken a general awareness of the significance of the Jesuit colleges of Campolide (1858–1910) and São Fiel (1863–1910) at the turn of the nineteenth century.

Forty years before the appearance of Egas Moniz's memoirs, renowned naturalist Theodore Dru Alison Cockerell (1866–1948) had already underlined the importance of the Portuguese Jesuit colleges and their collections when he published a plea in *Science* in favor of the Jesuits who had been exiled and deprived of their scientific instruments, books, and collections after the 1910 republican revolution. But why was an Anglo-American zoologist so concerned with the fate of the distant Portuguese naturalists? And why did a celebrated journal such as *Science* voice his concerns about the confiscation of the Jesuits' botanical and zoological collections and the closure of two seemingly unimportant boarding schools?

This book attempts to answer these questions. It argues that the priority the nineteenth-century Jesuit colleges placed on the teaching and practice of science was unusual both in context of Portugal itself and the broader history of science and education in Catholic schools during this period. The book interprets this commitment to science not only as a way of continuing a popular and effective centennial tradition but also as a response to the adverse anticlerical political and cultural Portuguese milieu in which the Society's restoration took place. In their efforts to achieve these twin goals—and while ultimately aspiring to the "greater glory of God"—the Portuguese Jesuits established scientific academies, observatories, laboratories, and museums in their colleges, and they also founded *Brotéria*, a scientific journal that would play a significant role in the progress of zoology, botany, plant breeding, biochemistry, and molecular genetics in Portugal. By putting the teaching and practice of science at the top of their educational agenda, and by projecting their public image as modern scientists and pedagogues, the Jesuits gradually regained their credibility in the eyes of the Portuguese royal family, the highest ranks of the nobility, the bourgeoisie, and the scientific community.

The Jesuits' commitment to science and education should accordingly be seen as an essential step in a broader campaign to recover their credibility as scientists and pedagogues, which would ultimately lead to the order's effective restoration in the Portuguese Empire, resulting, in turn, in the approval of new apostolic missions and colleges in Africa and East Asia. The colleges of Campolide and São Fiel consequently played a major role not only in the restoration of the Society but also in the diffusion of Ignatian spirituality, especially through the restoration of Marian congregations and the giving of the Spiritual Exercises. With the objective of providing an education grounded on the "alliance between religion and science," these colleges took a further step toward the formal scientific training of a new elite.[3]

What is particularly surprising about the Portuguese Jesuits' success in the provision of scientific teaching is that it was their informal education, and their participation in national and international scientific networks, that proved more relevant for their training as modern scientists and pedagogues than a formal scientific education. As we will see throughout this book, the Portuguese Jesuits learned and engaged in science by promoting and taking part in amateur expeditions to observe eclipses and collect new specimens of animals and plants, systematically collecting meteorological data, and attending international conferences. By teaching subjects directly related to their own research, the Jesuits were able to combine the teaching and practice of science in their colleges and involve students in diverse scientific tasks while simultaneously recreating the modern practices used at other scientific institutions.

Rather than promoting debates on the compatibility between science and religion, the journal *Brotéria* (1902–31) sought to present the Portuguese Jesuits' own scientific accomplishments and their direct involvement in the development of the botanical and zoological sciences, especially in the Portuguese Empire. In a period when networks of scientific correspondence played a particularly important role in the identification, description, and classification of novel species, *Brotéria* established itself as an international journal of taxonomy with works from renowned Portuguese and foreign botanists and zoologists. By focusing on the first three decades of *Brotéria*, this book will shed some light on the importance of the circulation of knowledge between different communities (collectors, amateurs, and experts) for the identification of

3 Colégio de Campolide, *Prospecto* (Lisbon: Typographia Universal, 1886), 1.

hundreds of new species of animals and plants pertaining to the European, African, South American, and Asian fauna and flora. As we will see, the creation and maintenance of networks of scientific correspondence, which transcended the boundaries of the Society of Jesus, not only contributed to the circulation of knowledge but also to *Brotéria*'s national and international scientific prestige.

As well as publishing taxonomical works, *Brotéria* also published a popular scientific series between 1907 and 1924. Directed at a wider, yet learned, readership, *Brotéria: Vulgarização científica* (Scientific popularization) published more than four hundred articles on a variety of topics, including physics, chemistry, agriculture, industry, medicine, and hygiene. By looking at the history of this popular scientific journal, the book underlines the importance the Portuguese Jesuits attached to reaching a wide readership, one broadly interested in science, technology, and medicine, as part of a strategy to finance their scientific work during this period. From the 1930s, the collaboration between the editor of *Brotéria* and the most important public laboratories played a particularly important part in the development of genetics and plant breeding during the early years of the Estado Novo (1933–74) and in reinforcing the Jesuits' political status in the new regime. Beginning in the 1960s, *Brotéria* went on to become one of the most important vectors in the promotion of biochemistry and molecular genetics at an academic level. Taking into account the centennial history of *Brotéria,* this book will argue that the strictly scientific purpose behind its foundation, its unusual longevity, the extensive taxonomical work it published, the role it played in the foundation and development of new research fields, and its editorial flexibility were among its most important features, both in the Jesuit context and in the broader history of scientific periodicals.

CHAPTER 1

Jesuit Science and Education: a Brief History

One of the most distinctive features of the Society of Jesus was its dedication to science and scientific education, especially when compared with other religious orders. The Jesuits' contributions to early modern science have been studied extensively and are now widely known, lending further credence to George Sarton's (1884–1956) dictum that "one cannot study the history of mathematics in the sixteenth and seventeenth centuries without coming across Jesuits at every corner."[1] Fifty years later, in his book on the history of electricity in the seventeenth and eighteenth centuries, John Heilbron (1934–) echoed this assertion by stating that "the single most important contributor to the support of the study of physics in the seventeenth century was the Catholic Church and, within it, the Society of Jesus."[2] Today, it is clear that Heilbron's and Sarton's statements can be extended to other scientific subjects, especially those connected to the mathematical and physical sciences such as astronomy, seismology, meteorology, and engineering.

The history of Jesuit science and education has been a topic of intense research for historians of science since the late 1990s.[3] By focusing on the Jesuits' educational and scientific work in the sixteenth to eighteenth centuries, this

1 George Sarton, "Preface to Volume 40: An Appeal for the Republication in Book Form of Father Bosmans' Studies on Belgian Mathematics in the Sixteenth and Seventeenth Centuries," *Isis* 40 (1949): 3–6, here 3.

2 John L. Heilbron, *Electricity in the 17th and 18th Centuries: A Study of Early Modern Physics* (Mineola, NY: Dover, 1999), 2.

3 The literature on Jesuit science is extensive, yet the first attempt at a global synthesis of the scientific history of the Society of Jesus from its foundation to the twentieth century was only published in 2015: Agustín Udías, *Jesuit Contribution to Science: A History* (Dordrecht: Springer, 2015). The historiography of the Jesuits' scientific practices is quite uneven, and most historians still pay more attention to the early modern period than to the scientific history of the Society of Jesus after its restoration in 1814. For the history of Jesuit science from 1540 to 1773, see especially John W. O'Malley et al., eds., *The Jesuits II: Cultures, Sciences, and the Arts, 1540–1773* (Toronto: Toronto University Press, 2006); Marcus Hellyer, *Catholic Physics: Jesuit Natural Philosophy in Early Modern Germany* (Notre Dame, IN: University of Notre Dame Press, 2005); Mordechai Feingold, ed., *Jesuit Science and the Republic of Letters* (Cambridge, MA: MIT Press, 2003); Feingold, ed., *The New Science and Jesuit Science: Seventeenth-Century Perspectives* (Dordrecht: Kluwer Academic, 2003); John W. O'Malley et al., eds., *The Jesuits: Cultures, Sciences and the Arts, 1540–1773* (Toronto: Toronto University Press, 1999); Ugo Baldini, *Legem impone subactis: Studi su filosofia e scienza dei gesuiti in Italia, 1540–1632* (Rome: Bulzoni, 1992).

 | DOI:10.1163/9789004382367_002

research has helped to challenge the Merton thesis and the traditional Whiggish narrative of the scientific revolution,[4] thus debunking the notion that there had been Protestant novelty and genius as opposed to Catholic backwardness and conservatism in early modern science.[5] Officially bound to teach Aristotelian Scholastic philosophy in their colleges, the Jesuits were often portrayed in the earlier literature as conservative clerics who firmly hindered the adoption of new philosophical models and scientific theories, such as heliocentrism, in early modern Europe.[6] But it is now clear that this portrait is far too naïve, particularly as it characterizes the Society as a rigid and uniform religious order while overlooking the significant individual and collective scientific and educational achievements of the Jesuits in Europe, East Asia, and South America.[7]

In the history of the Society of Jesus, there were always groups of men devoted to the teaching and practice of science. The creation of scientific courses and scientific academies within their colleges, the commitment to science for missionary purposes, and, in some cases, the pursuit of high-level scientific

4 Steven J. Harris, "Transposing the Merton Thesis: Apostolic Spirituality and the Establishment of the Jesuit Scientific Tradition," *Science in Context* 3 (1989): 29–65.

5 Analyzing the historiographical consequences of the traditional narrative of the scientific revolution is not the aim of this chapter. For an overview, see Sheila J. Rabin, "Early Modern Jesuit Science: A Historiographical Essay," *Journal of Jesuit Studies* 1, no. 1 (2014): 88–104; https://doi.org/10.1163/22141332-00101006. For modern narratives of the so-called scientific revolution, see Katharine Park and Lorraine Daston, eds., *The Cambridge History of Science*, vol. 3, *Early Modern Science* (Cambridge: Cambridge University Press, 2006); Steven Shapin, *The Scientific Revolution* (Chicago: University of Chicago Press, 1996); David C. Lindberg and Robert S. Westman, eds., *Reappraisals of the Scientific Revolution* (Cambridge: Cambridge University Press, 1990).

6 The Jesuits' acceptance of heliocentrism was one of the most controversial topics in some traditional accounts. For the most recent scholarship, see especially William Wallace, *Galileo and His Sources: The Heritage of the Collegio Romano in Galileo's Science* (Princeton: Princeton University Press, 1984); Christopher M. Graney, *Setting Aside All Authority: Giovanni Battista Riccioli and the Science against Copernicus in the Age of Galileo* (Notre Dame, IN: University of Notre Dame Press, 2015); and Alfredo Dinis, *A Jesuit against Galileo? The Strange Case of Giovanni Battista Riccioli's Cosmology* (Braga: Axioma–Publicações da Faculdade de Filosofia, 2017).

7 On the circulation of knowledge between Europe and China, see Florence C. Hsia, *Sojourners in a Strange Land: Jesuits and Their Scientific Missions in Late Imperial China* (Chicago: University of Chicago Press, 2009); Liam M. Brockey, *Journey to the East: The Jesuit Mission to China, 1572–1724* (Cambridge, MA: Harvard University Press, 2007). On the scientific activities of the Jesuits in the Americas, see Andrés I. Prieto, *Missionary Scientists: Jesuit Science in Spanish South America, 1570–1810* (Nashville, TN: Vanderbilt University Press, 2011), and Miguel de Asúa, *Science in the Vanished Arcadia: Knowledge of Nature in the Jesuit Missions of Paraguay and Río de la Plata* (Leiden: Brill, 2014).

practices makes it impossible to understand the scope and influence of this religious order without considering the science some of its members produced.[8]

1 A Distinctive Religious Order

The Society of Jesus is perhaps the most well-known, and controversial, religious order of the early modern period. Scholars and the general public have been drawn into the history of the Jesuits, especially in the period between the Society's official foundation in 1540 and the universal suppression in 1773 (i.e., the history of the so-called "old Society of Jesus"). Deeply informed by the numerous extant primary sources at local Jesuit archives all over the world, or by popular textbooks, novels, theater plays, and movies, both audiences usually have something to say about the religious order founded by Ignatius of Loyola (*c.*1491–1556) and his companions in 1540.

The fascination with this five-hundred-year-old institution gained a new impetus in 2013 when Pope Francis (1936–, r.2013–) was elected supreme head of the Catholic Church. As the first Jesuit elected at a papal conclave, Francis has attracted the attention of scholars and the media ever since. By using jargon so dear to the Jesuits—most notably discernment—and by citing the foundational documents of the Society of Jesus in his encyclicals and apostolic letters, Francis has played a crucial role in the diffusion of Ignatian spirituality and the Jesuit way of proceeding to the Catholic world.

The foundation of the Society was particularly molded by the life and conversion of Ignatius of Loyola.[9] After being seriously injured in his legs by a cannon ball during the Battle of Pamplona on May 20, 1521, the Basque nobleman went to the castle of Loyola where he underwent a series of excruciating operations that nonetheless failed to prevent him suffering from a lifelong limp.[10] During his recovery, Ignatius read *Flos sanctorum* (The lives of saints)

8 The expression "scientist" can be anachronistic when applied to scientific practitioners in the sixteenth and seventeenth centuries. This chapter uses it as a general category to refer to mathematicians, astronomers, natural philosophers, and other classical early modern scientific practitioners.

9 On the life of Ignatius and the early history of the Society of Jesus, see especially John W. O'Malley, *The First Jesuits* (Cambridge, MA: Harvard University Press, 1993); William Bangert, *A History of the Society of Jesus* (St. Louis, MO: Institute of Jesuit Sources, 1986).

10 Recounted to the Portuguese Jesuit Luís Gonçalves da Câmara (*c.*1520–75), his memoirs provide a vivid description of his life from 1521 to 1538. For the Latin version, see "Acta Patris Ignatii scripta a P. Lud. Gonzalez de Camara," in *Fontes narrativi de S. Ignatio de Loyola et de Societatis Iesu initiis*, vol. 1, *Narrationes scriptae ante annum 1557*, ed. Dionisio Fernández Zapico and Cándido de Dalmases (Rome: Monumenta Historica Societatis

by Jacopo da Voragine (*c.*1230–98) and *Vita Christi* (Life of Christ) by Ludolph of Saxony (*c.*1295–1378) and came to outline, from his own spiritual experience, what he would later call the "Rules for the Discernment of Spirits," a keystone of Jesuit spirituality and way of proceeding. Later, in Manresa, Loyola started working on his *Spiritual Exercises*,[11] a book derived from his own religious experience that represents his most treasured legacy to the Roman Catholic tradition. Giving the Spiritual Exercises was one of the most distinctive apostolic ministries of the Society of Jesus. As historian John W. O'Malley, S.J. puts it: "There is no understanding the Jesuits or the Society of Jesus without taking the *Exercises* into account."[12] References to the most renowned Ignatian meditations were constant in Jesuit texts and in their daily life.[13] The Exercises were given not only to Jesuit novices and other clerics but also to the public at large. Given their wide popularity, Ignatian terminology such as "discernment," "consolation," "desolation," and "indifference" soon entered the vocabulary of all laymen and women completing the Exercises, most notably aristocrats in princely courts in early modern Europe.[14]

Whereas the first three concepts are, in a certain sense, self-explanatory, the fourth requires a more detailed explanation, as it had a precise and distinctive

---

Iesu, 1943), 323–507. For an English translation, see Parmananda R. Divarkar, trans., *A Pilgrim's Testament: The Memoirs of St. Ignatius of Loyola*(St. Louis, MO: Institute of Jesuit Sources, 1995).

11 The *editio princeps* was printed in 1548 in Rome. Ignatius's notes were originally written in Spanish, with the text being translated into Latin by Andrés des Freux (Frusius [*c.*1515–56]). For the Latin version, see Arturo Codina, ed., *Exercitia spiritualia Sancti Ignatii de Loyola et eorum directoria* (Madrid: Rivadaneyra, 1919); José Calveras and Cándido de Dalmases, eds., *Monumenta Ignatiana: Sancti Ignatii de Loyola* Exercitia spiritualia; *Textuum antiquissimorum nova editio; Lexicon textus Hispani* (Rome: Institutum Historicum Societatis Iesu, 1969). For an English translation, see George E. Ganss, trans., *The* Spiritual Exercises *of Saint Ignatius* (St. Louis, MO: Institute of Jesuit Sources, 1992). For an introduction to the *Spiritual Exercises*, see especially Philip Endean, "The *Spiritual Exercises*," in *The Cambridge Companion to the Jesuits*, ed. Thomas Worcester (Cambridge: Cambridge University Press, 2008), 52–67, and Endean, "The *Spiritual Exercises*," in *The Cambridge Encyclopedia of the Jesuits*, ed. Thomas Worcester (Cambridge: Cambridge University Press, 2017), 757–62. When citing the *Spiritual Exercises*, it is customary to refer to the paragraph number attributed by Ignatius throughout the book.

12 John W. O'Malley, "The Distinctiveness of the Society of Jesus," *Journal of Jesuit Studies* 3, no. 1 (2016): 1–16, here 6.

13 See especially *Spiritual Exercises*, nos. 91–100, 36–48, 165–68, and 230–37.

14 On the meaning of these Ignatian maxims, see Brian O'Leary, "Consolation," in Worcester, *Cambridge Encyclopedia of the Jesuits*, 193–94; O'Leary, "Desolation," in Worcester, *Cambridge Encyclopedia of the Jesuits*, 228–29; Henry Shea, "Discernment," in Worcester, *Cambridge Encyclopedia of the Jesuits*, 234–35, and Shea, "Indifference," in Worcester, *Cambridge Encyclopedia of the Jesuits*, 392–93.

meaning in Ignatian spirituality. Unlike the traditional definition, aspiring to be indifferent was not seeking to be without interest or unconcerned but to "desire and choose only that which is more conducive to the end for which we are created," that is, "created to praise, reverence, and serve God our Lord, and by means of doing this to save their souls."[15] Being indifferent, for Ignatius, meant to freely seek and choose the divine will, as this reflected one's genuine desire of finding God in all things and of leading a Christian life oriented toward the greater glory of God, undoubtedly the most famous Ignatian motto.[16] Striving for the greater glory of God in all their labors was so dear to the Jesuits that the acronym AMDG (i.e., *Ad maiorem Dei gloriam*) was to be found on virtually every Jesuit manuscript, book, engraving, college, and church throughout the world in the sixteenth, seventeenth, and eighteenth centuries. Ironically, as it commonly served as an *ex-libris*, the most vociferous anti-Jesuit critics usually regarded it as a vestige of Jesuit action and frequently used this evidence in their arguments for censuring what they deemed to be Jesuit endeavors.

On September 27, 1540, Pope Paul III (1468–1549, r.1534–49) issued the bull *Regimini militantis ecclesiae* that approved the Formula of the Institute, the fundamental rule of the Society of Jesus.[17] Those who were to serve "as a soldier of God under the banner of the cross in our Society" were to "strive especially for the progress of souls in Christian life and doctrine and for the propagation of the faith by the ministry of the word, by spiritual exercises and works of charity, and specifically by the education of children and unlettered persons in Christianity."[18]

From its inception, the Society of Jesus was distinctive from the traditional monastic orders, such as the Augustinians, the Benedictines, the Cistercians, the Dominicans, and the Franciscans. As well as being devoted to the "salvation and perfection of the members' own souls," the end of the Society was to work for the "salvation and perfection of the souls of their neighbors," as stated

---

15 *Spiritual Exercises*, no. 23. On the meaning of Ignatian indifference as stated in the Principle and Foundation of the *Spiritual Exercises*, see Henry Shea, "Principle and Foundation," in Worcester, *Cambridge Encyclopedia of the Jesuits*, 648–49.

16 Robert E. Scully, "AMDG," in Worcester, *Cambridge Encyclopedia of the Jesuits*, 25–26, and Henry Shea, "Finding God in All Things," in Worcester, *Cambridge Encyclopedia of the Jesuits*, 300–1.

17 John W. Padberg, "Formula of the Institute," in Worcester, *Cambridge Encyclopedia of the Jesuits*, 304–6.

18 *Regimini militantis ecclesiae*, no. 1.

in the first printed version of the *Constitutions of the Society of Jesus* (1558).[19] Some of the most noticeable changes Ignatius introduced were largely practical, as they were intended to help Jesuits serve as ministers of the word and mercy of God in the world. Living in houses or colleges in the cities, and not in convents and monasteries, with no obligation to recite the liturgical hours in choir and wearing the simplest of garments were especially innovative, but they must be comprehended not merely as a break from the traditional way of life of monastic religious orders but as a way of serving the greater good, which in Ignatian terms was working for the propagation of faith in the world.

There were also significant changes in organization, as the Jesuits were not governed by provincial or general chapters but by a superior general who was elected by a general congregation for life and governed the Society from Rome with the help of his assistants and consultants. The Society was divided hierarchically and geographically into assistancies, provinces, vice-provinces, missions, and individual houses, including novitiates, professed houses, colleges, and universities. General directives from the Jesuit curia in Rome were sent to the provinces, where they were copied and re-sent to virtually every local superior. The flow of information went both ways, as all Jesuit houses, churches, colleges, and missions were to send annual letters (*litterae annuae*) to the curia. All Jesuit residences were also to send annual and triennial catalogs (*catalogi brevi et triennales*) containing detailed information about the current members of a specific community. Along with these official documents, Jesuits wrote thousands of private letters to superior general in Rome. Among the most glaring examples are the so-called *indipetae*, where Jesuits petitioned to be sent as missionaries to the Indies.[20] The annual letters, catalogs, and other documents were not only to be preserved in Rome but, according to their relative importance, were often copied and distributed to every Jesuit house in the

19 For a Latin version of the *Constitutions* of the Society of Jesus, see Arthur Codina, ed., *Monumenta Ignatiana: Sancti Ignatii de Loyola Constitutiones Societatis Iesu*, 3 vols. (Rome: Institutum Historicum Societatis Iesu, 1934–38). The edition used here is *The Constitutions of the Society of Jesus and Their Complementary Norms: A Complete English Translation of the Official Latin Texts*, ed. John W. Padberg, S.J., and trans. George E. Ganss et al. (St. Louis, MO: Institute of Jesuit Sources, 1996). The quotation pertains to the so-called "General Examen," or just "Examen," which constitutes a prelude to the *Constitutions*; "General Examen," in Padberg et al., *Constitutions of the Society of Jesus*, no. 3. For an introduction, see also Brian O'Leary, "Constitutions," in Worcester, *Cambridge Encyclopedia of the Jesuits*, 194–99.

20 Petitioning to the Indies was of the most distinctive features of the Society. The absolute numbers are staggering. It is estimated that fifteen thousand *indipetae* were written between 1540 and 1773. See Alison C. Fleming, "Letters, *Indipetae*," in Worcester, *Cambridge Encyclopedia of the Jesuits*, 463.

world. Ignatius himself should be given credit for this tradition—following his election as superior general on April 8, 1541, he promoted a global network of correspondence, having written or dictated several thousand letters.[21]

The Society's highly hierarchical organization and the extraordinary flow of information proved fundamental for establishing and maintaining long-distance networks of communication that were to play an important part in the circulation of knowledge in the early modern period.[22] The flow of people, texts, and objects between European centers and Jesuit peripheral missions provided the Society with a "unique institutional geography," and the "production and circulation of natural knowledge arose within and simultaneously helped sustain that institutional configuration."[23]

Together with the three traditional vows of obedience, poverty, and chastity, Jesuits also professed a vow of obedience to the pope, pledging to "obey at once, without subterfuge or excuse" whenever the pontiff wished to send them "among the Turks or to other infidels, even to the land they call India, or to any heretics or schismatics, or to any of the faithful."[24] Professing the fourth vow represented the final stage of a long period of spiritual and intellectual formation. After a two-year novitiate, where they completed a series of spiritual probations such as the thirty-day program of the Exercises, Jesuits professed simple vows of poverty, chastity, and obedience. When they had studied Latin grammar, rhetoric, and humanities to a sufficient level, scholastics studied philosophy for three years and theology for five years. Before studying theology, however, there was a period called regency, where Jesuits worked in spiritual or educational apostolates, generally teaching the lower classes of grammar, rhetoric, and humanities. After a period that typically lasted eleven to fourteen

21 The Archivum Romanum Societatis Iesu, the central archive of the Jesuits in Rome, conserves over six thousand letters written by Ignatius. See Mark A. Lewis, "Loyola, Ignatius of," in Worcester, *Cambridge Encyclopedia of the Jesuits*, 482–87, here 485.

22 Steven J. Harris, "Confession-Building, Long-Distance Networks and the Organization of Jesuit Science," *Early Science and Medicine* 1 (1996): 287–318; Harris, "Mapping Jesuit Science: The Role of Travel in the Geography of Knowledge," in O'Malley et al., *Jesuits: Cultures, Sciences and the Arts, 1540–1773*, 212–40, and Rivka Feldhay, "The Cultural Field of Jesuit Science," in O'Malley et al., *Jesuits: Cultures, Sciences and the Arts, 1540–1773*, 108–30.

23 Harris, "Mapping Jesuit Science," 215.

24 *Regimini militantis ecclesiae*, no. 3. The *Constitutions* also offer a similar formulation: "In addition to the three vows mentioned, the professed Society also makes an explicit vow to the present or future sovereign pontiff as the vicar of Christ our Lord. This is a vow to go anywhere His Holiness will order, whether among the faithful or the infidels, without pleading an excuse and without requesting any expenses for the journey, for the sake of matters pertaining to the worship of God and the good of the Christian religion." See, "General Examen," in Padberg et al., *Constitutions of the Society of Jesus*, no. 7.

years, Jesuits were ordained priests and were destined to a specific house, college, university, or mission, where they were to preach, give the Exercises, administer the sacraments, take care of the sick, and teach. Given the arduous and long process of formation, Jesuits were to experience a third year of probation before being fully admitted to the Society as professed members. During the tertianship, they would repeat the thirty-day silent retreat of the Spiritual Exercises and prepare themselves for solemnly professing their vows.[25] As Steven J. Harris puts it, the "overriding goal of this long apprenticeship was to prepare members, both morally and intellectually, for the difficult corporate tasks that would be assigned to them once they were admitted" as professed members.[26]

Unlike other religious orders, which nonetheless valued and promoted education, being a full member of the Society was "contingent upon satisfactory completion of advanced degrees," and that proved to be fundamental in the promotion of the Jesuit intellectual apostolates in Europe, East Asia, and South America.[27] While Ignatius did not admit a secular third order, as the Franciscans had done so successfully in the past, the Society was to accept coadjutors received to "aid the Society in either spiritual or temporal matters."[28] Lacking a formal education in humanities, the temporal coadjutors, also known as lay brothers, would simply profess the three vows of obedience, poverty, and chastity, becoming "cooks, tailors, gardeners, buyers, masons, carpenters, even architects."[29] Unlike lay brothers, spiritual coadjutors were priests.[30] But since they took "a somewhat mitigated course of studies" and did not profess the fourth vow, spiritual coadjutors could not participate "in the government of the Society at the highest level."[31]

---

25 Under Superior General Claudio Acquaviva (in office 1581–1615) it was determined that novices and tertians should undergo the full thirty-day program of the Spiritual Exercises, and that all Jesuits should do a shorter annual retreat lasting at least eight days; Endean, "Spiritual Exercises," 760.

26 Steven J. Harris, "Jesuit Ideology and Jesuit Science: Scientific Activity in the Society of Jesus, 1540–1773" (PhD diss., University of Wisconsin-Madison, 1988), 52–53.

27 Harris, "Jesuit Ideology and Jesuit Science," 51.

28 "General Examen," no. 13.

29 O'Malley, *First Jesuits*, 60. See also Jonathan Stott, "Brothers," in Worcester, *Cambridge Encyclopedia of the Jesuits*, 120–22.

30 Jonathan Stott, "Coadjutor, Temporal and Spiritual," in Worcester, *Cambridge Encyclopedia of the Jesuits*, 177.

31 Although, in principle, spiritual coadjutors could not hold superior offices, there were some exceptions; O'Malley, *First Jesuits*, 346.

2 Jesuit Science and Education

Traditional accounts of the Society's early history usually emphasize that education was not initially among Ignatius's priorities and interpret the successful foundation of the college of Messina by Jerónimo Nadal (1507–80) in 1548 as a turning point in the history of the Jesuits' commitment to education. In the following years, the Jesuits took the first steps in the creation of a global network of colleges and universities, responsible not only for the spiritual and intellectual formation of Jesuit scholastics but also for the education of lay students.[32] Even though teaching lay students was not a priority in the early years, schools soon "became the principal centers for all Jesuit ministry."[33] In 1556, when Ignatius died, there were at least thirty-nine operating or approved Jesuit schools; nine years later, in 1565, the Jesuits were running ninety-one schools in Europe.[34] The expansion of the network of colleges continued in the sixteenth, seventeenth, and eighteenth centuries, and in 1773, when the Society of Jesus was universally suppressed, the Jesuits were running more than eight hundred colleges and universities throughout the world.[35]

The support of the aristocracy and the rising need for Catholic education were two fundamental factors in the establishment of new colleges and universities, especially in early modern Europe.[36] Whereas the Jesuits initially tried to find appropriate patrons for the foundation of schools and universities, they would later wait for offers from potential donors. The most important requisite for the establishment of new institutions was to secure "continuous and guaranteed financial support" through a foundation contract.[37] Yet the support of the aristocracy in the foundation of colleges and the creation of boarding schools for educating nobles—especially in Italy, France, and the Germanic lands—was one of the most controversial aspects of Jesuit education, and it soon became one of the most frequent arguments in anti-Jesuit discourse. However, it must be emphasized that while educating the nobles was an important ministry for a few Jesuit teachers, "the vast majority of their

32 For an account of the Italian universities in the old Society, see Paul F. Grendler, *The Jesuits and Italian Universities, 1548–1773* (Washington, DC: Catholic University of America Press, 2017).

33 O'Malley, *First Jesuits*, 206.

34 Paul F. Grendler, "The Culture of the Jesuit Teacher, 1548–1773," *Journal of Jesuit Studies* 3, no. 1 (2016): 17–41, here 19–20.

35 O'Malley, *First Jesuits*, 239.

36 Harris, "Jesuit Ideology and Jesuit Science," 59.

37 Paul F. Grendler, "Jesuit Schools in Europe: A Historiographical Essay," *Journal of Jesuit Studies* 1, no. 1 (2014): 7–25, here 22.

students were not nobles."[38] Indeed, one of the most significant contributions the Society of Jesus made in this period was the creation of a universal system of secondary and university education, notably the "first free public education system that Europe or the rest of the world had seen."[39]

While early accounts claimed that the main goal of Jesuit education was to provide the necessary means for countering Protestantism, O'Malley argues that the main motives behind the foundation of the first schools in Europe included the Jesuits' desire to save souls and contribute to the common good by "producing good priests, good civic officials, and good citizens of every status."[40] Molded by the *modus Parisiensis*, the Jesuit education system incorporated several teaching practices that Ignatius and his companions had followed during their years of study in the University of Paris, namely the celebrated *disputationes*.[41] However, notwithstanding the importance of the *modus Parisensis* in shaping the methods in their colleges and universities, the Jesuits' educational system was also profoundly influenced by the Renaissance rediscovery of classical Greek and Latin texts.

Every Jesuit college and university in Europe, East Asia, or South America was to follow the same guidelines, texts, and methods, as the Jesuit educational system was meant to be universal.[42] Achieving uniformity in the curriculum and pedagogical practices was expounded in early texts, especially the *Constitutions* (1558) and the *Ratio studiorum* (1599),[43] foundational documents that prescribed meticulous rules for every aspect of education, including the subjects that were to be taught, the duration of classes, and the methods and books that should be followed. With regard to theology, the *Constitutions*

38 Grendler, "Jesuit Schools in Europe," 16–17.

39 Grendler, "Jesuit Schools in Europe," 7–8.

40 O'Malley, *First Jesuits*, 201–14, here 210.

41 Gabriel Codina Mir, *Aux sources de la pédagogie des jésuites: Le* Modus Parisiensis (Rome: Institutum Historicum Societatis Iesu, 1968).

42 For a historiographical essay on Jesuit pedagogy, see Claude Pavur, "The Historiography of Jesuit Pedagogy," in *Jesuit Historiography Online*, ed. Robert A. Maryks (Leiden: Brill, 2016); http://dx.doi.org/10.1163/2468-7723_jho_COM_194129.

43 For a critical edition in Latin of the *Ratio studiorum*, see László Lukács, ed., *Monumenta paedagogica Societatis Iesu: v; Ratio atque Institutio Studiorum Societatis Iesu (1586, 1591, 1599)* (Rome: Institutum Historicum Societatis Iesu, 1986). The critical edition used here will be *The* Ratio studiorum: *The Official Plan for Jesuit Education*, trans. Claude Pavur (St. Louis, MO: Institute of Jesuit Sources, 2005). On the *Ratio studiorum*, see also Cristiano Casalini and Francesco Mattei, "Acquaviva and the *Ratio studiorum*," in *The Acquaviva Project: Claudia Acquaviva's Generalate (1581–1615) and the Emergence of Modern Catholicism*, ed. Pierre-Antonio Fabre and Flavio Rurale (Chestnut Hill, MA: Institute of Jesuit Sources, 2017), 269–90, and Vincent J. Duminuco, ed., *The Jesuit* Ratio studiorum: *400th Anniversary Perspectives* (New York: Fordham University Press, 2000).

stipulated that the Scholastic doctrine of Thomas Aquinas (1224/25–74) should be followed.[44] The *Ratio studiorum* went even further in recommending that those studying Scholastic theology should consider Aquinas "their own particular authority," even though they should not be so "tied to Saint Thomas that they may not differ from that theology in any issue whatsoever," since it would be not right "that Jesuits be bound more tightly to Saint Thomas than the Thomists themselves are."[45] According to the *Constitutions*, the teaching of logic, as well as natural and moral philosophy and metaphysics, should follow Aristotle.[46] As in the case of theology, the *Ratio studiorum* provided more details on the exceptions to this general rule: when some doctrines contradicted "orthodox belief" or when the professor of philosophy came "across something that clashes with the teaching that educational institutions everywhere approve," the teaching "should definitely depart from Aristotle."[47]

Under Ignatius's leadership, it was established that Jesuit scholastics should study humanities, Latin, and Greek—and to a lesser extent Hebrew—logic, natural and moral philosophy, metaphysics, and theology.[48] Given that theology was seen as "the means most suited [to] aid our fellowmen to the knowledge and love of God and to the salvation of their souls," the Society's universities ought to be particularly committed to teaching it.[49] Following the medieval hierarchy of knowledge, the studies of languages, humanities, mathematics, natural sciences, philosophy, and metaphysics were deemed as handmaidens to the study of theology.[50] Although Ignatius recognized that teaching "how to read and write" would be a "work of charity," Jesuits would not ordinarily be devoted to teaching those "elementary subjects."[51] Similarly, the study of medicine and law was not usually undertaken at the Society's universities and colleges as it was deemed "more remote from our Institute."[52]

Even though scholarship usually focuses on Ignatius's legacy, the generalates of Diego Laínez (1512–65, in office 1556–65), Francisco de Borja (1510–72, in office 1565–72), Everard Mercurian (1514–80, in office 1573–80), and Claudio Acquaviva (1543–1615, in office 1581–1615) were also crucial for defining the

44 *Constitutions of the Society of Jesus*, no. 464.
45 *Ratio studiorum*, no. 175.
46 *Constitutions of the Society of Jesus*, no. 470.
47 *Ratio studiorum*, no. 207.
48 *Constitutions of the Society of Jesus*, no. 351.
49 *Constitutions of the Society of Jesus*, no. 446.
50 *Constitutions of the Society of Jesus*, nos. 447–51.
51 *Constitutions of the Society of Jesus*, no. 451.
52 *Constitutions of the Society of Jesus*, no. 452.

Jesuit apostolates and for the standardization of Jesuit education.[53] As Paul F. Grendler notes, it was under Laínez's direction that schools were "elevated to the most important ministry" of the Society, and it was under Acquaviva's mandate that the final version of the *Ratio studiorum* was printed and adopted in all Jesuit colleges and universities.[54]

The core curriculum of Jesuit schools and universities was undeniably focused on the humanities, philosophy, and theology, yet the teaching of mathematics and natural sciences was nevertheless significant, especially when compared with what was by then the common practice. Although other religious orders also dedicated themselves to education, the Jesuits' dedication to "worldly subjects, such as literature, theater and astronomy" was "systemic, not occasional."[55]

The interest in incorporating mathematics in the Jesuit curriculum dates back to the foundation of the first Jesuit schools, namely the colleges of Messina and Rome. For the former, Nadal proposed a two-year course on mathematics in 1548 and a three-year course in 1552. As for the latter, the three-year philosophy course encompassed five semesters of mathematics from 1551 onward.[56] The scientific curriculum of the Collegio Romano—arguably the most important institution run by the Jesuits with regard to the teaching and practice of mathematics and the natural sciences in the early modern period—was followed as an educational model by every school and university until the widespread implementation of the *Ratio studiorum*.[57] According to the 1599 version of the *Ratio*, the three-year philosophy course should contemplate at least a forty-five-minute weekly mathematics class in its second year of studies. Most

---

53 Some attention has been paid to significant primary sources other than the *Ratio* and the *Constitutions*. For the period between 1540 and 1616, see Cristiano Casalini and Claude Pavur, eds., *Jesuit Pedagogy, 1540–1616: A Reader* (Chestnut Hill, MA: Institute of Jesuit Sources, 2016). For the history of Jesuit education under the generalate of Everard Mercurian, see Thomas M. McCoog, ed., *The Mercurian Project: Forming Jesuit Culture 1573–1580* (St. Louis, MO: Institute of Jesuit Sources, 2004). For the history of the Society of Jesus under Acquaviva, see Fabre and Rurale, *Acquaviva Project*.

54 Grendler, "Culture of the Jesuit Teacher," 12.

55 O'Malley, "Distinctiveness of the Society of Jesus," 14.

56 Bernardo Mota, "O estatuto da matemática em Portugal nos séculos XVI e XVII" (PhD diss., University Lisbon, 2008), 134–35. On the teaching of mathematics in Jesuit colleges in this period, see also Antonella Romano, *La Contre-Réforme mathématique: Constitution et diffusion d'une culture mathématique jésuite à la Renaissance (1540–1640)* (Rome: École Française de Rome, 1999), and Romano, "Teaching Mathematics in Jesuit Schools: Programs, Course Content, and Classroom Practices," in O'Malley et al., *Jesuits II*, 355–70.

57 On the history of Collegio Romano, see Ricardo García Villoslada, *Storia del Collegio Romano dal suo inizio (1551) alla soppressione della Compagnia di Gesù (1773)* (Rome: Pontificia Università Gregoriana, 1954).

importantly, if a student was "suitable for these studies and inclined towards them," he should attend private classes.[58]

This emphasis on private instruction was particularly significant for the scientific formation of Jesuits and lay students. As Mordechai Feingold has suggested, it seems that "formal teaching mattered less than instruction outside the classroom," as private lessons provided the perfect educational environment for learning and practicing more sophisticated mathematics (and natural sciences).[59] The inclusion of mathematics in the philosophy course reflected the importance the Jesuits attached to mathematics in the late sixteenth and seventeenth centuries. Whereas, in the medieval university, the teaching of arithmetic, geometry, astronomy, and music—the classical mathematical disciplines that composed the *quadrivium*—were taught before the philosophy course, in the Jesuit colleges they were taught together with physics and metaphysics.[60]

The teaching of advanced scientific topics in Jesuit colleges and universities encompassed subjects such as mechanics, geography, astronomy, cosmology, anatomy, optics, electricity, meteorology, and cosmography, and it was especially conducted at the academies. Composed by a "group of committed students, chosen from the entire student body," the academies would gather under the direction of a Jesuit in order to "hold certain special exercises pertaining to studies."[61] Although they were dedicated to different subjects—grammar, rhetoric, and humanities, or philosophy and theology—the Jesuit academies followed the same organization.[62] The most well-known Jesuit academy in the early modern period is probably the mathematics academy of the Collegio Romano.[63] Founded in 1581 by Christopher Clavius (1538–1612), one of the most prominent mathematicians of the day, the Collegio Romano academy trained some of the most celebrated Jesuit mathematicians and astronomers in the early modern period, including Matteo Ricci (1552–1610), Carlo Spinola (1564–1622), Christoph Grienberger (1561–1636), Orazio Grassi (1583–1654),

58 *Ratio studiorum*, no. 38.

59 Mordechai Feingold, "Jesuits: Savants," in Feingold, *Jesuit Science and the Republic of Letters*, 1–45, here 32.

60 Peter Dear, "Jesuit Mathematical Science and the Reconstitution of Experience in the Early Seventeenth Century," *Studies in History and Philosophy of Science* 18 (1987): 133–75, here 135.

61 *Ratio studiorum*, no. 481.

62 The officers of the academy included a rector, two counsellors, and a secretary; *Ratio studiorum*, no. 487.

63 Ugo Baldini, "The Academy of Mathematics of the Collegio Romano from 1553 to 1612," in Feingold, *Jesuit Science and the Republic of Letters*, 47–98.

Jan Ciermans (Wremann) (1602–48), Giovanni Paolo Lembo (*c.*1570–1618), Paul Guldin (1577–1643), and Otto van Maelcote (1572–1615).[64]

After the Society's official restoration in 1814, the Jesuits resumed their scientific activities, especially in their colleges and universities, where they founded natural history museums, chemistry laboratories, and physics cabinets. However, their most significant endeavor was arguably the creation of a network of astronomical observatories and seismological stations.[65] From Rome to the Philippines, from Shanghai to Cuba, and from Madagascar to Argentina, the network was virtually worldwide. These observatories, which were dedicated to astronomy, meteorology, seismology, and geomagnetism, made significant contributions to the study of eclipses, planetary transits, tropical hurricanes, and earthquakes. Some individual Jesuits played a pivotal role in the emergence of new subjects, such as astrophysics in the case of Italian astronomer Angelo Secchi (1818–78), and paleontology, in the case of French scholar Pierre Teilhard de Chardin (1881–1955), probably the most well-known Jesuit scientist in the modern era.[66] Less well known, but equally significant, was the work done by Jesuit naturalists in early twentieth-century India, including Ethelbert Blatter (1877–1934), Hermenegildo Santapau (1903–70), and Koyapillil M. Mathew (1930–2004).

Although all educational institutions were to follow the same guidelines, the individual and collective scientific endeavors of the Jesuits were diverse and heavily dependent on the local and daily contexts of their missions, colleges, and universities.[67] The empirical sciences, including Aristotelian natural philosophy, the classical sciences, and mixed mathematics, were usually seen as the "forms of early modern scientific activity most compatible with the spirit and purpose of the Society" given their practical utility in the three main stages of Jesuit activity, that is, the apostolates in education, at court, and in the foreign missions.[68] Nevertheless, it must be stressed that, although scientific

64 On Clavius and the teaching of mathematics at Jesuit universities and schools, see Dennis C. Smolarski, "The Jesuit *Ratio studiorum*, Christopher Clavius, and the Study of Mathematical Sciences in Universities," *Science in Context* 15 (2002): 447–57; Romano Gatto, "Christoph Clavius' *Ordo servandus in addiscendis disciplinis mathematicis* and the Teaching of Mathematics in Jesuit Colleges at the Beginning of the Modern Era," *Science and Education* 15 (2006): 235–58.

65 Agustín Udías, *Searching the Heavens and the Earth: The History of Jesuit Observatories* (Dordrecht: Kluwer Academic Publishers, 2003).

66 Agustín Udías, "Jesuit Contribution to Science 1814–2000: A Historiographical Essay," in *Jesuit Historiography Online*, ed. Robert A. Maryks (Leiden: Brill, 2016); doi: 10.1163/2468-7723_jho_COM_192552.

67 Harris, "Mapping Jesuit Science," 214.

68 Harris, "Jesuit Ideology and Jesuit Science," 225.

knowledge was sometimes "employed as a means to achieve the Society's (ultimately religious) ends," the scientific practices of the Jesuits were "not coldly exploitative nor purely instrumental."[69]

When considering the Jesuits' involvement in science and education, especially in the missions, it is important to look beyond the handmaiden metaphor—notably perpetuated in the histories of the Jesuits in China. The execution of worldly tasks, including the scientific apostolates at court and foreign missions, must instead be understood in the broader context of Jesuit ideology. The execution of tasks such as teaching at colleges and universities, observing the heavens, drawing accurate maps, describing indigenous flora and fauna, and partaking in the circulation of knowledge between distant—and often incommensurable—communities was a reflection of a practical spirituality, molded by Ignatius and firmly rooted in the Jesuits' way of proceeding. The commitment to worldly apostolates was the preferred way for Jesuits to work for their own salvation.

3 The Portuguese Assistancy, 1540–1759

The Portuguese assistancy (Assistentia Lusitaniae) was composed of seven provinces and vice-provinces (Portugal, Brazil, Maranhão, Goa, Malabar, Japan, and China) and was undoubtedly the largest and most scattered administrative territory under the Jesuits' jurisdiction.[70] Between the arrival of Francis Xavier (1506–52) and Simão Rodrigues (1510–79) in Lisbon in June 1540 and their expulsion in September 1759, the Jesuits from the Portuguese assistancy were especially committed to the propagation of the Christian faith and doctrine by means of spiritual works such as preaching, giving the Spiritual

69 Harris, "Jesuit Ideology and Jesuit Science," 288.

70 For the history of the Portuguese assistancy, see especially Francisco Rodrigues, *História da Companhia de Jesus na assistência de Portugal*, 7 vols. (Porto: Livraria Apostolado da Imprensa, 1931–50), and Dauril Alden, *The Making of an Enterprise: The Society of Jesus in Portugal, Its Empire and Beyond, 1540–1750* (Stanford: Stanford University Press, 1996). Rodrigues's monumental collection is still one of the most useful sources for the religious and educational history of the Portuguese Jesuits. Alden's account not only updates Rodrigues's work but also sheds some light on the main controversies regarding the economic and political activities of the Portuguese Jesuits in the colonies. Mainly concerned with political, social, economic, and diplomatic issues, *Making of an Enterprise* deals with the creation, expansion, organization (governance, recruitment, and fiscal administration), and vicissitudes of the Portuguese assistancy. Written with no apologetic or anti-Jesuit agenda, it represents the most detailed and balanced history of the Portuguese assistancy from 1540 to 1750.

Exercises, and administering the sacraments, and by means of worldly labors, most notably education.[71]

Supported by King João III (1502–57, r.1521–57) and his successors for their work as preachers, teachers, and ministers of the mercy of God, the Jesuits established several missions, churches, houses, novitiates, schools, and universities throughout the Portuguese Empire.[72] Before the 1759 expulsion, they ran one university and thirty-seven schools, being responsible for the education of tens of thousands of students in Portugal, Brazil, Azores, Madeira, Angola, and Macao.[73] Most colleges were founded in urban areas, and they served various purposes as they were "not only preparatory schools but also dormitories for resident staff and sojourners, as well as the administrative centers of dependent houses, missions, and estates."[74]

The tradition of establishing schools and universities, first in Portugal and then in the colonies, dates back to the first years of the Society. On October 18, 1553, five years after the foundation of the college in Messina, the Jesuits began teaching Latin, Greek, and rhetoric at the Colégio de Santo Antão in Lisbon.[75]

---

71 The 1759 expulsion led to the destruction or scattering of most manuscripts and printed sources pertaining to the Portuguese assistancy. The most important extant sources can be found in the following archives and libraries: Arquivo Nacional Ultramarino (Lisbon); Arquivo Nacional da Torre do Tombo (Lisbon), especially the collection Armário Jesuítico e Cartório dos Jesuítas; Biblioteca da Ajuda (Lisbon), particularly the collection Jesuítas na Ásia; Biblioteca Nacional de Portugal (Lisbon); and Archivum Romanum Societatis Iesu (Rome), especially the collections *Lus.*, *Goan.*, *Bras.*, and *Jap. Sin.* For the sources pertaining to Biblioteca Nacional that concern the 1759 expulsion, see the catalog: Jorge Couto, ed., *A expulsão dos jesuítas dos domínios portugueses: 250.° aniversário* (Lisbon: Biblioteca Nacional de Portugal, 2009). For the sources pertaining to the collection Jesuítas na Ásia, see Francisco da Cunha Leão, ed., *Jesuítas na Ásia: Catálogo e guia*, 2 vols. (Macao: Instituto Cultural de Macau, 1998). Some other sources can be found at the following libraries: Biblioteca Pública da Universidade de Coimbra (Coimbra), Biblioteca Pública de Évora (Évora), and Biblioteca da Academia das Ciências de Lisboa (Lisbon). For an accurate description of Jesuit sources pertaining to the Portuguese assistancy, see Alden, *Making of an Enterprise*, 683–89.

72 The history of the Portuguese Jesuit missions in Brazil, East Asia, and Africa exceeds the aim of this chapter. For a historiographical account of Jesuit missions, see Ronnie Po-chia Hsia, "Jesuit Foreign Missions: A Historiographical Essay," *Journal of Jesuit Studies* 1, no. 1 (2014): 47–65.

73 This network was particularly impressive considering that the Portuguese assistancy had been facing a shortage of manpower to work in the missions and colleges since its involvement in the Portuguese Padroado.

74 Alden, *Making of an Enterprise*, 654.

75 On the history of the Lisbon college, see especially Henrique Leitão, *A ciência na "Aula da Esfera" no Colégio de Santo Antão, 1590–1759* (Lisbon: Comissariado Geral das Comemorações do V Centenário do Nascimento de S. Francisco Xavier, 2007), and Leitão, ed., *Sphaera*

In 1553, Santo Antão had three hundred pupils, and by 1554 the Jesuits already taught six hundred students in Lisbon. The growth of the student body was exponential in the first decades: by 1591, the Jesuits of Santo Antão were responsible for the education of 2,500 students.[76] Along with Santo Antão, the early Jesuits also ran the Colégio das Artes in Coimbra (1555–1759) and the University of Évora (1559–1759).[77]

Thus, as historian of science Henrique Leitão notes, the Jesuits "radically altered the cultural and educational scenario" in Portugal.[78] They established the first organized system of secondary education, founded important schools outside the major urban areas, and taught, for more than two centuries, regular courses of mathematics and natural philosophy.[79] Although the teaching of natural philosophy was somewhat irregular in the sixteenth and seventeenth centuries, it was always part of the philosophy curriculum at the University of Évora and the colleges of Coimbra, Lisbon, and Braga, in Portugal, and the colleges of São Salvador da Baía and Rio de Janeiro, in Brazil, Goa, and Cochin, in India, and Macao, in China. At the Colégio das Artes, in Coimbra, and at the University of Évora, the Jesuits taught natural philosophy at a higher university level, as this would enable students to attend the higher faculties of the university and pursue their studies of theology, civil and canon law, and medicine.[80] The Colégio das Artes was also especially acknowledged in the Jesuit educational network for the edition of a series of commentaries on the works of Aristotle. Written by Manuel de Góis (1543–97) and Sebastião do Couto (1567–1639), the *Commentarii collegii Conimbricensis* (Coimbra commentaries [1592–1606]) were widely used as textbooks throughout the seventeenth

mundi: A ciência na "Aula da Esfera"; Manuscritos científicos do Colégio de Santo Antão nas colecções da BNP (Lisbon: Biblioteca Nacional de Portugal, 2008).

76 Leitão, *Sphaera mundi*, 20.

77 Founded in 1548 by King João III, the Colégio das Artes was a propaedeutic college to the University of Coimbra. The crown entrusted it to the Jesuits in 1555.

78 Henrique Leitão, "A Periphery between Two Centers? Portugal in the Scientific Route from Europe to China (16th and 17th Centuries)," in *Travels of Learning: A Geography of Science in Europe*, ed. Ana Simões, Ana Carneiro, and Maria Paula Diogo (Dordrecht: Kluwer Academic Publishers, 2003), 19–46, here 27.

79 Leitão, "Periphery between Two Centers," 27–28.

80 Luís Miguel Carolino and Henrique Leitão, "Natural Philosophy and Mathematics in Portuguese Universities, 1550–1650," in *Universities and Science in the Early Modern Period*, ed. Mordechai Feingold and Victor Navarro Brotóns (Dordrecht: Springer, 2006), 153–68, here 154. On the teaching of philosophy at the University of Évora, see also Domingos Maurício, "Obra científico-literária e pedagógica da Universidade de Évora," *Brotéria* 69 (1959): 377–92; João Pereira Gomes, "Últimas actividades filosóficas na Universidade de Évora," *Brotéria* 69 (1959): 393–404; and Pereira Gomes, *Os professores de filosofia da Universidade de Évora* (Évora: Câmara Municipal de Évora, 1960).

century, being—along with Clavius's *In sphaeram Ioannis de Sacro Bosco commentarius* (Commentaries on Ioannis de Sacrobosco's *Sphere* [1596])—the most influential Jesuit commentaries in the early modern period.[81] It must be added that although they followed Aristotle in their classes and writings, the Jesuits taught natural philosophy in a speculative way, and they progressively included the latest astronomical discoveries and the new cosmologies in the curricula of the colleges of Évora and Coimbra.[82]

With regard to the teaching of mathematics, the most important institution the Jesuits ran in the Portuguese Empire was the Colégio de Santo Antão. From the college's foundation in 1553, the Jesuits taught some courses of elementary mathematics, mostly based on *De sphaera* (The sphere) by Johannes de Sacrobosco (*c*.1195–*c*.1256).[83] In December 1573, and given the importance of having regular classes in mathematics in sixteenth-century Portugal, Cardinal Henrique of Portugal (1512–80, r.1578–80) signed a contract with the Jesuits granting an annual endowment to the Colégio de Santo Antão on the condition that they establish a public course of mathematics.[84] In 1590, in response to this appeal from the crown, the Jesuits finally created the celebrated Aula da Esfera, a public course of mixed mathematics that was regularly taught in Santo Antão until 1759.[85] Throughout its centennial history, the Aula da Esfera was taught by distinguished mathematicians and astronomers, some of whom had been

81 The most recent work on the *Cursus Conimbricensis* is Cristiano Casalini, *Aristotle in Coimbra: The* Cursus Conimbricensis *and Education at the College of Arts* (New York: Routledge, 2017). See also Mário Santiago de Carvalho, *Psicologia e ética nel* Cursus Conimbricensis (Rome: Anicia, 2015); Jesué Pinharanda Gomes, *Os Conimbricenses* (Lisbon: Instituto de Cultura e Língua Portuguesa, 1992); Alfredo Dinis, "Tradição e transição no curso Conimbricense," *Revista portuguesa de filosofia* 47 (1991): 535–60; and Dinis, "O comentário Conimbricense à *Física* de Aristóteles," *Brotéria* 134 (1992): 398–406.

82 Carolino and Leitão, "Natural Philosophy and Mathematics in Portuguese Universities," 164.

83 Francisco Rodrigues, *A formação intellectual do jesuíta: Leis e factos* (Porto: Livraria Magalhães e Moniz, 1917), 76–78.

84 Leitão, *Sphaera mundi*, 21. Son of King Manuel I (1469–1521, r.1495–1521) and brother to King João III, Henrique was inquisitor general from 1539, regent of Portugal between 1562 and 1568, and succeeded to the throne after his grand-nephew, King Sebastião (1554–78, r.1557–78), disappeared in the Battle of Alcácer Quibir, in Morocco (1578).

85 Henrique Leitão, "Jesuit Mathematical Practice in Portugal, 1540–1759," in Feingold, *New Science and Jesuit Science*, 229–47, here 234. The first professor of the Aula da esfera was João Delgado (1553–1612). For a study, see Luís Carolino, "João Delgado S.J. e a *Quaestio de certitudine mathematicarum* em inícios do século XVII," *Revista brasileira de história da matemática* 6 (2006): 17–49. On the teaching of mathematics at the Jesuits colleges in Portugal, see also Domingos Maurício, "Os jesuítas e o ensino das matemáticas em Portugal," *Brotéria* 20 (1935): 189–205; Luís de Albuquerque, "A 'Aula da Esfera' do Colégio de Santo Antão no século XVII," *Anais da Academia portuguesa de história* 21 (1972): 337–91.

taught by Clavius or his successors, most notably Grienberger, Lembo, Johann Chrysostomus Gall (1586–1642), Cristoforo Borri (1583–1632), and João Delgado (1553–1612).[86] Since this class on the sphere was expected to contribute to the instruction of maritime pilots, the courses were lectured in the vernacular. Until 1640, the Jesuits taught cosmography, nautical astronomy, and navigation, with an emphasis on the uses of instruments such as the astrolabe and the quadrant, and the construction of nautical charts and globes. Later, the class on the sphere included some more theoretical topics, namely geometry, theoretical astronomy, and algebra.[87] But it also included controversial subjects such as astrology and chiromancy.[88] Although they generally followed the pedagogical guidelines of the *Ratio studiorum*, the Portuguese colleges also had some original features, especially with regard to the teaching of the scientific disciplines, namely the unusual emphasis on mixed mathematical sciences, teaching in the vernacular, and the inclusion of controversial topics in Lisbon, and the design of large panels of *azulejos* (glazed tiles) for teaching Euclidean geometry in Coimbra.[89]

The presence of talented European mathematicians, astronomers, and natural philosophers in the main colleges of the Portuguese assistancy played a significant role in the introduction and circulation of new scientific methods, observations, objects, and theories in the vast Portuguese Empire.[90] One of

86 Leitão, "Jesuit Mathematical Practice in Portugal," 235.

87 Carolino and Leitão, "Natural Philosophy and Mathematics in Portuguese Universities," 162.

88 Although astrology and all divinatory arts had a long history of condemnations in the Catholic world, at least four professors taught astrology at Santo Antão: Delgado, Francisco da Costa (1567–1604), Simon Fallon (*c.*1604–42), and Luís Gonzaga (1666–1747). For more details on the teaching of astrology in the Aula da Esfera, see: Henrique Leitão, "Entering Dangerous Ground: Jesuits Teaching Astrology and Chiromancy in Lisbon," in O'Malley et al., *Jesuits II*, 371–89.

89 These tiles were applied on the walls of the mathematics room at the Jesuit college in Coimbra, and most of them displayed geometrical diagrams from André Tacquet's (1612–60) popular edition of the *Elements* of Euclid. From about two hundred *azulejos* pertaining to the Jesuit college of Coimbra, only thirty are known today; Henrique Leitão and Samuel Gessner, "Euclid in Tiles: The Azulejos of the Jesuit College in Coimbra," *Mathematische Semesterberichte* 61 (2014): 1–5, and Henrique Leitão, "Azulejos que testemunham uma tradição de ensino científico," in *Azulejos que ensinam*, ed. António Leal Duarte and Carlota Simões (Coimbra: Museu Nacional Machado de Castro–Centro de Matemática da Universidade de Coimbra, 2007), 16–33.

90 For the history of the scientific activities of the Jesuits in the Portuguese East Asian missions, see especially Francisco Rodrigues, "Mathematicos portugueses na China," *Revista de história* 12 (1923): 81–118; Rodrigues, *Jesuítas portugueses astrónomos na China, 1583–1805* (Porto: Tipografia Porto Médico, 1925); Ugo Baldini, "As assistências ibéricas da Companhia de Jesus e a actividade científica nas missões asiáticas (1578–1640): Alguns

the clearest examples of this concerns Galileo Galilei's (1564–1642) telescopic observations in 1609. A distinguished manufacturer of telescopes and an astronomer at the Collegio Romano, Giovanni Paolo Lembo built the telescope the Jesuits used to confirm Galileo's observations in 1610, following the request of Cardinal Roberto Bellarmino (1542–1621).[91] Five years later, Lembo came to Lisbon and taught mathematics at Santo Antão for two years. In the Lisbon college, he observed the phases of Venus—arguably the first telescopic observations made in Portugal—confirming Galileo's earlier findings. He then discussed his observations with his pupils, and taught them how to build a telescope.[92] In 1615, while Lembo was teaching mathematics in the Aula da Esfera, Portuguese Jesuit Manuel Dias (1574–1659) published *Tianwenlüe* (Epitome of questions on the heavens [1615]), the first text describing Galileo's telescopic observations to a Chinese readership.[93]

At the turn of the seventeenth century, Superior General Tirso González (1624–1705, in office 1687–1705) urged the Portuguese Jesuits to reform the teaching of mathematics. Given the importance of mathematics and astronomy in the East Asian missions, especially at the Imperial Observatory in Beijing, and the shortage of Portuguese mathematicians, his deep desire was to create a consistent and independent national school of mathematics.[94] One

---

aspectos culturais e institucionais," *Revista portuguesa de filosofia* 54 (1998): 195–245; and Baldini, "The Portuguese Assistancy of the Society of Jesus and Scientific Activities in Its Asian Missions until 1640," in *História das ciências, Portugal e o Oriente*, ed. Luís Saraiva (Lisbon: Fundação Oriente, 2000), 49–104.

91 The famous letter confirming the observations was signed by Clavius, Grienberger, van Maelcote, and Lembo.

92 Henrique Leitão, "Longemira: Os primeiros telescópios em Portugal," *Gazeta de física* 33 (2010): 17–21, and Ugo Baldini, "Giovanni Paolo Lembo's Lessons in Lisbon: A Partial Content Analysis," in *Proceedings of the International Conference History of Astronomy in Portugal: Institutions, Theories, Practices*, ed. Luís Saraiva (Lisbon: Sociedade Portuguesa de Astronomia, 2014), 123–81.

93 In April 1601, after completing his philosophy studies in Coimbra (1596–1600), he departed to the East Asian missions. After a sojourn in Macao (1604–10), he headed to Shaozhou (1610–12) and then to Beijing (1613–1616), where he finished this summary of European cosmography and astronomy; Henrique Leitão, "The Contents and Context of Manuel Dias' *Tianwenlüe*," in *History of Mathematical Sciences: Portugal and the East Asia, III; The Jesuits, the Padroado and East Asian Science (1552–1773)*, ed. Luís Saraiva and Catherine Jami (Singapore: World Scientific, 2008), 99–121, and Rui Magone, "The Textual Tradition of *Tianwenlüe*," in Saraiva and Jami, *History of Mathematical Sciences*, 123–38.

94 On this, see especially Ugo Baldini, "L'insegnamento della matematica nel Collegio di S. Antão a Lisbona, 1590–1640," in *A Companhia de Jesus e a missionação no Oriente*, ed. Nuno da Silva Gonçalves (Lisbon: Fundação Oriente–Brotéria, 2000), 275–310, and Baldini, "The Teaching of Mathematics in the Jesuit Colleges of Portugal, from 1640 to Pombal," in *The Practice of Mathematics in Portugal*, ed. Luís Saraiva and Henrique Leitão (Coimbra: Acta

of the objectives behind this reformation was to prevent the Portuguese from asking German and Italian mathematicians and astronomers to teach mathematics in the Portuguese colleges for a few years before allowing them to depart to the East Asian missions. From 1700 onward, as a result of the ordinances, the Aula da Esfera was taught only by Portuguese Jesuits, most notably by Manuel de Campos (1681–1758), Inácio Vieira (1678–1739), and Eusébio da Veiga (1718–98).[95]

Given the importance of astronomy and cartography in the missions for the precise definition of borders, the Portuguese crown turned to the Jesuits for the appointment of talented astronomers and cartographers. As the number of mathematicians in the Portuguese assistancy was never sufficient to cover the vast territory of the East Asian and Brazilian missions, the curia continued to send foreign Jesuits to work in the Portuguese Empire, despite González's ordinances. One of the most significant examples was the arrival of Italian Jesuits Giovanni Battista Carbone (1694–1750) and Domenico Capassi (1694–1736) in Lisbon in 1722. Carbone and Capassi had been sent to work as cartographers in the Brazilian state of Maranhão at the request of King João V (1689–1750, r.1706–50). In the first years following their arrival, they founded two astronomical observatories, at the royal court and at the College of Santo Antão, and made important astronomical observations that were published in *Philosophical Transactions*, the official journal of the Royal Society of London.[96] Whereas Capassi and the Portuguese Jesuit Diogo Soares (1684–1748) were eventually sent to Brazil in 1729 under the title of royal mathematicians or royal astronomers, Carbone stayed in Lisbon and became one of the closest advisors to King João V.[97]

The history of the Jesuits' scientific and educational activities changed dramatically in the 1750s. Following the devastating earthquake of November 1,

---

Universitatis Conimbrigensis, 2004), 293–465, particularly the appendices by Baldini and Leitão, 629–758.

95 Eusébio da Veiga was the last professor of mathematics of the Aula da Esfera. Between 1750 and 1755, he was also responsible for the Jesuit observatories in Lisbon. For a biography, see António Júlio Trigueiros, "Eusébio da Veiga, o último professor da Aula da Esfera: Reconstituição biográfica de um jesuíta do século XVIII," *Lusitania sacra* 32 (2015): 147–78.

96 Leitão, "Jesuit Mathematical Practice in Portugal," 242–43.

97 The most recent and detailed study of Carbone's scientific and diplomatic activities in Portugal is Luís Tirapicos, "Ciência e diplomacia na corte de D. João V: A acção de João Baptista Carbone, 1722–1750" (PhD diss., University of Lisbon, 2017). For an account of Capassi's cartographical work in Brazil, see especially André Ferrand de Almeida, *A formação do espaço brasileiro e o projecto do novo Atlas da América portuguesa (1713–1748)* (Lisbon: Comissão Nacional para as Comemorações dos Descobrimentos Portugueses, 2001).

1755, which shattered the city of Lisbon and caused the deaths of tens of thousands, the Jesuits fell into almost absolute disgrace.[98] Accused of making illegal profits from commerce in Brazil, disregarding the frontiers between the Portuguese and Spanish colonies in South America, supporting popular riots, and plotting to murder the king, the Jesuits were consecutively banned from court, prohibited from preaching, hearing confessions, and teaching, and were finally expelled from the Portuguese Empire in 1759. After the expulsion, the Jesuits were presented as the main obstacle to the country's progress and modernization. Charges of educational backwardness and scientific obscurantism were widely diffused in the following decades and proved extremely resilient to the point of deeply affecting the restoration of the Jesuits a century later.

98 The estimates of casualty tolls caused by the earthquake and tsunami of All Saints' Day range from ten thousand to one hundred thousand in Lisbon alone; Álvaro Santos Pereira, "The Opportunity of a Disaster: The Economic Impact of the 1755 Lisbon Earthquake," *Journal of Economic History* 69 (2009): 466–99, here 468.

CHAPTER 2

The Pombaline Expulsion and the Building of Anti-Jesuitism

When examining the expulsion of the Jesuits from Portugal and the rise of anti-Jesuitism in the eighteenth century, historians usually focus on the role played by Sebastião José de Carvalho e Melo, better known by his later title, the marquis of Pombal (1699–1782).[1] Yet although he was undoubtedly the leading figure behind the expulsion, a closer look into the years that preceded the Jesuits' proscription and exile reveals that his early actions against the Jesuits were heavily dependent on a series of circumstantial events, starting with the appointment of a secular visitor.[2]

The traditional accounts of the 1759 expulsion usually focus on two of Pombal's main accusations against the Jesuits: their disregard of the Treaty of Madrid (1750) and their involvement in illegal trading in Brazil, and their role as moral instigators of the plot to murder the king. However, the Pombaline accusation of obscurantism and scientific illiteracy also played a central role in the history of anti-Jesuitism in Portugal, mainly due to its wide acceptance and

1 Anti-Jesuitism has been a pervasive subject in Jesuit studies. The most recent and detailed books on anti-Jesuitism are Michel Leroy, *Le mythe jésuite: De Béranger à Michelet* (Paris: Presses Universitaires de France, 1992); Geoffrey Cubitt, *The Jesuit Myth: Conspiracy, Theory and Politics in Nineteenth-Century France* (Oxford: Clarendon Press, 1993); Sabina Pavone, *The Wily Jesuits and the* Monita secreta: *The Forged Secret Instructions of the Jesuits; Myth and Reality* (St. Louis, MO: Institute of Jesuit Sources, 2005); and Pierre-Antoine Fabre and Catherine Maires, eds., *Les antijésuites: Discours, figures et lieux de l'antijésuitisme à l'époque modern* (Rennes: Presses Universitaires de Rennes, 2010). This chapter revises and expands the arguments presented in Francisco Malta Romeiras, Henrique Leitão, "The Role of Science in the History of Portuguese Anti–Jesuitism," *Journal of Jesuit Studies* 2 (2015): 77–99; Francisco Malta Romeiras, "Jesuit Historiography in Modern Portugal," in *Jesuit Historiography Online*, ed. Robert Maryks (Leiden: Brill, 2016); and Francisco Malta Romeiras, "Francisco Saldanha da Gama: The Last Visitor of the Portuguese Assistancy," in *With Eyes and Ears Open: The Role of Visitor in the Society of Jesus*, ed. Thomas McCoog (Leiden: Brill, 2019), 172–90.

2 José Caeiro, *De exilio provinciae Lusitaniae Societatis Iesu libri quinque* (ANTT, *Manuscritos da Livraria* 2600–2; ARSI, *Lus.* 93, 94/1–2) is the most detailed Jesuit account of the 1759 expulsion. In the 1990s, Júlio de Morais and José Leite edited and translated it into Portuguese; José Caeiro, *História da expulsão da Companhia de Jesus da província de Portugal*, 3 vols. (Lisbon: Verbo, 1991–99). On the imprisonment and exile of the Portuguese Jesuits, see also Anselmo Eckart, *Historia persecutionis Societatis Iesu in Lusitania* (Nuremberg, 1779–80), ARSI, *Lus.* 96 (reprinted and edited as *Memórias de um jesuíta prisioneiro de Pombal*, trans. Joaquim Abranches and Ana Maria Lago da Silva [Braga: Apostolado da Imprensa, 1987]).

 | DOI:10.1163/9789004382367_003

longevity. This argument was not only directly relevant for the expulsion of the Jesuits in the eighteenth century but was also a keystone of the anti-Jesuit propaganda that eventually led to the expulsion of the Society of Jesus from Portugal in the twentieth century.

The consecutive expulsions, the suppression, and, more recently, the restoration of the Society of Jesus have proven to be topics of intense research in the field of Jesuit studies.[3] Because he expelled the Jesuits from the Portuguese Empire in 1759 and deployed an unprecedented array of political, diplomatic, and economic actions against the Society that influenced the succeeding expulsions and the order's suppression in 1773, the marquis of Pombal has long been considered the archetype of eighteenth-century anti-Jesuitism. From the 1750s, this powerful secretary of state to King José (1714–77, r.1750–77) promoted his relatives—including his younger brothers Paulo de Carvalho e Mendonça (1702–70) and Francisco Xavier de Mendonça Furtado (1701–69)—to the most prominent ecclesiastical, diplomatic, and political positions in the Portuguese Empire. They played a direct role in Pombal's battles with the Jesuits, and their activities greatly increased the wealth of the Carvalho family.[4]

Francisco Xavier de Mendonça Furtado was appointed governor of Grão-Pará and Maranhão on April 19, 1751. He was responsible for the implementation of the Treaty of Madrid, which redrew the boundaries between the Portuguese and Spanish colonies in the New World, and for the Jesuits' eventual expulsion from Brazil in 1759.[5] Despite not being officially nominated inquisitor general, Paulo de Carvalho became responsible for the government of the Portuguese Inquisition in 1760 after the arrest of the inquisitor general José de Portugal (1720–1801) and his brother António de Portugal (1714–1800).[6] There were several explanations for their arrest and imprisonment in the convent of Buçaco until 1777, but according to the most commonly accepted version, Pombal and

3 For a global perspective, see especially Jeffrey D. Burson and Jonathan Wright, eds., *The Jesuit Suppression in Global Context: Causes, Events, and Consequences* (Cambridge: Cambridge University Press, 2015), and Robert A. Maryks and Jonathan Wright, eds., *Jesuit Survival and Restoration: A Global History, 1773–1900* (Leiden: Brill, 2014).

4 The literature on the marquis of Pombal is immense. For an overview of his life and policies, see especially Kenneth Maxwell, *Pombal, Paradox of the Enlightenment* (Cambridge: Cambridge University Press, 1995), and Nuno Gonçalo Monteiro, *D. José: Na sombra de Pombal* (Lisbon: Temas e Debates, 2008).

5 On Francisco Xavier de Mendonça Furtado, see Ângela Domingues, "Francisco Xavier de Mendonça Furtado," in *Dicionário da história da colonização portuguesa no Brasil*, ed. Maria Beatriz Nizza da Silva (Lisbon: Verbo, 1994), cols. 359–62, and Miguel Dantas da Cruz, "Francisco Xavier de Mendonça Furtado," in *e-Dicionário da terra e do território no império português*, ed. José Vicente Serrão, Márcia Motta, and Susana Münch Miranda (Lisbon: CEHC–IUL, 2015); doi: 10.15847/cehc.edittip.2015v003 (accessed January 29, 2019).

6 José and António de Portugal were legitimized sons of King João V (1689–1750, r.1706–50) and were commonly known as "children of Palhavã."

the inquisitor José de Portugal had a tempestuous discussion over the publication of the regalist work *De potestate regia* (On royal powers), which abruptly ended when the inquisitor general tore off Pombal's wig and flung it in his face. After the imprisonment of the inquisitor general, Paulo de Carvalho e Mendonça became the leading figure of the Portuguese Inquisition,[7] which then proceeded with the prosecution of the Jesuit Gabriele Malagrida (1688–1751), burnt at the stake in an *auto-de-fé* on September 21, 1761.[8]

Historians generally emphasize the role of Pombal and his brothers in the Jesuits' downfall, mentioning only briefly the activities of two other decisive figures: Francisco de Almada Mendonça (*c.*1720–83), Pombal's first cousin and minister plenipotentiary to the Holy See, and Cardinal Francisco Saldanha da Gama (1713–76), who was appointed visitor and reformer of the Portuguese assistancy on April 1, 1758. Although he was acting as a papal legate, Saldanha da Gama acted on behalf of the crown and directed a series of repressive actions against the Jesuits that would culminate in their expulsion.[9]

1 The Pombaline Expulsion

Accused of making illegal profits from commerce with the missionary province of the state of Grão-Pará and Maranhão, where Francisco Xavier de Mendonça Furtado was governor, several Jesuits were banished from Brazil on September 19, 1757.[10] Two days later, in a separate incident, the Jesuit confessors to the royal family were banned from court on account of having allegedly supported a popular riot in Porto against the foundation of the Companhia Geral da Agricultura e das Vinhas do Douro (General company of agriculture of wines of Douro). Created by Pombal in September 1756 with the objective of ensuring the quality of the wine exported to England, the government wine company created a demarcated wine region and sought to control the price range of the wine exported, which in turn led to protests from some winegrowers

7 See Monteiro, *D. José: Na sombra de Pombal*, 184–88.

8 Throughout the book, I have included some biographical information for the most relevant Jesuits in modern Portugal—date and place of birth (b.), date of entering the novitiate (e.), and date and place of death (d.). Gabriele Malagrida: b. December 5, 1688, Menaggio (Italy); e. October 23, 1711, Genoa (Italy); d. September 21, 1761, Lisbon (Portugal) (*DHCJ*, 3:2481; *Sommervogel*, 5:394–95).

9 Francisco Malta Romeiras, "Francisco Saldanha da Gama: The Last Visitor of the Portuguese Assistancy," in *With Eyes and Ears Open: The Role of Visitors in the Society of Jesus*, ed. Thomas McCoog (Leiden: Brill, 2019), 172–90.

10 For a more detailed account of the anti-Jesuit campaign in Brazil, see Dauril Alden, "The Gang of Four and the Campaign against the Jesuits in Eighteenth-Century Brazil," in O'Malley et al., *Jesuits II*, 707–24.

and publicans. On February 23, 1757, the publicans organized a popular riot, the "Revolta dos Taberneiros" (Publicans' riot). As a result, twenty-six individuals were put to death and three hundred more suffered various other fates (confiscation of property, deportation, and public lashes). Although there was no evidence to support the charges, the Jesuits were accused of instigating the riot and were immediately expelled from court.[11]

Following the banishments, Pombal instructed Almada Mendonça to inform the pope of these incidents as proof that the Portuguese Jesuits were in need of reformation. Almada Mendonça initiated this diplomatic mission on October 8, 1757, but it took him six months to persuade Pope Benedict XIV (1675–1758, r.1740–58) to take action.[12] The pope granted the petition only after he had heard the serious accusations as reported in *Relação abreviada* (Abbreviated report [Lisbon, 1757]).[13] Published on December 3, the feast of Francis Xavier, this anti-Jesuit booklet accused the Jesuits of disobeying the Treaty of Madrid and of governing and profiting from a powerful Guaraní republic comprising thirty-one villages and more than one hundred thousand indigenous slaves.[14] Despite being printed as an anonymous report, it is now indisputable that *Relação abreviada* was actually written by Pombal, who ensured its widespread circulation in Europe and throughout the Portuguese Empire: the booklet was translated into French, Spanish, Italian, German, and Latin, with twenty thousand copies being distributed all over the Portuguese Empire.[15]

On March 9, 1758, Almada e Mendonça argued that the *Relação abreviada* demonstrated the miserable condition of the Jesuits in the Portuguese Empire and thus asked the pope for either a thorough reformation or total suppression

11 Rui Ramos, Bernardo Vasconcelos e Sousa, and Nuno Gonçalo Monteiro, *História de Portugal*, 8th ed. (Lisbon: Esfera dos Livros, 2009), 366–68.

12 Stefan Gatzhammer, "Antijesuitismo europeu: Relações político-diplomáticas e culturais entre a Baviera e Portugal (1750–1780)," *Lusitania sacra* 5 (1993): 159–250, here 169–71; Monteiro, *D. José: Na sombra de Pombal*, 127–28.

13 Fortunato de Almeida, *História da Igreja em Portugal*, 8 vols. (Coimbra: Imprensa Académica, 1910–22), 5:290.

14 *Relação abbreviada da republica, que os religiosos jesuitas das provincias de Portugal, e Hespanha, estabelecerão nos dominios ultramarinos das duas monarquias, e da guerra, que nelles tem movido, e sustentado contra os exercitos hespanhoes e portuguezes; Formada pelos registos das secretarias dos dous respectivos principaes commissarios, e plenipotenciarios; E outros documentos authenticos* (Lisbon, 1757).

15 José Eduardo Franco, "Os catecismos antijesuíticos pombalinos: As obras fundadoras do antijesuitismo do marquês de Pombal," *Revista lusófona de ciência das religiões* 7, no. 8 (2005): 247–68, here 249. Arquivo Nacional da Torre do Tombo (hereafter ANTT), Biblioteca Nacional de Portugal (hereafter BNP), and Biblioteca da Brotéria (hereafter BB), in Lisbon, hold several copies of *Relação abreviada*. Archivum Romanum Societatis Iesu (hereafter ARSI) holds at least two copies; *Hist. Soc.* 270, fols. 2^r–64^v, and *Hist. Soc.* 184, fols. 6^r–11^v.

of the Society.[16] Although the pope declared that such grave charges should be addressed to the Jesuit curia in Rome, Almada e Mendonça insisted that the pontiff take immediate action without consulting the Jesuits. Most likely, he hoped to take advantage of the *sede vacante* in the Society: Superior General Alessandro Luigi Centurione (1686–1757, in office 1755–57)[17] had died on October 2, 1757, and his successor, Lorenzo Ricci (1703–75, in office 1758–73), was not elected until May 21, 1758.[18] Thus, without any consultation with the Jesuit curia, the pope nominated a visitor. As soon as Benedict XIV had decided in favor of a visitor, he sent Cardinal Domenico Passionei (1682–1761), a Jansenist sympathizer and papal secretary of briefs, to the Portuguese embassy where the cardinal received a draft of the brief confirming the visitor's appointment.[19] Finally, on April 1, 1758, the Holy See officially designated Cardinal Saldanha da Gama as the Society's visitor and reformer.

Satisfied with the success of his diplomatic mission, Almada e Mendonça informed Pombal of the appointment and requested two diamond rings for Cardinals Passionei and Alberico Archinto (1698–1758), Benedict's secretary of state, to whom he owed the promulgation of the brief.[20] The brief *In specula supremae dignitatis* entrusted Saldanha da Gama with the visitation of the churches, professed houses, novitiates, colleges, hospitals, and missions of the Portuguese assistancy, and it also conferred on him ample powers, including the authority to dismiss rectors and superiors from their offices without the right of appeal to a higher court.[21] Following the reading of the brief, the

16 ARSI, *Lus.* 110, fols. 199^{r}–199^{v}.

17 Alessandro Luigi Centurione: b. August 29, 1686, Genoa (Italy); e. May 14, 1703, Chieri (Italy); d. October 2, 1757, Castel Gandolfo (Italy) (*DHCJ*, 2:1655).

18 Lorenzo Ricci: b. August 1, 1703, Florence (Italy); e. November 16, 1718, Rome (Italy); d. November 24, 1775, Rome (Italy) (*DHCJ*, 2:1656).

19 The draft had been written by António Rodrigues, a priest and personal secretary to Francisco de Almada Mendonça; Almeida, *História da Igreja em Portugal*, 5:290–91. On the influence of Jansenism in the suppression, see also Dale K. van Kley, "Jansenism and the International Suppression of the Jesuits," in *The Cambridge History of Christianity*, vol. 7, *Enlightenment, Reawakening and Revolution 1660–1815*, ed. Stewart J. Brown and Timothy Tackett (Cambridge: Cambridge University Press, 2006), 302–28.

20 ANTT, *Ministério dos Negócios Estrangeiros* liv. 126, fols. 12^{r}–16^{v}.

21 ARSI, ASV, and BNP hold at least one copy of the brief *In specula supremae dignitatis*; see ARSI, *Hist. Soc.* 248, fols. 304^{r}–309^{v}; ASV, *Segretaria di Stato Portogallo* 180, fols. 190^{r}–193^{v}; BNP, *Pombalina* 458. The brief was also included in the renowned *Collecção dos negócios de Roma*, which compiled the most important official documents concerning the 1759 expulsion. Copies can be found in ANTT, *Armário jesuítico* liv. 1., doc. 7; BNP, *Pombalina* 457. Originally published in 1759, this Pombaline compilation was reprinted in the late nineteenth century; see *Collecção dos negocios de Roma no reinado de El-Rei Dom José I, ministerio do marquez de Pombal*, 4 vols. (Lisbon: Imprensa Nacional, 1874–75).

Portuguese Jesuits wrote to their companions in Rome who were completely unaware of the nomination of Cardinal Saldanha da Gama.[22]

On May 15, 1758, Saldanha da Gama forbade the Jesuits from engaging in "secular businesses."[23] He compared Jesuit missionaries to the infamous vendors in the temple, admonished them for making a profit out of their commerce, and challenged them to pursue apostolic poverty as "their single patrimony." Under the risk of *latae sententiae* excommunication, the Jesuits were given three days to disclose all their mercantile activities. The pastoral letter's severity was puzzling because neither Saldanha da Gama nor his delegates had investigated the allegations contained in *Relação abreviada*. However, it would appear that Saldanha da Gama issued the decree without conducting a due inquiry out of fear of "a manifest and pungent scandal," and it would in any case have been extremely difficult to investigate the accusations against the Jesuits in such a short period of time because the mercantile activities in dispute concerned practices in the province of Brazil and the vice-province of Maranhão. As Jesuit José Caeiro (1712–91)[24] sharply put it, Saldanha da Gama had become "the accuser, the witness, and the judge."[25] In the following weeks, the Jesuits were also prohibited from hearing confessions and preaching in the archdiocese of Lisbon.[26]

On September 3, 1758, while allegedly returning from a rendezvous with his mistress, the marchioness Teresa Távora de Lorena (1723–?), King José was seriously injured in an assassination attempt. The event was kept secret for three months, and the Holy See was not notified of the suspected conspiracy behind

22 See Francisco Rodrigues, *Jesuitophobia: Resposta serena a uma diatribe* (Porto: Typographia Lusitania, 1917), 231; Almeida, *História da Igreja em Portugal*, 5:293–94.

23 *Pastoral do cardeal-diácono Saldanha como visitador apostólico e reformador geral da Companhia de Jesus em Portugal, proibindo a ingerência dos jesuítas em "negócios seculares," dada em 15 de Maio de 1758* (Lisbon, 1758). ANTT, BNP, ARSI, and Arquivo Histórico do Patriarcado de Lisboa (hereafter AHPL) hold copies of this provision. It was also included in the compilation *Collecção dos negócios de Roma* and *De exilio provinciae Lusitaniae Societatis Iesu*: *Collecção dos negócios de Roma*, 1:53–59; Caeiro, *História da expulsão da Companhia de Jesus*, 2:157–62; ARSI, *Hist. Soc.* 248, fols. 310r–329v; ARSI, *Hist. Soc.* 184, fols. 30r–35v; BNP, *Pombalina* 456; ANTT, SP 3575 (61).

24 José Caeiro: b. April 14, 1712, Reguengos de Monsaraz (Portugal); e. May 3, 1726, Évora (Portugal); d. November 10, 1791, Frascati (Italy) (*DHCJ*, 1:596),

25 Caeiro, *História da expulsão da Companhia de Jesus*, 2:99.

26 *Edital do cardeal patriarca de Lisboa José Manoel da Câmara de Atalaia de 7 de Junho de 1758 em que proíbe os jesuítas de confessar e pregar no seu patriarcado* (Lisbon, 1758): *Collecção dos negócios de Roma*, 1:59; Caeiro, *História da expulsão da Companhia de Jesus*, 2:163; ANTT, SP 3558 (10); ASV, *Segretaria di Stato Portogallo* 114, fol. 53r (in Portuguese); ASV, *Segretaria di Stato Portogallo* 114, fol. 53v (in Italian); ARSI, *Hist. Soc.* 184, fol. 37r.

the attempt until December 19.[27] As a result of the so-called Távora Affair, various members of this noble family were indicted and brutally executed on January 13, 1759 for conspiracy to assassinate the king.[28] In 1759, allegations of Jesuit involvement spread like wildfire, with the order's members being accused of promoting moral laxity and of instigating the conspiracy.[29]

In the aftermath of the royal assassination attempt, Pombal emerged from the shadows and openly directed a number of repressive actions against the Jesuits: the college of Santo Antão, in Lisbon, was invaded in a fruitless search for tobacco and other hidden goods (December 23, 1758); all Jesuits were put under house arrest under penalty of mortal sin (January 12, 1759); their assets were confiscated (January 19, 1759); and, finally, they were prohibited from teaching (June 28, 1759).[30] On September 3, 1759, exactly one year after the assassination attempt, the Jesuits were expelled from Portugal and all its dominions.[31]

Hence, while the visitation had been commissioned by the crown to address the Jesuits' alleged challenges to the implementation of the Treaty of Madrid and the accusations of illegal trading, it was the Society's alleged participation in the regicide conspiracy that actually triggered the expulsion. The 1,480[32]

27 ASV, *Segretaria di Stato Portogallo* 178, fols. 300^{v}–301^{r}.

28 Teresa's closest relatives were indicted and executed for conspiring in the assassination of King José, namely her mother-in-law, Marchioness Leonor de Távora (1700–59), her sibling and father-in-law, Marquis Francisco de Assis de Távora (1703–59), and her husband, Marquis Luís Bernardo de Távora (1723–59). José de Mascarenhas da Silva e Lencastre, duke of Aveiro (1708–59), and Jerónimo de Ataíde, count of Atouguia (1721–59), executed on the same day, were also related to the young marchioness. The latter was married to Leonor and Francisco's eldest daughter, Mariana de Távora (1722–?), and the former was married to one of Teresa's sisters, Leonor Tomásia (1719–c.1761). On the Távora affair, see Monteiro, *D. José: Na sombra de Pombal*, 135–66.

29 On the expulsion of the Jesuits from the Portuguese Empire, see Emanuele Colombo and Niccolò Guasti, "The Expulsion and Suppression in Portugal and Spain: An Overview," in Burson and Wright, *Jesuit Suppression*, 117–38, and António Júlio Trigueiros and Mariagrazia Russo, eds., *I gesuiti dell'assistenza Lusitana esiliati in Italia (1759–1831)* (Padua: Libraria Editrice Universitá di Padova, 2013).

30 Almeida, *História da Igreja em Portugal*, 5:303; ASV, *Segretaria di Stato Portogallo* 115, fols. 106^{r}–109^{v}; ASV, *Segretaria di Stato Portogallo* 116, fols. 22^{r}–33^{v}.

31 *Lei dada para a proscrição, desnaturalização e expulsão dos regulares da Companhia de Jesus nestes reinos e seus domínios* (Lisbon, 1759); *Collecção dos negocios de Roma no reinado de El-Rei Dom José I, ministerio do marquez de Pombal*, 4 vols. (Lisbon: Imprensa Nacional, 1874–75), 1:114–16; ANTT, *Colecção de Leis*, folder 6, document 20; BNP, *Pombalina* 453; ARSI, *Hist. Soc.* 184, fols. 148^{r}–151^{v}.

32 It is difficult to arrive at an accurate and definitive number, since there are some discrepancies between different primary (and secondary) sources. According to António Júlio Trigueiros, in 1759 there were a total of 1,480 Jesuits in the Portuguese assistancy; António Júlio Trigueiros, "*Per nascere poca terra, per morire tutto il mondo*: Il fallimentare

Jesuits of the Portuguese assistancy essentially suffered two different fates during those years: 1,105 were expelled (seventy-five percent) and died in the Papal States and 222 were incarcerated (fifteen percent). Until the dismissal of Pombal, eighty Jesuits died in prison; forty died on the journeys from Brazil and Goa to Lisbon; and five were deported to Africa.[33]

2 The Arguments of the Anti-Jesuit Campaign

Although anti-Jesuitism was born with the very foundation of the Society of Jesus and had a long and convoluted history, it is perhaps safe to say that it reached its zenith in eighteenth-century Portugal.[34] Pombal's direct role in the Jesuits' expulsion from the Portuguese Empire in 1759, his indirect action in their expulsion from other European countries, and his decisive influence in Clement XIV's (1705–74, r.1769–74) suppression of the Society of Jesus in 1773 are well known and well attested. Pombal set up an anti-Jesuit campaign of unprecedented strength, scope, and virulence. Although the full force of this campaign was felt most of all in Portugal, its effects rapidly spread to all of Europe and even to regions outside Europe. By bringing together old and new accusations, and, most of all, by investing immense resources and careful planning in his propaganda campaign, Pombal became a major contributor to the image of the Jesuits as sinful, decadent, ambitious, deceitful, reactionary, and mediocre.

Pombal's violent animosity toward the Jesuits and his campaigning against the Society are well known, but a more detailed inspection of the arguments

progetto di distruzione di una identità," in Trigueiros and Russo, *I gesuiti dell'assistenza Lusitana esiliati in Italia (1759–1831)*, 19–23, here 20. According to Nuno Gonçalves, there were 1,698 Jesuits in the Portuguese territories in 1759; Nuno da Silva Gonçalves, "Pombal e a restauração da Companhia de Jesus," *Brotéria* 174 (2014): 319–38, here 321. *Synopsis historiae Societatis Jesu*, c. 321, which Silva Gonçalves cites, indicates that, in 1749, there were 1,754 Jesuits in Portugal and its overseas missions. And finally, Dauril Alden accounts for 1,760 Jesuits in the Portuguese territories in 1756; Alden, *Making of an Enterprise*, 676. As Trigueiros reached his estimate after identifying all of the expatriated Jesuits, and the work was based on a thorough analysis of the catalogs, 1,480 would appear to be a more conservative approximation to the number of Jesuits in the Portuguese assistancy, at least in 1759. There are several factors that can account for these discrepancies (such as the difficulty in interpreting homonymies, for instance), but this discussion largely exceeds the aim of this chapter.

33 Trigueiros, "Per nascere poca terra," 20.

34 For the history of anti-Jesuitism in the Portuguese Empire, see José Eduardo Franco, *O mito dos jesuítas em Portugal, no Brasil e no Oriente (séculos XVI a XX)* (Lisbon: Gradiva, 2006).

used is required, since, as it turns out, at least one of the accusations enjoyed a surprising longevity and ultimately contributed to shaping the image of the Society in the following centuries. The sophisticated arsenal Pombal deployed against the Society of Jesus included political, diplomatic, and economic actions. The core of his campaigning, however, was a massive publishing offensive. Under his orders and supervision, hundreds of printed works of all sorts condemning the Jesuits and their apostolates (pamphlets, reports, books, treatises, theatrical plays, etc.) were published and distributed all over Europe. This was a gigantic operation, well planned and sustained over roughly two decades, most likely the widest and most ambitious propaganda campaign in Europe up to the eighteenth century.[35]

The most important and most widely diffused books published in this period were the so-called anti-Jesuit catechisms: *Relação abreviada* (1757), *Erros ímpios e sediciosos* (Wicked and seditious offenses [1759]), *Dedução cronológica e analítica* (Chronological and analytical deduction [1767–68]), and *Compêndio histórico do estado da Universidade de Coimbra* (Historical compendium of the state of the University of Coimbra [1771]).[36] These works were translated and widely available in Europe and Asia. *Relação abreviada*, as already mentioned, was translated into French, German, Spanish, and Italian. Besides being translated into French, Spanish, and Italian, *Dedução cronológica e analítica* was also translated into Chinese because Pombal wanted it to be as "universal as the missionary labors of the Jesuits."[37] Anti-Jesuitism was not only vigorously reinforced with these materials; it was also profoundly influenced by Pombal's arguments. This campaign and Pombal's diplomatic efforts were so significant during this period that, as already said, they also played an important part in the Jesuits' expulsion from other European countries and in the Society's universal suppression in 1773.[38]

---

35 Such is the opinion of the authors of the most complete and detailed accounts of Pombal's campaign: José Eduardo Franco and Christine Vogel, "Um acontecimento mediático na Europa das Luzes: A propaganda antijesuítica pombalina em Portugal e na Europa," *Brotéria* 169 (2009): 349–506; Christine Vogel, *Der Untergang der Gesellschaft Jesu als europäisches medienereignis (1758–1773): Publizistische Debatten im Spannungsfeld von Aufklärung und Gegenaufklärung* (Mainz: Verlag Philipp Von Zabern, 2006). Vogel's monograph has been translated into Portuguese; see *Guerra aos jesuítas: A propaganda antijesuítica do marquês de Pombal em Portugal e na Europa*, trans. Cristina Osswald (Lisbon: Temas e Debates–Círculo de Leitroes, 2017).

36 Franco, "Os catecismos antijesuíticos pombalinos," 247–68.

37 Franco and Vogel, "Um acontecimento mediático," 351.

38 Pombal's influence and role in the expulsion of the Jesuits from different countries in Europe and in the suppression of the Society is well documented. For a comparative study of the Jesuit expulsions, see Bertrand Roehner, "Jesuits and the State: A Comparative

The campaign designed by the marquis of Pombal was built upon three major theses. The first claimed that the Jesuits had uncontrolled power, especially overseas, where they were accused of building a powerful and independent Guaraní republic in Portuguese South America. As well as being accused of disobeying the Treaty of Madrid, the Jesuits were also censured for profiting from illegal trading in Brazil. This line of reasoning, which was mostly conveyed by *Relação abreviada*, triggered the nomination of Saldanha da Gama as visitor and reformer of the Portuguese assistancy in 1758. Nevertheless, the accusation of profiting illegally from commerce was not new.[39] In June 1755, the crown established the Companhia do Comércio do Grão-Pará e Maranhão, a state-controlled trade company in the state of Grão-Pará and Maranhão (roughly corresponding to today's northern Brazil). The company was granted a monopoly on exports and sea commerce, and its foundation led to the first episodes of open conflict between Pombal and the Jesuits. The conflict rapidly escalated because—as the Jesuits claimed—the monopoly of the Companhia do Grão-Pará e Maranhão delivered a death blow to the Jesuit missions and colleges in Brazil, the economic survival of which was dependent on the trade in Brazilian goods. It bears remembering that Pombal was deeply involved in this quest, not only because his brother Francisco Xavier was governor of Grão-Pará and Maranhão from 1751 but also because he held shares in the company.[40]

The second thesis was that the Jesuits had been involved in the attempted murder of King José on September 3, 1758. Although the Távoras were ultimately held responsible, the Jesuits were also believed to have been involved in the

Study of Their Expulsions (1590–1990)," *Religion* 27 (1997): 165–82. On the Portuguese case, see also António Júlio Trigueiros, "'O negócio jesuítico' e o papel da política regalista portuguesa," *Brotéria* 169 (2009): 149–67. On the influence of the Pombaline literature on the politicians who expelled the Jesuits from Spain, see Enrique Giménez López, "Portugal y España ante la extinción de los jesuitas," in *Los jesuitas españoles expulsos: Su imagen y su contribución al saber sobre el mundo hispánico en la Europa del siglo XVIII*, ed. Manfred Tietz (Madrid: Iberoamericana, 2010), 337–57; Giménez López, "El antijesuitismo en la España de mediados del siglo XVIII," in *Fénix de España: Modernidad y cultura propia en la España del siglo XVIII (1737–1766); Homenaje a Antonio Mestre Sancís*, ed. P. Fernández Albaladejo (Madrid: Marcial Pons, 2006), 283–326; Mar García Arenas, "La Compañía de Jesús en la *Deducción Cronológica y Analítica* pombalina," *Revista de historia moderna: Anales de la Universidad de Alicante* 21 (2003): 315–48.

39 On the commercial activities of the Jesuits in the Portuguese Empire, see Serafim Leite, *História da Companhia de Jesus no Brasil*, 10 vols. (Lisbon: Livraria Portugália, 1938–50), 4:153–210; Alden, *Making of an Enterprise*, 528–51.

40 The shares had been purchased on behalf of his second wife, Leonor Ernestina, countess of Daun (1721–89); see Domingos Maurício, "Pombal e a Companhia do Grão-Pará e Maranhão," *Brotéria* 80 (1965): 743–50.

plot.[41] As well as triggering the Jesuits' expulsion, this accusation also led to the infamous execution of Malagrida, the confessor of the marchioness of Távora, who was judged to have personally participated in the assassination attempt.[42] This argument was strongly promoted by Pombal in the booklet *Erros ímpios e sediciosos*.[43]

The third thesis was that the Jesuit educational system was profoundly deficient and that the Jesuits were illiterate and scientifically backward. This accusation made explicit a set of critiques of Jesuit education and teaching methods that had been gaining momentum since the early decades of the eighteenth century. Eighteenth-century enlightened philosophers and scientists such as Frei Manuel do Cenáculo (1724–1814), António Ribeiro Sanches (1699–1783), José Anastácio da Cunha (1744–87), and Luís António Verney (1713–92) had also been particularly critical of the Jesuits' teaching methods, which they deemed obsolete.[44] Similar criticism had erupted in other countries in Europe,

---

41 These public executions, carried out with extreme barbarity, sent shock waves through Europe and revealed the darker side of Pombal's personality. See Marquês de Alorna, *As prisões da Junqueira durante o ministério do marquês de Pombal*, 2nd ed. (Lisbon: Typographia Universal, 1882 [1857]); César da Silva, *A execução dos Távoras* (Lisbon: João Romano Torres, 1921); Manuel João Gomes, ed., *O processo dos Távoras: A expulsão dos jesuítas* (Lisbon: Editora Afrodite, 1974). Nuno Gonçalo Monteiro, "Pombal's Government: Between Seventeenth-Century Valido and Enlightened Models," in *Enlightened Reform in Southern Europe and Its Atlantic Colonies, c.1750–1830*, ed. Gabriel Paquette (Farnham: Ashgate, 2009), 321–38.

42 Malagrida was burned at the stake in an *auto de fé* on September 20, 1761. On Malagrida, see Paulo Mury, *História de Gabriel Malagrida da Companhia de Jesus*, trans. Camilo Castelo-Branco (Lisbon: Livraria Editora de Mattos Moreira, 1875); *Processo do padre Gabriel Malagrida*, ANTT, Tribunal do Santo Ofício, Inquisição de Lisboa, proc. 8064. See also footnote 8 for his biographical data.

43 Dom José I de Portugal, *Erros impios e sediciosos, que os religiosos da Companhia de Jesus ensinarão aos reos, que forão justiçados, e pertenderão espalhar nos póvos destes reynos* (Lisbon: Officina de Miguel Rodrigues, impressor do Eminentíssimo Senhor Cardeal Patriarca, 1759); *Collecção dos negócios de Roma*, 1:85–99; ANTT, *Arquivo do Arquivo*, Avisos e Ordens mç. 2, doc. 5; BNP, *Impressos Reservados* RES. 1040//13 A.; ASV, *Segretaria di Stato Portogallo* 115, fols. 145^{r}–161^{v}; ARSI, *Hist. Soc.* 184, fols. 124^{r}–143^{v}.

44 Luís António Verney, *Verdadeiro método de estudar, para ser útil à república, e à igreja: proporcionado ao estilo, e necesidade de Portugal* (Valença: Antonio Balle, 1746) (reprint, Porto: Domingos Barreira, 1984); Manuel do Cenáculo, *Memórias históricas e appendix segundo a disposição da colecção das disposições do superior provincial para a observância e estudos da Congregação da Ordem Terceira de São Francisco* (Lisbon: n.p., 1794); António Ribeiro Sanches, *Cartas sobre a educação da mocidade* (Cologne: n.p., 1760) (reprint, Porto: Domingos Barreira, n.d.); José Anastácio da Cunha, *Notícias literárias de Portugal: 1780*, trans. Joel Serrão (Lisbon: Seara Nova, 1971). *O verdadeiro método de estudar* and Verney's opposition to the Jesuits' teaching methods have been extensively studied in the past. See especially the works by António Alberto Banha de Andrade, Francisco Vieira Pires, José

yet Pombal sharpened what was mostly a vague animosity into a very specific accusation: not only were the Jesuits culturally and scientifically incompetent and obsolete; they had also become the main obstacle to "progress" and to Portugal's scientific development. By cleverly playing with the perceived notion of Portugal's "decadence" from its sixteenth-century imperial grandeur, Pombal identified the Jesuits as the guilty party. He was, thus, brilliantly adjusting to Enlightenment perceptions that equated modernity with scientific progress. The consequence followed with implacable logic: the Jesuits' impressive network of colleges was, after all, the main obstacle to progress and the root of all of Portugal's woes; its extraction from the country was not only necessary but even desired.

The first two theses were the most decisive triggers for the expulsion. The second argument was actually the proximate cause, as the Jesuits were expelled exactly one year after the royal assassination attempt—highly symbolic timing. But these two arguments were also circumstantial, built around a specific set of historical events. It was the third one, broader and more abstract, that turned out to be the more durable and, in a sense, the one that most deeply affected the image of the Society of Jesus in Portugal and conditioned its future.

After the expulsion, Pombal was especially interested in promoting the accusation of Jesuit educational backwardness and scientific illiteracy. With the Society now off stage, he was virtually unhindered in his freedom to develop in painstaking detail a devastating narrative showing that the consequences of the Jesuit presence in Portugal had been disastrous. By doing so, he expected to provide a theoretical background that would justify his decision to expel the Society while at the same time announcing in shining prose his unlimited dedication to the country's progress, modernization, and development. What he could not have predicted was that he was putting in place a critique that would prove to be extremely resilient to the point of affecting the return of the Society a century later.

Pombal's gradual construction of a complex and detailed critique of the Society emerges from a closer inspection of the four major anti-Jesuit works he promoted. *Relação abreviada* and *Erros ímpios e sediciosos* were relatively short and focused works; they did not address educational or cultural issues, much less any scientific ones.[45] After the expulsion, however, Pombal's arguments changed considerably in content, detail, and dimension: *Dedução cronológica e analítica* and *Compêndio histórico* were extremely ambitious works, much

---

Vitorino de Pina Martins, Joaquim Ferreira, Luís Cabral de Moncada, Maria Luísa Borralho, Manuel Curado, and Pedro Calafate.

45 Franco and Vogel, "Um acontecimento mediático," 354–61.

more intricate and longer than the first two anti-Jesuit treatises. *Dedução cronológica e analítica* was a monumental enterprise, originally published in three volumes. With a total of 1,387 pages, it was, by far, the longest, most complex, and most influential piece of anti-Jesuit propaganda ever published in Portugal.[46] Its objective was to provide an ample and detailed picture of the history of Portugal, demonstrating the catastrophic effects of the arrival and the presence of the Society of Jesus from the mid-sixteenth century onward. While the first part dealt with the political history of the Portuguese Empire, and implied that the Jesuits were responsible for the monarchy's institutional decadence, the second part accused them of also being responsible for the decline of the church's institutions.[47]

One of the most pervasive themes put forward was the accusation that the "fateful arrival" of the Jesuits was the single event most responsible for Portugal's eighteenth-century "generalized idiocy."[48] The Jesuits were accused of destroying the best schools—namely the celebrated Colégio das Artes, which had been founded in Coimbra in 1547—and of ruining the University of Coimbra, which they "subdued, prostituted, and darkened."[49] The allegations ran wild, and the Jesuits, who had "no weapons apart from their impostures," were actually accused of establishing a "tyrannical empire" that had been "a terrible scourge for the crown, the arts, the armory, the commerce and the agriculture for more than two centuries in these realms and all their dominions."[50]

Despite being much less extensive and ambitious than *Dedução cronológica e analítica*, *Compêndio histórico* was nevertheless crucial in constructing a specific cultural and educational argument directed against the Jesuits.[51] The book's five hundred pages sought to document and explain the alleged

---

46 Although the work indicates José Seabra da Silva as its author, it is clear today that Carvalho e Melo personally oversaw its production; see Franco and Vogel, "Um acontecimento mediático," 363–64.

47 Franco and Vogel, "Um acontecimento mediático," 361–68.

48 José Seabra da Silva, *Deducão chronológica e analítica na qual se manifestão pela sucessiva serie de cada hum dos reynados da monarquia portuguesa, que decorrerão desde o Governo do Senhor Rey D. João III até o presente, os horrorosos estragos, que a Companhia denominada de Jesus fez em Portugal, e todos os seus domínios por hum plano, e systema por ella inalteravelmente seguido desde que entrou neste Reyno, até que foi delle proscripta, e expulsa pela justa, sabia, e providente Ley de 3 de Setembro de 1759*, 3 vols. (Lisbon: Officina de Miguel Manescal da Costa, 1767–68), 2:iv.

49 *Dedução cronológica e analítca*, 2:v.

50 *Dedução cronológica e analítca*, 1:i.

51 Junta de Providência Literária, *Compendio historico do estado da Universidade de Coimbra no tempo da invasão dos denominados jesuitas e dos estragos feitos nas sciencias e nos professores, e directores que a regiam pelas maquinações, e publicações dos novos estatutos por eles fabricados* (Lisbon: Regia Officina Typographica, 1771).

decadence of the University of Coimbra, the oldest and most important Portuguese university, with the central accusation being that the Jesuits had ruined the university and "buried the Portuguese monarchy in the darkness of ignorance."[52] Throughout *Compêndio histórico*, the Jesuits were accused of having corrupted the teaching of theology, canon and civil law, medicine, physics, chemistry, pharmaceutics, botany, and anatomy.[53]

To counterbalance the anti-Jesuit campaign, Portuguese Jesuits wrote several letters and other apologetic works.[54] Among these accounts, the most detailed were *De exilio provinciae Lusitaniae Societatis Iesu libri quinque* (The expulsion of the Society of Jesus from the Portuguese province) and *De exilio provinciarum transmarinarum assistentiae Lusitanae Societatis Iesu* (The expulsion of the Society of Jesus from the overseas provinces of the Portuguese assistancy), both by José Caeiro and *Historia persecutionis Societatis Iesu in Lusitania* (History of the persecution of the Society of Jesus in Portugal), by Anselmo Eckart (1721–1809).[55] In *De exilio provinciae Lusitaniae,* Caeiro absolved King José while criticizing his minister Pombal, whom he considered responsible for the Jesuits' imprisonment and suppression. For Caeiro, the enthronement of King José on the feast day of Ignatius of Loyola boded well for the future of his reign, but he failed to meet his expectations when he appointed Pombal as his secretary of state. After underlining the national acceptance of the Jesuits and attesting that they were personally esteemed by some members of the royal family, Caeiro described the succession of events that led to the ban of the Society of Jesus, emphasizing the role Pombal played in each of those events. Similarly, *De exilio provinciarum transmarinarum* focused on the ultramarine empire and provided a detailed account of the exile from

52 *Compêndio histórico*, ix.

53 The decadence of the university, and particularly the decline in the teaching of medicine, was attributed to the Jesuits' arrival and the "poisonous root" of "Scholastic physics." *Compêndio histórico*, ix–xii.

54 See, for instance, the following anti-Jesuit "exhortative letter "and its "compulsory response": *Carta exhortatoria aos padres da Companhia de Jesus da provincia de Portugal* (n.p: n.p, 1754?), Biblioteca da Brotéria (hereafter BB), Bib-Controvérsia Pombalina 1 (misc.), and Francisco de Pina e de Melo, *Resposta compulsoria à carta exhortatoria para que se retrate o seu author das calumnias que proferìo contra os reverendissimos padres da Companhia de Jesus da província de Portugal* (Montemor-o-Velho: n.p, 1755), BB, Bib-Controvérsia Pombalina 2 (misc.).

55 Caeiro's book became the most "canonical" description of the expulsion, according to the Jesuits themselves. On José Caeiro, see Júlio de Morais, *Historiador desconhecido: José Caeiro, grande escritor da época Pombalina* (Braga: Livraria Cruz, 1939).

the provinces of Brazil, Maranhão, and Goa.[56] In *Historia persecutionis Societatis Iesu in Lusitania*, Eckart also reported the major events that led to the expulsion of the Jesuits.[57] While describing his captivity at Forte de Almeida and Forte de São Julião da Barra, Eckart reported the annual arrivals of his brethren, the occasional releases, and the successive deaths in confinement, from his imprisonment in 1759 to his release in 1777.[58] In short, the aim of these latter works was to provide a detailed history of the 1759 expulsion, to proclaim its unfairness, and to emphasize the pivotal role played by Pombal. However, as they were poorly distributed until the twentieth century, they proved to be less influential than their authors had anticipated and, most importantly, much less influential than the anti-Jesuit treatises.

3 Pombal's Educational Reform and Its Outcome

After eliminating the Society of Jesus and its schools in Portugal, Pombal proceeded to institute his own educational system. The statutes and teaching

56 José Caeiro, *De exilio provinciarum transmarinarum assistentiae Lusitanae Societatis Iesu*, ARSI, *Lus.* 94–93; *Jesuítas do Brasil e da Índia: Na perseguição do marquês de Pombal (século XVIII)*, trans. Manuel N. Martins (Baía: Escola Tipográfica Salesiana, 1936). Caeiro also wrote an apology in Portuguese that remains unpublished; see José Caeiro and Timóteo de Oliveira, *Apologia da Companhia de Jezus nos reynos e dominios de Portugal: Na qual se mostra evidentemente a sua innocencia e se convencem, se fazem ver com os olhos, e tocar com as mãos as innumeraveis mentiras, falsidades, e calumnias, com que neste reino se maquinou a sua ruina; Obra unicamente dirigida a huma justa defense, para restauração da sua fama, e para evitar no juizo da posteridade o escandalo de tão falsas, como enormess accuzações*, ARSI *Lus.* 95.

57 Eckart's account was translated into Portuguese and published as *Memórias de um jesuíta prisioneiro de Pombal* in 1987 (see footnote 31). *Historia persecutionis Societatis Iesu* was translated into French by Auguste Carayon, S.J. (1813–74) and published in the collection *Documents inédits concernant la Compagnie de Jésus*, 23 vols. (Poitiers: H. Oudin, 1863–86) with the title *Les prisons du marquis de Pombal ministre de S.M. le roi du Portugal (1759–1777)*. It was also translated into Spanish with the title *Historia de la persecución de la Compañía de Jesús*, but this manuscript remained unpublished.

58 It bears remembering that pro-Jesuit works were not exclusive to Jesuit (and Catholic) scholarship. In his history of the 1759 expulsion, for example, the German Protestant Christoph Gottlieb von Murr (1733–1811) was highly critical of Pombal's actions in the execution of the noble family Távora and the expulsion of the Jesuits; see Christoph Gottlieb von Murr, *Geschichte der Jesuiten in Portugal unter der Staatsverwaltung des Marquis von Pombal* (Nuremberg: Felsechrischen Buchhandlung, 1787–88) (reprint João Bautista Hafkemeyer, trans, *Historia dos jesuitas no ministerio do Marques de Pombal: Extrahida de manuscriptos por Christovão Theophilo de Murr* [Porto Alegre: Typographia do Centro, 1923]).

programs of the Pombaline reform unquestionably displayed a genuine desire for scientific and educational modernity, deeply informed by Enlightenment ideals.[59] However, it is also beyond question that any sensible evaluation of Pombal's reformist efforts must balance what was stated in the letter of laws, statues, and curricula with what was in fact realized. Any serious attempt at assessing Pombal's educational policies must therefore go beyond a mere analysis of the normative *corpus* and look into the decisions that were actually taken and the events that really took place. Such an inquiry is frequently not pursued, and historians often seem content to extol the unquestionable modernity of Pombal's official documents and declared intentions.

Before the expulsion in 1759, the Society of Jesus was in charge of one university, in Évora, and thirty-seven colleges in the Portuguese Empire.[60] With regard to pre-university education, that is, the colleges, the first point to note is that Pombal deliberately avoided using the extensive infrastructure of the Jesuit colleges: after 1759, only five of the former Jesuit schools were reused for educational purposes. This clearly signaled a break with the Jesuit past, but it was not a cost-effective decision. The changes in the total number of students are even more surprising. In 1759, the Jesuits were responsible for the education of about twenty thousand pre-university students in Portugal. Precise numbers are perhaps impossible to determine, but this order of magnitude is surely the correct one.[61] In the expulsion's immediate aftermath, and as a historian of education puts it, this number dropped abruptly, creating "an almost

59 There are some fine scholarly studies of Pombal's educational policies. See António Alberto Banha de Andrade, *A reforma pombalina dos estudos secundários (1759–1771): Contribuição para a história da pedagogia em Portugal* (Coimbra: Universidade de Coimbra, 1981); Manuel Antunes, ed., *Como interpretar Pombal? No bicentenário da sua morte* (Lisbon: Brotéria, 1982); Rómulo de Carvalho, *História do ensino em Portugal: Desde a fundação da nacionalidade até ao fim do regime de Salazar-Caetano* (Lisbon: Fundação Calouste Gulbenkian, 1986); *História da Universidade em Portugal*, 3 vols. (Coimbra: Universidade de Coimbra, 1997); Ana Cristina Araújo, ed., *O marquês de Pombal e a Universidade* (Coimbra: Universidade de Coimbra, 2000).

60 In 1759, the Jesuits directed sixteen colleges in Portugal, one in Madeira, three in Azores, twelve in Brazil, one in Angola, three in India, and one in Macao. See Joaquim Veríssimo Serrão, *História de Portugal*, 18 vols. (Lisbon: Verbo, 1977–2010), 6:254. See also Manuel Antunes, "O marquês de Pombal e os jesuítas," in Antunes, *Como interpretar Pombal?*, 25–142; António Júlio Trigueiros, "A expulsão e exílio dos jesuítas de Évora em 1759," in *Universidade de Évora (1559–2009): 450 anos de modernidade educativa*, ed. Sara Marques Pereira and Francisco Lourenço Vaz (Lisbon: Chiado Editora, 2012), 357–78.

61 The estimate was established by António Leite, "Pombal e o ensino secundário," in Antunes, *Como interpretar Pombal?*, 165–81, here 171.

total void in the field of educational activities."[62] The Jesuit school network was not replaced by anything remotely comparable; in absolute number of students, Pombal's secondary education system was only a very small fraction of what the Jesuits had been able to achieve. To put matters into their proper—and bleak—perspective, statistics show that the number of twenty thousand pre-university students was only reached again in Portugal in the 1930s, that is, more than 170 years after the Society had departed.[63] According to the official data, the number of enrollments in *private* and *public* schools was only matched again in 1931. However, it would take four more years for the public educational system to be comparable with the Jesuit educational network in eighteenth-century Portugal, for it was only in 1935 that public schools alone hosted more than twenty thousand students. A more detailed analysis must also take demographic growth into consideration. In 1758, Portugal had a population of around 2,533,000, and by the 1930s this had increased to 6,802,429 inhabitants. Thus, in proportional terms, the population of pre-university students in the 1930s was still very far from its value in 1759. Besides the massive decline in the number of students, historians of education agree that there was also a clear decline in quality, caused by an over-simplification of the courses' contents, a reduction in their length, a reduction in the number of schools where Latin was taught, and finally the substitution of skilled professors with unprepared teachers.[64]

The most emblematic of all the schools created by the Pombaline educational reform was the Real Colégio dos Nobres, a college accepting only young noblemen that was established in 1761 in Lisbon.[65] The college was not only radically class-selective; it was also very small, hosting a maximum of one hundred students. Again, to put matters into perspective, the most important Jesuit college in Lisbon, the Colégio de Santo Antão, usually had more than two thousand students each year, in some periods reaching around three thousand. The modern and ambitious statutes of the Real Colégio dos Nobres determined that, in addition to intense schooling in the humanities (with classes of Latin,

62 Rómulo de Carvalho, "As ciências exactas no tempo de Pombal," in Antunes, *Como interpretar Pombal?*, 215–32, here 215.

63 For more details on the statistical data, see Jorge Ramos do Ó, *Ensino liceal (1836–1975)* (Lisbon: Ministério da Educação, 2009), 40, 45; Teresa Ferreira Rodrigues et al., *História da população portuguesa. Das longas permanências à conquista da modernidade* (Porto: CEPESE–Afrontamento, 2008), 253, 511.

64 Andrade, *A reforma pombalina*, 897.

65 *Estatutos do Collegio Real de Nobres da corte e cidade de Lisboa* (Lisbon: Officina de Miguel Rodrigues, 1761), 775. See also Rómulo de Carvalho, *História da fundação do Colégio Real dos Nobres de Lisboa* (Coimbra: Atlântida–Livraria Editora, 1959).

Greek, French, Italian, English, rhetoric, poetry, etc.), pupils could also opt for training in scientific disciplines. The scientific course, with a rich variety of disciplines (arithmetic, geometry, trigonometry, algebra, infinitesimal analysis, mechanics, hydraulics, optics, astronomy, military and civil architecture, and experimental physics) was, in effect, the truly emblematic novelty of the college. The reality, however, was that it was nothing but an unmitigated failure: in its whole history, only seven students took the scientific disciplines. The leading expert in the history of this college notes the complete "uselessness of scientific education" at the Real Colégio dos Nobres.[66]

As for the reform of the university, the picture is also dramatic. The new statutes of the University of Coimbra (1772) undoubtedly reflected a genuine desire for modernity, higher academic standards, and increased attention to scientific matters. To focus only on the scientific aspects, a new faculty of mathematics was created, the experimental teaching of science was promoted, and modern facilities such as a botanical garden, an astronomical observatory, a physics cabinet, a chemistry laboratory, and a natural history museum were planned. But, once again, a more attentive look at the reality reveals a rather depressing result, as the statistics bluntly show. In the five decades *before* the reform, between 1724 and 1771, an average of 2,827 new students enrolled each year at the University of Coimbra. *After* the reform, in the years between 1772 and 1820, this number dropped to an average of 452 new students per year—a dramatic reduction of eighty-four percent, which explains why some scholars label the phenomenon an educational "catastrophe." Though this is already an impressive figure, it does not directly convey the impact of the extinction of the University of Évora, which the Jesuits managed from 1559 to 1759. Since 320 new students enrolled at the University of Évora in 1758, the total decrease in university enrollments amounts to a staggering eighty-six percent. Any scrupulous comparison between the intentions of Pombal's reform, as stated in its foundational documents and texts, on the one hand, and the stark reality, on the other, cannot fail to notice the shocking imbalance between what was proposed and what was actually achieved. As one of the most distinguished historians of education in Portugal puts it: "The monumentality of the project [i.e., Pombal's educational reform] was rather staged."[67]

Although it is known that rhetorical exaggerations typically characterize Pombal's documents, some explanation must be sought in order to understand the drastic imbalance between programs and reality. First of all, despite the great resources he had at his disposal, Pombal was operating in adverse

66 Carvalho, "As ciências exactas no tempo de Pombal," 217.

67 Carvalho, *História do ensino em Portugal*, 466.

economic circumstances. The country had been severely affected by the earthquake of 1755 and was facing serious financial difficulties that certainly hindered his plans. Second, and perhaps more importantly, Pombal's programmatic texts can easily be misunderstood by the modern reader. Despite his constant rhetoric about modernization, development, and education, Pombal never intended to establish a system of generalized education. On the contrary, to him, as to the Enlightenment intellectuals who had inspired his ideas, education was mostly an elite affair.[68] The foundation of the Real Colégio dos Nobres was based on a doctrine of a Portuguese physician, Ribeiro Sanches, who defended the proposition that only noblemen should receive an education. He believed that, if commoners were allowed to study, some smaller offices would disappear because literate children would not want to pursue their fathers' labors. Hence the apparent "paradox" of Pombal's educational reform—the impulse for modernization coupled with a drastic reduction of students—might not be that paradoxical after all.

Upon her father's death, Queen Maria I (1734–1816, r.1777–1816) immediately barred Pombal from her inner circle and dismissed him and his closest associates from their offices. Thereafter, Maria released the Jesuits incarcerated at the fortresses of Almeida and São Julião da Barra, the surviving members of the Távora family, the bishop of Coimbra, and around eight hundred political prisoners.[69] From 1777 until 1782, the fallen secretary of state resided away from court in his estate in Pombal, where he died on May 8, a few days before turning eighty-three. Despite his political downfall, in one respect Pombal clearly achieved his objectives: after him, the Society's image was hopelessly associated with educational obsolescence, scientific obscurantism, and the obstruction of progress.

---

68 The most important work defending this view is *Cartas sobre a educação da mocidade* (see footnote 44).

69 Maxwell, *Pombal*, 152–54.

CHAPTER 3

Carlos Rademaker and the Restoration of the Society of Jesus in Portugal

On July 21, 1773, Clement XIV issued the brief *Dominus ac Redemptor* suppressing the Society of Jesus throughout the world. Consecutively expelled from Portugal (1759), France (1762–65), Naples, Parma, and Spain (1767), thousands of Jesuits "found themselves in a new, discomfiting category: that of the ex-Jesuit."[1] Following the suppression, some Jesuits continued to operate in Quebec, Prussia, and, especially, in the Russian Empire, where their colleges and missions thrived with the vigorous support of Catherine II (1729–96, r.1762–96) and her successors.[2] On March 7, 1801, less than a year after the election of Pius VII (1742–1823, r.1800–23), the ex-Jesuits living in the Russian Empire regained their ecclesiastical status, as the pontiff gave them permission to "join together and be united in one body or congregation of the Society of Jesus." Granted in recognition of their labors and zeal in the salvation of souls "in an area almost deprived of the Gospel," this consent represented the first step toward the restoration of the Society of Jesus.[3] Three years later, on July 13, 1804, Pius VII extended this permission to the Kingdom of the Two Sicilies and on August 7, 1814—shortly after returning from a five-year period of captivity in France—issued the bull *Sollicitudo omnium ecclesiarum* that restored the Society in "all other states and jurisdictions," a deep desire "since the very beginning of our pontificate."[4] The pontiff granted Superior General Tadeusz Brzozowski (1749–1820, in office 1814–20) all the faculties to freely admit and accept those seeking to live as Jesuits, according to the previously approved Formula of the Institute of 1550, and reinstated the distinctive ministries

1 Jonathan Wright and Jeffrey D. Burson, "Introduction: Towards a New History of the Eighteenth-Century Jesuit Suppression in Global Context," in Wright and Burson, *Jesuit Suppression in Global Context*, 1–10, here 2.

2 Daniel L. Schlafly Jr., "General Suppression, Russian Survival, American Success: The 'Russian' Society of Jesus and the Jesuits in the United States," in Wright and Burson, *Jesuit Suppression in Global Context*, 201–15.

3 "Sollicitudo omnium ecclesiarum," in *Promising Hope: Essays on the Suppression and Restoration of the English Province of the Society of Jesus*, ed. Thomas McCoog (Rome: Institutum Historicum Societatis Iesu, 2003), 323–30, here 327. The bull was translated by Joseph K. Drane, S.J.

4 "Sollicitudo omnium ecclesiarum," 329.

 | DOI:10.1163/9789004382367_004

of the Society. Besides hearing confessions, preaching, and administering sacraments, the Jesuits were given permission to establish new novitiates, seminaries, and colleges. In a continent devastated by the Napoleonic Wars (1803–15), the Society's restoration reflected Pius's aspiration of rebuilding the church and regaining influence in Europe. Aware of the controversial scope of his action, he admonished the "nobility of princes and temporal lords, and also our venerable brother archbishops and bishops, and others in any seat of honor" to accept the Society's restoration and to receive the Jesuits "kindly and, as is becoming, with charity."[5]

Given the idiosyncrasies of the Portuguese case, the crown formally rejected Pius's arguments, and the effective restoration of the Society of Jesus only took place forty-four years after the bull's promulgation. The return of the Jesuits to the Portuguese Empire turned out to be contingent on a series of events, including the liberal wars between 1828 and 1834, the expulsion of the Jesuits from Italy in 1848, and the support of the crown and the aristocracy in the foundation of the first residencies and colleges from 1858 onward. It should also be noted that their presence was tacitly accepted by opposing political parties, as the Jesuits' counsel was sought to solve key diplomatic and educational matters, including the organization of the overseas missions.

1 The Ephemeral French Mission

On September 12, 1814, the state journal *Gazeta de Lisboa* (Lisbon gazette) reported the restoration of the Society of Jesus by publishing in vernacular the bull of the restoration.[6] The official announcement, which was publicly acclaimed in Lisbon by the apostolic delegate Vincenzo Macchi (1770–1860), the highest ranks of the nobility, the bourgeoisie, and the common people, soon reached Brazil, where the Portuguese court and the royal family had been based since 1808.[7] Less than a year after its promulgation, the secretary of state for foreign affairs issued a formal response from Rio de Janeiro rejecting the

5 "Sollicitudo omnium ecclesiarum," 329–30.

6 *Gazeta de Lisboa* 215 (September 12, 1814).

7 Acácio Casimiro, "O govêrno de D. João VI e a restauração da Companhia de Jesus (1814–1815)," *Brotéria* 31 (1940): 469–75, here 470. In the aftermath of the first Napoleonic invasion, in November 1807, the court and the royal family fled to Brazil, where they resided from 1808 to 1821. On the history of the Napoleonic invasions, see Rui Ramos, "Invasões francesas, tutela inglesa e monarquia brasileira (1807–1820)," in Ramos, Vasconcelos e Sousa, Monteiro, *História de Portugal*, 439–56.

bull.[8] The notice emphasized the astonishment felt upon its publication and the intransigence in admitting any negotiations on the national restoration of the Society of Jesus.[9] As the law of expulsion of 1759 and the brief of suppression of 1773 were deemed to be in full force, the surviving Jesuits from the old Portuguese assistancy were prevented from returning to Portugal.[10]

By the 1820s, the political context had changed considerably. In 1821, after thirteen years of living in Brazil, King João VI (1767–1826, r.1816–22) was compelled by the courts to return to Portugal, where he solemnly swore in the constitution that would put an end to the absolute regime. Despite staying in Rio de Janeiro to serve as regent, Pedro de Alcântara (1798–1834), the heir to the throne, soon embraced the cause of independence and became the first emperor of Brazil (r.1822–31). Upon his father's death on March 10, 1826, Pedro also became the sovereign of Portugal. But as the union of the two kingdoms was neither desirable nor possible, he soon abdicated in favor of his daughter Maria (1819–53, r.1826–28 and 1834–53) and promised her hand to his younger brother Miguel (1802–66, r.1828–34). Exiled in Vienna since 1824, Miguel returned to Portugal and, contrary to what had been agreed, was crowned king on July 7, 1828. He then rejected the Constitutional Charter of 1826—which his brother Pedro had written—and reinstated an absolute regime, thus beginning a six-year civil war.[11]

It was in this particular political context that the Jesuits returned to Portugal. A few months after being crowned, King Miguel asked Nuno Álvares Pereira de Melo, duke of Cadaval (1799–1837), to bring the Jesuits back to restore their colleges and ministries.[12] On March 15, 1829, the Portuguese mission was officially founded as a branch of the French province, with Philippe Delvaux (1787–1865) appointed its superior.[13] Three years after their return, the Jesuits regained the Colégio das Artes in Coimbra (January 9, 1832) and the Colégio do Espírito Santo in Évora (September 10, 1832). However, with the arrival of the

8 "Officio do marquez de Aguiar para José Manuel Pinto de Sousa, ministro em Roma (1 de Abril de 1815)," in *Supplemento á collecção dos tratados, convenções, contratos e actos publicos celebrados entre a corôa de Portugal e as mais potencias desde 1640*, ed. Júlio Firmino Júdice Biker (Lisbon: Imprensa Nacional, 1879), 18:76–77.

9 "Officio do marquez de Aguiar," 76.

10 Out of the 1,100 Jesuits exiled to Italy in 1759, roughly thirty were still alive in 1814. See Gonçalves, "Pombal e a restauração da Companhia de Jesus," 329.

11 On the history of the civil war, see Rui Ramos, "Ruptura constitucional e guerra civil (1820–1834)," in Ramos, Vasconcelos e Sousa, Monteiro, *História de Portugal*, 457–89.

12 Francisca Branco Veiga, "O breve regresso da Companhia de Jesus no reinado de D. Miguel (1829–1834)," *Brotéria* 179 (2014): 387–400, here 390.

13 Philippe Delvaux: b. February 1, 1787, Rochefort (Belgium); e. July 20, 1819, Montrouge (France); d. February 21, 1865, Quimper (France) (*DHCJ*, 2:1080).

liberal troops on July 24, 1833, the seventeen French Jesuits working in Coimbra were imprisoned and were only released after eleven months in captivity due to the intervention of the French ambassador. Whereas the Jesuits living in Lisbon, including Delvaux, were deported to England and Italy in August 1833, the Jesuits of Coimbra only boarded to Genoa on July 7, 1834.[14] During their ephemeral stay in Portugal, the Jesuits were accused by their opponents of being ultramontane, conservative, and obsolete clerics, and of being active supporters of the *ancien régime*.[15]

On May 30, 1834, in the immediate aftermath of the civil war, the crown banned the male religious orders and confiscated their convents, monasteries, and other properties.[16] The influence of the clergy decreased significantly in the following years. The regular orders, which represented around forty percent of the clergy, simply disappeared. And between 1820 and 1840, the number of secular ministers decreased from twenty-four thousand to the staggering figure of ten thousand.[17] Beginning in 1834, the government micromanaged every aspect related to the organization of the church, from the functioning of lay congregations to the nomination of priests. As for the nobility, the divide between both factions increased with the ban of thirty-one conservative peers from the upper house of Parliament, the Câmara dos Pares do Reino, in 1826. The substitution of the old order by a new elite—constituted by free thinkers, liberal officers, and moguls of industry and commerce—accompanied economic, political, and religious turmoil in the 1830s and 1840s.[18]

---

14 In that same year, an anti-Jesuit booklet printed in Rio de Janeiro foretold the epilogue of the French mission: F.E.A.V., *Manifestação dos crimes e atentados cometidos pelos jesuítas: Em todas as partes do mundo, desde a sua fundação até à sua extinção* (Rio de Janeiro: Typ. de Miranda & Carneiro, 1833). On the establishment of the French mission, see also Auguste Carayon, *Lettres inédites du R.P. Joseph Delvaux sur le rétablissement des jésuites en Portugal: 1829–1834* (Paris: L'Écureux, 1866).

15 José Eduardo Franco, "Da extinção à restauração dos jesuítas: Construção e reprodução do mito dos jesuítas de matriz pombalina no século XIX," *Brotéria* 179 (2014): 401–28.

16 Written by the minister of justice, Joaquim António de Aguiar (1792–1884), and signed by King Pedro I/IV (1798–1834, r.1822–31/r.1826), the bill stipulating the expulsion extinguished all male religious orders immediately and the female orders conditionally. Prohibited from accepting novices, female orders were supposed to be extinguished when the last nun died. In fact, as most orders secretly continued to accept novices, female orders were not officially expelled from Portugal until 1910.

17 Ramos, "A revolução liberal (1834–1851)," in Ramos, Vasconcelos e Sousa, Monteiro, *História de Portugal*, 491–519, here 495.

18 The tobacco monopoly played a significant role in the rise of a new political and economic elite. See Maria Filomena Mónica, "Negócios e política: Os tabacos (1800–1890)," *Análise social* 27 (1992): 461–79.

Some stability, however, was achieved in the 1850s. Following the Saint-Simonian program of António Fontes Pereira de Melo (1819–87), improvements to the country's infrastructure (including the building of railways and new roads, and the heavy investments in telegraphy), a rise in industrial and commercial activities, and increasing levels of productivity in the primary sector led to steady economic growth.[19] Yet even though the economy grew forty to sixty percent between 1850 and 1910, the literacy rate remained incredibly low (twenty-one percent), especially when compared with other European countries.[20] Thus, given their regular commitment to education, the gradual return of the religious orders in the 1850s seemed to represent an opportunity to correct this trend.

2 Carlos Rademaker and the Restoration of the Portuguese Province

The Society of Jesus was effectively restored in Portugal in 1858 by Carlos João Rademaker, S.J. (1828–85) (figure 1).[21] Born to José Basílio Rademaker (1789–1856) and Charlotte Lecusson Verdier (dates unknown) in Lisbon, Carlos spent his childhood in Italy, where his father served as minister plenipotentiary to King Miguel.[22] In Chieri, a small city in the outskirts of Turin, Carlos learned the first letters with his father and private tutors. Starting in 1836, he studied rhetoric in Turin with Giovanni Bosco (dates unknown)—a relative of the founder of the Salesians—and in 1842 he was awarded a commendation for his excellent rhetorical skills. Between 1842 and 1846, he studied philosophy and law in Genoa and Turin and, despite his father's disapproval, entered the Society of Jesus on October 28, 1846. In 1848, after nineteen years living in Italy, the sudden expulsion of the Jesuits from Turin forced Carlos to leave the novitiate and to return to Portugal with his family. As his desire to be a Jesuit persisted,

19 Fontes Pereira de Melo was arguably the most influential politician in Portugal between 1851 and 1890. He held public offices in several governments and was the man responsible for the industrial modernization of the country. He was the first minister of public works after the political movement called the regeneration (1851–56) and was appointed prime minister on three occasions (1871–77, 1878–79, 1881–86). For a biography, see Maria Filomena Mónica, *Fontes Pereira de Melo* (Lisbon: Alêtheia, 2009).

20 Ramos, "A regeneração e o Fontismo (1851–1890)," in Ramos, Vasconcelos e Sousa, Monteiro, *História de Portugal*, 521–48, here 526.

21 Carlos João Rademaker: b. July 6, 1818, Lisbon (Portugal); e. October 26, 1846, Chieri (Italy); d. June 6, 1885, Lisbon (Portugal) (*DHCJ*, 4:3276; *Sommervogel*, 6:1370–1).

22 For a biography of Carlos Rademaker, see Nuno Olaio, "Carlos João Rademaker (1828–1885): Percurso do restaurador da Companhia de Jesus em Portugal," *Lusitania sacra* 12 (2002): 65–119.

FIGURE 1 Carlos Rademaker, S.J. (1828–85)
APSI

he wrote to Provincial Francesco Pellico (1802–84) asking him to continue his studies in the Society.[23] In his reply, Pellico suggested that Rademaker should postpone his decision and invested his confessor with the authority to dismiss him from his vows, if he wished.[24]

Back in Portugal, Rademaker continued his correspondence with Italian Jesuits in the hope of being readmitted to the Society. During this period, the encouragement of Giuseppe Maria Vigitello (1799–1859), an Italian Jesuit exiled in Lisbon, and the support of the Lavradio family were crucial not only for his ordination but also for the restoration of the Portuguese province.[25] António de Almeida Portugal, marquis of Lavradio (1794–1874), had been aide-de-camp to King Miguel during the civil war and was a close friend of the Rademaker family. Companion in exile to José Rademaker, with whom he had embarked in Genoa in 1848, Lavradio accommodated Vigitello in Lisbon and probably introduced him to young Carlos. Following Vigitello's counsels, Carlos studied theology and was finally ordained a priest on September 20, 1851, three years after leaving the novitiate in Turin. Due to Vigitello's diplomatic efforts, Rademaker was admitted to the province of Turin in 1851 and, four years later, joined the Spanish province.

In 1856, after his father's passing, Rademaker invested part of his inheritance in the Instituto da Caridade, a charitable institution he had been co-directing since 1853,[26] because he viewed it as a means to restore the Society, as he told Provincial Domingo Olascoaga (1808–60) in September 1856.[27] As well as enjoying a good reputation in Lisbon, the Instituto da Caridade already had two prospective collaborators, two Portuguese priests who had manifested the desire to join the Society and were willing to help Rademaker in the education of orphans and in the foundation of new pious works. If the Spanish provincial could spare two Jesuits, he argued, the restoration of the Society in Portugal would finally take place, for he had the official support of the British consul and, more importantly, of the governor of Lisbon.[28] Although he was a liberal

23 Francesco Pellico: b. February 12, 1802, Saluzzo (Italy); e. November 12, 1834, Chieri (Italy); d. April 29, 1884, Chieri (Italy) (*Mendizábal*, 5948; *Sommervogel*, 6:455).

24 Olaio, "Carlos João Rademaker," 71.

25 Giuseppe Maria Vigitello: b. March 17, 1799, Turin (Italy); e. November 5, 1816, Venice (Italy); d. November 21, 1859, Turin (Italy) (*Mendizábal*, 2416; *Sommervogel*, 7:744–45).

26 His father left him the considerable amount of twelve million *réis*. See Manuel Borges Grainha, *História do colégio de Campolide da Companhia de Jesus* (Coimbra: Imprensa da Universidade, 1913), 10. In 2019, that would roughly correspond to €270,000.

27 Domingo Olascoaga: b. January 16, 1808, Vidania, Guipúzcoa (Spain); e. July 30, 1825, Spain; d. December 25, 1860, Loyola (Spain) (*Mendizábal*, 2557).

28 Olaio, "Carlos João Rademaker," 79.

war hero, Luís de Mello Breyner (1807–76), the count of Sobral and governor of Lisbon (in office 1851–52 and 1856–58), personally supported the restoration of Society. In a period when the gap between absolutists and liberals was still sharp, this support seemed puzzling. Since he was Lavradio's nephew, it is highly probable that the contact between Sobral and Rademaker was mediated by Lavradio himself.

In 1857, eleven years after being admitted as a novice, Rademaker traveled to Loyola, where he professed his first vows as a Jesuit, and then to Madrid to seek permission to restore the Society of Jesus in Portugal.[29] With the official consent of Superior General Peter Jan Beckx (1795–1887, in office 1853–87), he began searching for a suitable location to transfer the Instituto da Caridade and to reinstate the Portuguese province.[30] It did not take him long to find such a place. In Campolide, in the outskirts of Lisbon, there was a considerable estate for sale, the Quinta da Torre. As he wrongly thought that it had once belonged to the Society of Jesus before the 1759 expulsion, he contacted its owner, the famous poet João de Lemos (1819–90), and invested one-third of his father's inheritance in its acquisition (four million *réis*).[31]

Although it was officially founded on the feast of Luigi Gonzaga (1568–91), patron saint of youth (June 21, 1858), the Colégio de Campolide was dedicated to the Immaculate Conception (figure 2). The meaning of this dedication was twofold. First, the dogma of the Immaculate Conception was relatively recent. While it had been part of the Catholic tradition for centuries, it was only in 1854 that Pius IX (1792–1878, r.1846–78) defined it as a dogma of the Catholic faith. Second, the kingdom of Portugal had been dedicated to the Immaculate Conception by King João IV (1604–56, r.1640–56) in 1640. After regaining independence from Spanish rule, King João placed his crown before a statue of Our Lady of the Immaculate Conception and proclaimed her the true queen of Portugal. From that moment on, the Portuguese sovereigns were never crowned again as a way of paying homage to the spiritual queen of Portugal. By dedicating the first college of the restored Society to the Immaculate Conception, Rademaker seemed to be adopting a clear strategy to please both the Holy See and the crown. Given the order's convoluted history in Portugal, such an option was probably carefully planned.

---

29 Ambrósio Pina, *Carlos Rademaker, 1828–1885: Restaurador dos jesuítas em Portugal no século XIX* (Porto: Livraria Apostolado da Imprensa, 1967), 165.

30 Peter Jan Beckx: b. February 8, 1795, Zichem (Belgium); e. October 29, 1819, Hildesheim (Germany); d. March 4, 1887, Rome (Italy) (*DHCJ*, 1:1671–75).

31 Grainha, *História do colégio de Campolide*, XIV. In 2018, four million *réis* would roughly correspond to €90,000.

FIGURE 2 Colégio de Campolide in 1908
APSI

After the foundation of the Colégio de Campolide, the Jesuits were officially restored in Portugal, first as a mission of the Spanish province (1858–79) and then as an independent province (1880–1910). During the first years following the restoration, the creation of new institutions, apostolates, and missions owed much to the support of the aristocracy, the government, and the royal family. In June 1860, José de Menezes da Silveira e Castro, marquis of Valada (1826–95), bought a property in Barro, Torres Vedras, for four hundred thousand *réis* and offered it to the Society. As well as establishing a new college for the orphans of the former Instituto da Caridade, the Colégio de Nossa Senhora dos Anjos, Rademaker also founded a novitiate in the new property.[32] Once again, the friendship between the Jesuit and a highborn aristocrat was probably mediated by Lavradio. Besides being his nephew, Valada was also Lavradio's brother-in-law, for the marchioness of Lavradio was Valada's half-sister. But while the support of Lavradio and his family was perhaps to be expected, given the political and religious ties that connected him and Rademaker, the support of liberal aristocrats was utterly surprising.

In contrast to what had happened with the colleges, the Jesuits were granted explicit authorization from the government of the marquis of Loulé (1804–75) to establish a novitiate. Loulé had been aide-de-camp to King Pedro in the civil war and was, by marriage, great-uncle to King Pedro V (1837–61,

32 Olaio, "Carlos João Rademaker," 80–82.

r.1853–61). Nominated prime minister three times (in office 1856–59, 1860–45, and 1869–70) and leader of the Partido Histórico (Historical party), Loulé was one of the most influential liberal politicians of the 1850s and 1860s. His support in the foundation of a novitiate was, therefore, extremely significant. In the 1860s, the political situation and, particularly, the government, seemed to be favorable to the Jesuits. Unlike the Daughters of Charity of Saint Vincent of Paul—expelled in 1862 after four years of political debate over the legitimacy of their presence—the Jesuits "had nothing to fear," for even though the government was "completely progressive," Rademaker considered Loulé "a gentleman of good qualities."[33] In fact, shortly after the expulsion of the Daughters of Charity, the Jesuits were asked to take over a small college for the education of orphans previously directed by this female congregation. Established in 1852 in Louriçal do Campo, in the outskirts of Castelo Branco by the Franciscan friar Agostinho da Anunciação (1808–74), the Colégio de São Fiel was directed by the Daughters of Charity until 1862 (figure 3). After the expulsion, Anunciação and Isabel Maria de Bragança (1801–76), renowned benefactor of the college and regent of Portugal (r.1826–28), traveled to Rome and obtained official permission to hand the college to the Portuguese Jesuits.

As in the old Society, the popularity of the Jesuits among the aristocracy was crucial for their restoration. One of the most important expressions of Rademaker's popularity in this period was arguably an invitation from King Pedro V himself. Given the Jesuit's talent, the monarch invited him to preach the sermon of Good Friday in the royal chapel on March 21, 1861.[34] With the explicit approval of the government and the tacit endorsement of the sovereign, Rademaker's acceptance among the more liberal sectors of the aristocracy was gaining impetus. After preaching the sermon in the royal chapel, he was invited by José Mascarenhas, marquis of Fronteira (1802–81), to preach the eulogy of his brother Carlos(1803–61) in the Church of São Domingos de Benfica.[35] Throughout the sermon, Rademaker often referred to the piety of

33 Joaquim Campo Santo, "Padre Carlos Rademaker," *Novo mensageiro do Coração de Jesus* 23 (1903): 409–10. On the expulsion of the Daughters of Saint Vincent of Paul in 1862, see Artur Vilares, *As congregações religiosas em Portugal, 1901–1926* (Lisbon: Fundação Calouste Gulbenkian-Fundação para a Ciência e Tecnologia, 2003).

34 Olaio, "Carlos João Rademaker," 90.

35 Peer of the realm and aide-de-camp to King Pedro V, Carlos Mascarenhas (1803–61) had been an important military officer in the civil war. Carlos Rademaker, *O discurso fúnebre nas solenes exéquias do Exmo. Sr. D. Carlos Mascarenhas* (Lisbon: Imprensa Nacional, 1861) (reprinted in Marquês de Fronteira, *Memórias do marquês de Fronteira e d'Alorna D. José Trazimundo Mascarenhas Barreto ditadas por ele próprio em 1861* [Coimbra: Imprensa da Universidade, 1928–32], 5:348–60).

FIGURE 3 Colégio de São Fiel in 1903
APSI

Carlos, who had passed away on May 3, 1861. As well as relating his heroic deeds to the "faith in the religion of his ancestors," Rademaker argued that "the nobleness of the Catholic spirit" had always inspired Mascarenhas's soul. As his intimate friend (and probable confessor), the Jesuit recalled that the liberal war hero repeatedly told him that there was "no happiness without religion and faith."[36] Giving credit to his rhetorical skills, however, Rademaker saved the best for last. When referring to his deathbed, the Jesuit compared Mascarenhas to his great-grandfather, the marquis of Alorna (1726–1802), who had been held in captivity for eighteen years owing to his ties with the Távora family. Just like his "unfortunate grandfather" in the dungeons of Junqueira, Rademaker argued, Mascarenhas realized that he would only find comfort in his faith.[37] Subtly, the Jesuit ended his sermon with a reference to Pombal and the Távora Affair, the event that triggered the 1759 expulsion. The sermon was nonetheless well received by the liberal aristocracy, and Rademaker was invited to preach a more important eulogy a few months later, the eulogy of King Pedro V in November 1861.[38]

36 Rademaker, *O discurso fúnebre*, 357–58.
37 Rademaker, *O discurso fúnebre*, 358.
38 Olaio, "Carlos João Rademaker," 91.

Besides having regained, at least in part, their popularity among the liberal and conservative elites, the restored Jesuits in Portugal were sought by the crown to solve the problems over the Padroado do Oriente. On February 21, 1857, after decades of chaotic functioning and desertification, the Holy See established the borders and determined the organization of the Padroado, entrusted to the Portuguese crown since the fifteenth century.[39] The concordat delegated to the Portuguese government the task of recruiting and sending ministers to the East Asian missions. As he had not yet persuaded any other priests to become missionaries, the minister of the navy Carlos Bento da Silva (1812–91) turned to Rademaker and asked him to send Jesuits to the Portuguese missions. In 1861, with Rademaker's consent, the Jesuits then directed the most important institution in the formation of missionaries since the turn of the eighteenth century, the Real Colégio das Missões Ultramarinas in Cernache do Bonjardim.[40] One year later, the royal college sent the first group of Jesuit missionaries to Macao to direct the Seminário de São José and a school for external students. The missions in East Asia were thriving once again, and, in 1881, the Jesuits turned to Africa, establishing several missions in Zambésia (Mozambique), where they stayed until 1910.[41]

Although the Jesuits' return seemed to have been met with a reasonable degree of social peace, especially considering the 1759 and 1834 expulsions, there were also some dissident voices. Manuel Borges Grainha (1862–1925), a former student of the Jesuits and perhaps the most vociferous anti-Jesuit opponent in this period, was especially critical of successive governments for implicitly, and sometimes explicitly, approving the Society's restoration and for entrusting the overseas missions to the order.[42] Even though Loulé was ultimately

39 For the history of the Padroado do Oriente, the work by António Rego is still useful. See António Rego, *O Padroado português do Oriente: Esboço histórico* (Lisbon: Agência Geral das Colónias, 1940). For a more recent account, see Manuel Braga da Cruz, "O Padroado português no Oriente," *Didaskalia* 33 (2003): 239–55.

40 Olaio, "Carlos João Rademaker," 91–92.

41 Francisco Augusto Correia, *O método missionário dos jesuítas em Moçambique, 1881–1910: Um contributo para a história da missão da Zambésia* (Braga: Livraria Apostolado da Imprensa, 1992).

42 Among other works, Borges Grainha published the *História do colégio da Companhia de Jesus*, which remains one of the most important sources on the history of this Jesuit college. After the Jesuits' expulsion in 1910, he recollected and translated into vernacular the *annuae* and *triannuae litterae* found in Campolide. With the objective of further tainting the image of the college, he published this book in 1913. Besides including the *litterae* from 1858 to 1910, he also added an extensive preface (76pp.) where he criticized the Jesuits, their relationship with the Lisbon elites, and their teaching methods. Whereas the *litterae* allow for a historical reconstruction of the college, the preface is one of the most vivid examples of the anticlerical and anti-Jesuit milieu in early twentieth-century Portugal.

responsible for their return, Borges Grainha also blamed the ministers and secretaries of state that governed in the following years, including Prime Minister Fontes Pereira de Melo (in office 1871–77, 1878–79, and 1881–86), Minister of Foreign Affairs José Mendes Leal (1820–86, in office 1869), and Prime Minister João Carlos Saldanha de Oliveira Daun (1790–1876, in office 1835, 1846–49, 1851–56, and 1870).[43] Whereas the critique of the more conservative Pereira de Melo was to be expected, the inclusion of Saldanha and Mendes Leal—two staunch freemasons and anticlerical politicians—in Borges Grainha's charges was somewhat surprising. As well as adding a deeper level of complexity to the political and religious context, the critique put forward by Borges Grainha also revealed how challenging it was to be publicly opposed to the return of a religious order that was resourcefully solving the diplomatic and political problems of the day.

In 1862, having succeeded in the restoration of the Society in Portugal, Rademaker embarked on a journey to Spain and Italy to conclude his studies and religious formation. After studying theology for a year in León (1862–63), he went to Rome for his tertianship (1863–64). With an atypical formation for a Jesuit in this period, he would only profess the final vows in the Society on August 15, 1868, that is, twenty-two years after being accepted as a novice in Chieri. After returning to Portugal in 1864, Rademaker became progressively involved in the popular missions in the regions of Algarve, Alentejo, Estremadura, Douro, Minho, Beira Alta, and Trás-os-Montes.[44] In the middle of an anticlerical political campaign in Porto, in May 1872, the press censured him for establishing lay congregations in several cities and villages, and for persuading young women to become nuns. Given the delicate state of affairs, his superiors asked him to leave the country. During his exile for thirteen years in Spain, Italy, and France (1872–84), he taught at Jesuit colleges, gave the Spiritual Exercises, and organized several popular missions, his principal ministry after the restoration. Poor health eventually compelled him to return to Campolide in 1884, where he passed away on June 6, 1885.[45] Despite all the controversies, when Rademaker died the Society of Jesus seemed to be flourishing in Portugal. Twenty-seven years after the foundation of the Colégio de Campolide, the Portuguese province was constituted by 161 Jesuits, including forty-seven priests, sixty-nine scholastics, and forty-five lay brothers.[46]

43 Grainha, *História do colégio de Campolide*, XXXVIII–IX.
44 Olaio, "Carlos João Rademaker," 93–95.
45 Olaio, "Carlos João Rademaker," 102–5.
46 See Table 1 in the appendix.

3 The Jesuit Colleges and the Rise of Anticlericalism

The first years after the foundation of the Colégio de Campolide were crucial for the restoration of the Society of Jesus and for the formation of the first generation of Jesuits. Shortly after the official opening of the college, in June 1858, there were already four candidates who wanted to be admitted into the Society. As superior of the Portuguese mission, Rademaker sent them to novitiates in Loyola, Spain, and Toulouse, France.[47] As the first Jesuit institution established in nineteenth-century Portugal, Campolide soon became a popular boarding school for the Lisbon elites. "Following the affections of the royal family, high clergy, party chairs, and state bureaucrats," in Borges Grainha's words, aristocrats and the bourgeoisie "seeking ennoblement" sent their children to Campolide regardless of their ideology or political stance.[48] The student body of this Jesuit college was thus composed of both liberal and conservative aristocrats, as well as the offspring of distinguished anticlerical politicians, republicans, and freemasons. Intriguingly, in 1910, when the Jesuits were expelled for the third time from Portugal and the overseas colonies, their opponents accused them of secretly establishing colleges for the education of the nobility and the bourgeoisie.[49]

In 1862, while Rademaker was finishing his theological studies, John Meagher (1823–1901) was nominated superior of the Portuguese mission.[50] Son to Jeremiah Meagher (dates unknown), British vice-consul in Lisbon since 1821, John governed the Portuguese mission until the arrival of Francis Xavier Fulconis (1814–80) in 1863.[51] Under Fulconis's government (1863–80), the Jesuit colleges continued to prosper. During the 1870s and 1880s, the elites continued to support the Jesuits. When they started to build a church in Campolide (1879–84), the Jesuits managed to have more than half of their expenses paid by prominent aristocrats, including the count of Mesquitella, the duchess of Cadaval, the count of Praia e Monforte, and the marquis and marchioness of Monfalim.[52]

---

47 Grainha, *História do colégio de Campolide*, 11.

48 Grainha, *História do colégio de Campolide*, xli–xlii.

49 For more details on the rise of anti-Jesuitism in nineteenth-century Portugal, see Chapter 5.

50 John Meagher: b. May 5, 1823, Lisbon (Portugal); e. September 7, 1842, England; d. March 21, 1901, Stonyhurst (England) (*Mendizábal*, 9927).

51 Francis Xavier Fulconis: b. November 14, 1814, L'Escarène (France); e. November 19, 1830, Rome (Italy); d. April 21, 1880, Nice (France) (*Mendizábal*, 5227).

52 According to the *História do colégio de Campolide*, the main benefactors were Josefina Bertrand (four million *réis*), the count of Mesquitella (three million *réis*), the count of Praia e Monforte (1,500,000 *réis*), the duchess of Cadaval (1,680,000 *réis*), the marquis and

The reputation of the colleges in this period played a vital part in the restoration of the Society, for even renowned anti-Jesuit politicians "did not dare to say anything against the college [of Campolide]," according to the *litterae annuae*.[53] During these years, the rector and other professors of a public high school in Lisbon, the Liceu de Lisboa, praised the Jesuits' teaching methods, "especially in physics and mathematics."[54] The increasing popularity of the Jesuits and their pedagogy led to a steady growth of the student body in Campolide from forty-three students in 1865 to 160 in 1880.[55] The inversion of this trend in the 1880s coincided with the centennial celebration of Pombal's death in the Parliament (1882) and with the subsequent rise of anti-Jesuitism in the public sphere.[56] Besides organizing conferences and rallies, influential anti-Jesuits promoted the publication of weekly periodicals exclusively dedicated to diffusing their anticlerical ideologies and countering the Society's restoration in Portugal.[57] Yet despite this campaign, during the 1890s, enrollment at Campolide practically tripled, reaching a staggering total of 333 students in 1897.[58] This surge, however, soon came to an end. In the 1900–1 school year, there was again a decline of the student body in Campolide. According to the annual letters sent to Rome, the explanation for having only 250 students was threefold: the Jesuits blamed "the envy of our enemies," "the unfortunate outcome of the official exams," and, especially, "the tempest against the religious orders."[59]

Although they did not mention it explicitly, the Jesuits were referring to the most controversial religious matter of the day, the so-called Calmon Affair. On February 17, 1901, the Brazilian consul José Calmon (1839–1913, in office 1892–1901) and his daughter Rosa (*c.*1866–?) were the protagonists of a scandal in the Church of Trindade in Porto. Rosa had aspired to become a Sister of

marchioness of Monfalim (one million *réis*), the Tavares family (nine hundred thousand *réis*), António Menéres (nine hundred thousand *réis*), and Mr. Mariz Sarmento (four hundred thousand *réis*). Considering that some donors remained anonymous, this means that the Jesuits secured at least 13,380,000 *réis* for building the church. As the total cost was twenty-three million *réis*, the donations represented at least fifty-eight percent of the investment. See Grainha, *História do colégio de Campolide*, xliv.

53 Grainha, *História do colégio de Campolide*, 56.

54 Grainha, *História do colégio de Campolide*, 59.

55 See Table 2 in the appendix.

56 For more details, see Chapter 5.

57 The most important bulletins published in this period were: *A semana de Loyola* (1884–86), *O anti-jesuíta* (1894–95), *O liberal de Gaya* (1899–1900), and *O jesuíta* (1901). For more details, see, especially, António Araújo, *Jesuítas e antijesuítas no Portugal republicano* (Lisbon: Roma Editora, 2004), 10.

58 See Table 2 in the appendix.

59 Grainha, *História do colégio de Campolide*, 125.

Saint Dorothy since 1899, but Calmon was firmly opposed to the religious vocation of his daughter. On that day, Calmon shouted that Rosa was being kidnapped and requested help, claiming that she was grabbing the lateral grate of the church and was refusing to leave with her parents. In reality, however, Rosa had in fact had been held in captivity by her family for the past two years. After reaching out to her father's hierarchical superior in Lisbon in October 1899, she tried to escape her family's house in 1900, but police officers caught her at a train station near Porto and forced her to return home. From that moment on, Calmon's family home was permanently guarded by law officers. Fearing that his daughter might try to escape again, José asked prominent physicians to testify that Rosa suffered from a mental disorder—"a religious passion"—so that he could legally forbid her from becoming a nun.

The case soon left the legal sphere and entered into the public and diplomatic spheres. Whereas, according to Portuguese law, the consul could not prohibit his thirty-two-year-old daughter from becoming a nun, Calmon argued that as he was a Brazilian diplomat he was not subject to the Portuguese Constitution.[60] In early March, as the controversy quickly escalated, the Brazilian government summoned Calmon and transferred him (and his family) to a new diplomatic post.

After the turmoil in the Church of Trindade, the division between both factions became even sharper, resulting in a series of anticlerical revolts throughout the country. Pressured by the rise of anticlericalism in the Parliament and the press, the government was forced to take action. Since it was no longer possible to ignore the gradual return of the religious orders, the government asked civil governors to make official inquiries on March 10, 1901 that revealed there were several convents and monasteries working regularly throughout the country, as was publicly known. The solution found by the government of Ernesto Hintze Ribeiro (1849–1907, in office 1893–97, 1900–4, and 1906) was to promulgate a decree on April 18. Paradoxically, however, the decree had the particular merit of frustrating both Catholics and anticlerical thinkers alike. On the one hand, it prohibited the admission of novices, the profession of solemn vows, and, in general, the practice of monastic life. But on the other it regimented the existence of religious congregations as long as they remained committed to charity, education, and the overseas missions.[61] The only condition for legalizing such religious orders was that they had to provide their statutes to the state within six months. The Jesuits used this opening to

60 Isabel Corrêa da Silva, "A Rosa brasileira que incendiou a questão religiosa em Portugal: O caso Calmon (1899–1901)," *Tempo* 20 (2014):1–20; doi:10.1590/TEM-1980-542X-2014203626.

61 Silva, "A Rosa brasileira," 16.

their advantage and created the Associação Fé e Pátria (Association of faith and fatherland).[62]

The decision to create this association went back and forth in the first months after the promulgation of the decree. After being asked by Superior General Luis Martín García (1846–1906, in office 1892–1906) to share its thoughts on the statutes of the Associação Fé e Pátria, a small committee of Portuguese Jesuits advised him not to allow it.[63] After being notified of the decision, Provincial Luís Campo Santo (1853–1913, in office 1891–1903) urged the committee to write again to the Jesuit curia, bearing in mind that the alternative of not creating an association that submitted to the temporal power was the dissolution of the Society of Jesus in Portugal.[64] Shortly after receiving a positive response from the Jesuit curia, the statutes were handed over to Hintze Ribeiro on September 24, 1901 and were finally published together with the statutes of other religious orders on October 18, 1901.[65]

The Associação Fé e Pátria claimed to be a Catholic institution dedicated to education in Portugal and to the "propagation of faith and civilization in the missions" of "Zambésia, India, Cochin, Macao, and Timor."[66] As the decree prohibited the profession of vows and the existence of novitiates, both the novitiate and the philosophate were officially presented as institutions devoted to the formation of future missionaries. Although the newly created association directed three boarding schools (Campolide, São Fiel, and Guimarães) and two religious institutes (Barro and Setúbal), the statutes explicitly stated that the perpetuation of their educational work was contingent on the permission of the buildings' legal owners. The Jesuits, conscious of the anticlerical controversies over the promulgation of the law, deemed it wiser to detach the property of the buildings from the ministry of teaching, for if anything would happen to the newly created association, the buildings would remain theirs.

The concern with the colleges' property, however, was not new. From 1869 onward, the strategy of the Jesuits in the Portuguese province was to transfer their most important buildings to a group of English Jesuits in order to protect

62 On the convoluted history of the foundation of the Associação Fé e Pátria, see especially Correia, *O método missionário dos jesuítas em Moçambique*, 101–4.

63 Luis Martín García: b. August 19, 1846, Burgos (Spain); e. October 13, 1864, Loyola (Spain); d. April 18, 1906, Rome (Italy) (*DHCJ*, 1:1676–82).

64 Luís Campo Santo: b. June 29, 1853, Guimarães (Portugal); e. October 12 1868, Barro (Portugal); d. December 12, 1913, Brussels (Belgium) (*Mendizábal*, 13146).

65 "Estatutos da Associação Fé e Pátria," *Diário do governo* (October 18, 1901): 784–85.

66 "Estatutos da Associação Fé e Pátria," 784.

their property "from any invasion by modern enemies."[67] The first contract with the English Jesuits, which was signed on March 3, 1869, concerned an ancient Benedictine convent in Rua do Quelhas, in the center of Lisbon. The future headquarters of the Portuguese province (1880–1910), the convent had been purchased a few years earlier thanks to a generous donation from one of the youngest daughters of the count of Penamacor, Maria da Assunção de Saldanha e Castro (*c.*1840–72), when she joined the Congregation of the Sacred Heart of Jesus.[68] Following the example of the Benedictine convent, the Jesuits sold the colleges of Campolide and São Fiel to their English brethren on March 21 and November 13, 1873 respectively.[69] As the main objective of transferring the property of the colleges was to secure diplomatic protection, selecting a group of English clerics was probably a wise decision given that diplomatic relations between Portugal and England dated back to the fourteenth century. Fearing that an official deed of purchase would not suffice, for their critics could disregard it and accuse them of having made a fictitious sale, the Jesuits hoisted the Union Jack in Campolide during the solemn feasts of the college.[70] As the strategy seemed to work, there was a continuous effort to transfer the official property of the colleges to foreign Jesuits. In 1897, following the death of their last collaborator in England the year before, Portuguese Jesuits again transferred the buildings and estate of Campolide to a new group that included four Portuguese and one English.[71] In the case of São Fiel, the property was transferred twice, first to a group of five Portuguese in 1894 and then to the Magalhães brothers in 1909.[72] As the eldest brother was Brazilian and the youngest was Portuguese, the rector of São Fiel urged the latter to be naturalized as a British citizen, as this could improve the diplomatic protection afforded to the college. Lastly, fearing the two brothers might die without

67 Grainha, *História do colégio de Campolide*, 64–65. The trades were made with three English Jesuits: Georges Lambert (1821–82), Ignatius Cory Scoles (1834–96), and Henry Foley (1811–91).

68 Grainha, *História do colégio de Campolide*, xxvi–xxvii. The convent was sold for 2,800,000 *réis*, which roughly corresponds to €63,000 in 2019.

69 Grainha, *História do colégio de Campolide*, xxi; xxvi–xxvii. Campolide was sold for 5,500,000 *réis* and São Fiel was sold for two million *réis*. In 2019, those values would roughly correspond to €124,000 and €45,000.

70 Grainha, *História do colégio de Campolide*, 65.

71 The official owners of Campolide from 1897 onward were António Cordeiro, S.J. (1847–1917), José Joaquim de Magalhães, S.J. (1853–1924), José Dias Silvares, S.J. (1857–1929), and Joseph Bramley, S.J. (1862–1937).

72 The official owners of São Fiel between 1909 and 1910 were Arnaldo Magalhães, S.J. (1870–1953) and José Maria Magalhães, S.J. (1894–1914).

an official will, the rector of São Fiel designated a German Jesuit teaching at São Fiel as their official heir.[73]

In the early 1900s, the Jesuits' endeavors in Portugal were generally well received. However, the path to achieve their effective restoration was not straightforward. Unlike other European sovereigns, the Portuguese ruler did not welcome the promulgation of *Sollicitudo omnium ecclesiarum* in 1814. Whereas Portugal had been home to the first province of the Society of Jesus in the sixteenth century, and had been the first country to expel the Jesuits two hundred years later, in the nineteenth century it was among the last nations to officially recognize the restoration and accept their return. In a country still torn by the civil war between conservatives and liberals, the restoration of the Jesuits was contingent on a succession of events, beginning with the return of Carlos Rademaker from Turin in 1848. With the benefits from Rademaker's paternal inheritance and the support of conservative nobles, including Lavradio and his closest relatives, the Jesuits were able to found new colleges, a novitiate, a philosophate, and a new church. The stability of their position in nineteenth-century Portugal owed much to the implicit, and sometimes explicit, support of the more liberal sectors of the society. As well as preaching to liberal audiences on occasions of paramount importance, where he personally gained royal favor and the support of liberal aristocrats, the counsel of Rademaker and his first companions was also sought by the government to solve pressing diplomatic and political problems, including the training of missionaries and the organization of the ultramarine missions. Liberal politicians and intellectuals, recognizing the importance of integrating the circuits of power, sent their children to study in Campolide and São Fiel.

Despite the rise of anticlericalism in the 1880s and 1900s, the Jesuits' colleges were thriving thanks to the provision of an excellent education in the humanities and, especially, in the natural sciences. Due to the successful implementation of a hands-on experimental teaching style in their physics, chemistry, botany, and zoology classes, Campolide and São Fiel soon became the most distinguished boarding schools of the day. By resourcefully investing in the promotion of science and scientific education in their colleges, the Jesuits disputed the Pombaline accusations of obscurantism and gradually regained their credibility as scientists and educators.

73 Grainha, *História do colégio de Campolide*, xxx–xxxi.

CHAPTER 4

For the Greater Credibility: Science and Education in Modern Portugal

The Society of Jesus directed two types of institutions in each of its provinces: houses destined for the religious formation of its members (novitiates, philosophates, and theologates) and colleges for external students. In Portugal, the most important colleges the Jesuits directed after the restoration were undoubtedly the Colégio de Campolide (Lisbon, 1858–1910) and the Colégio de São Fiel (Louriçal do Campo, 1863–1910) (figure 4). Both colleges offered a primary and a secondary education course to pupils between seven and sixteen years old.[1] Nevertheless, the socio-economic origins of the students were somewhat different. The Colégio de Campolide, founded in the country's capital, was mainly attended by children of prominent politicians and highborn aristocrats, being the preferred boarding school for both republican and monarchist parliamentarians, freemasons and Catholics alike, particularly because it was seen as an important way of integrating the circuits of power. In contrast, the Colégio de São Fiel was attended by a lower rank of aristocracy, a countryside bourgeoisie, and by orphans and poor children from Louriçal do Campo and its outskirts. Despite the undeniable importance of geography, there was another aspect that also explains the socio-economic differences between the students of the two colleges—while Campolide was founded by Carlos Rademaker with the benefits from his father's inheritance, São Fiel was not founded by the Portuguese Jesuits but by the Franciscan friar Agostinho da Anunciação.[2]

When they assumed direction of São Fiel in 1863, the Jesuits respected the founder's determination to provide an education for local orphans and poor children. With the increasing demand for this college, they began to accept new students, many of whom were the children of local elites. While the orphans'

1 On the requirements to enroll at the Jesuit colleges, the prospectus of Campolide specified that new students should be between seven and twelve years old. The extant sources regarding São Fiel do not mention the age requirements for a new student to enroll. However, when the Jesuits took over the direction of São Fiel in 1863, the rector adopted the school rules and regulations followed at Campolide. See Colégio de Campolide, *Prospecto* (Lisbon: Typographia Universal, 1886), 1–2. This chapter is a revised version of: Francisco Malta Romeiras, "For the Greater Credibility: Jesuit Science and Education in Modern Portugal (1858–2002)," *History of Science* 56 (2018): 97–119.

2 For more details on the foundation of São Fiel, see Chapter 3.

 | DOI:10.1163/9789004382367_005

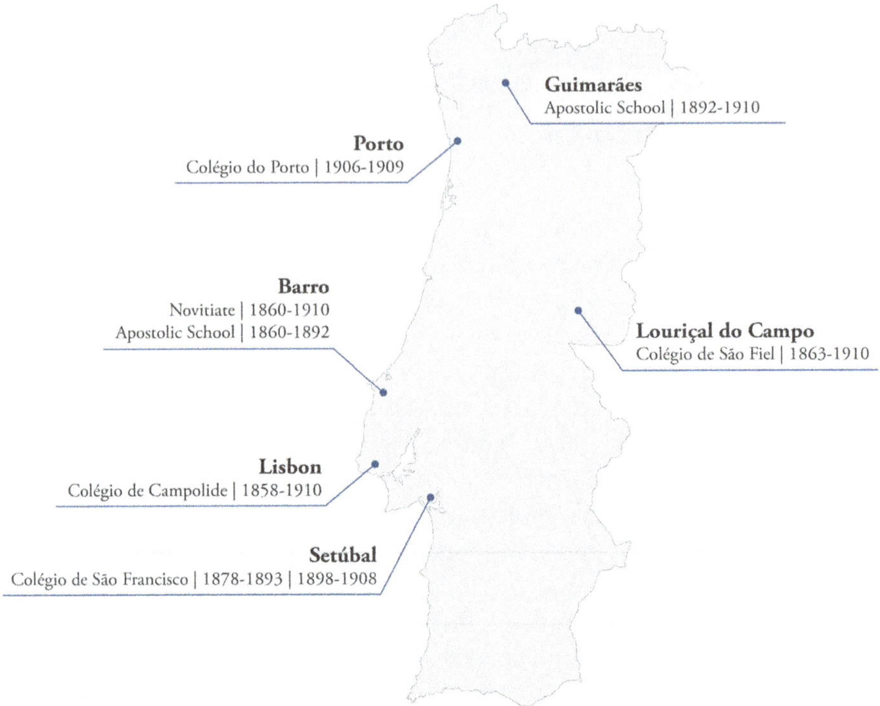

FIGURE 4 Jesuit colleges in Portugal (1858–1910)

expenses were paid for via a trust fund created and preserved by Anunciação, the new students paid a monthly tuition fee. There was still another distinction regarding the paying students, as the wealthiest pupils paid a higher fee to attend the college. This naturally implied that the socio-economic origins of the students were more diverse in São Fiel than in Campolide. Because the monthly fee to attend Campolide was twice the fee to attend São Fiel, some students also came from Lisbon to Louriçal Campo.[3]

Notwithstanding the differences between both colleges, the tuition fee represented a break with the customary practice in the old Society of Jesus. Before the universal suppression, in accordance with the *Constitutions* and the

3 In 1880, the tuition fee in Campolide was fifteen thousand *réis* (roughly €337 in 2019). In São Fiel, in 1880, out of the 138 students, sixty-eight paid a monthly fee of eight thousand *réis*, fifty-six paid 6,500 *réis* of tuition, and fourteen attended the college for free. Eight thousand and 6,500 *réis* would roughly correspond to €180 and €146 in 2019. See Joaquim Augusto de Sousa Refóios, *O collegio de São Fiel no Louriçal do Campo e o de Nossa Senhora da Conceição na Covilhã: Apontamentos sobre o jesuitismo no districto de Castello-Branco* (Coimbra: Imprensa da Universidade, 1883), 17.

Ratio studiorum, it was free to attend Jesuit colleges. In 1853, given the general finances of the Society, the Twenty-Second General Congregation authorized the colleges to charge tuition provisionally. However, what was supposed to be a transitory measure became a permanent one.[4]

Despite targeting quite different sectors of society, Campolide and São Fiel were both equally dedicated to science.[5] At both colleges, the Jesuits created modern chemistry and physics laboratories, organized substantial botanical, zoological, and geological collections, promoted scientific field trips to observe eclipses and collect novel specimens of animals and plants, and engaged in original research work in physics, botany, and zoology. As well as using hands-on experimental teaching methods in their natural science classes, the Portuguese Jesuits also made significant efforts to recreate the daily practices of modern scientific institutions, with the objective of providing both teachers and students "abundant means to further perfect themselves, accompany the progress of sciences, and contribute to its development with personal works."[6]

There are a number of reasons why the Portuguese Jesuits' use of experiments when teaching the natural sciences proved so successful, including the quality of their natural history collections, the modernity of their physics instruments, and the Jesuits' teaching skills. However, the success of the Portuguese colleges ultimately depended on a number of financial, political, and sociological factors. Between 1858 and 1910, the colleges underwent a significant number of improvements, including the construction of larger rooms, dormitories, infirmaries, and recreational areas, the creation and modernization of sanitation and water systems (including hot water), electrical installation, and the foundation of laboratories and natural history museums.[7] The creation and constant improvements of the chemistry and physics laboratories, the enlargement of the zoological and botanical collections, and the acquisition of scientific books and instruments were all crucial for the success of the experimental teaching of the natural sciences.

---

4 Grendler, "Jesuit Schools in Europe," 24–25.

5 Moniz, *A nossa casa*, 254.

6 António de Oliveira Pinto, "O Instituto de sciencias naturaes do collegio de Campolide," *O nosso collegio* 5 (1908–9): 99–113, here 100–1.

7 From 1885 to 1906, the Jesuits spent at least fifty-one million *réis* in Campolide (roughly €1,145,000 in 2019. These improvements were especially relevant since they enabled a steady growth of the student body. While, in the mid-1860s, the Colégio de Campolide only housed about forty students, by 1910 it housed nearly three hundred alumni. As for São Fiel, the data also corroborate this hypothesis, as this college housed eighty students in the 1860s, and more than 360 students in 1910.

Notwithstanding their importance to the annual budget of the colleges, the tuition fees and occasional donations only partly explain the heavy investments made. What distinguished the finances of Jesuit colleges from other private and public institutions was the amount the professors were paid. In this period, the professors of Campolide and São Fiel were all Jesuits. As such, they were prevented from being paid a salary, in accordance with their vow of poverty. Because wages are among the highest costs of any educational institution, and this expense was absent from their accounts, the colleges of Campolide and São Fiel managed to be self-sustainable and were often profitable.[8] The official requirement to fully invest their profits, combined with the desire to promote the teaching and practice of science, thus gave rise to investments that led to the creation of laboratories and natural history museums, the purchase of scientific books and instruments, and even the acquisition of full libraries and collections. In the 1891–92 school year, for instance, the natural history museum of Campolide was enriched by the acquisition of a complete museum that had been established by the renowned Italian naturalist Domenico Vandelli (1735–1816) for the marquis of Angeja (1716–88), one of the most notorious eighteenth-century aristocratic collectors.[9] By the turn of the century, the eagerness of the Portuguese Jesuits to project their public image as modern scientists and pedagogues, combined with the tacit acceptance of their return by the royal family and the highest ranks of nobility, the steady increase of admissions to their colleges, and the appropriate financial conditions, appeared to be *una mirabil congiuntura* (a marvelous conjuncture) for Jesuit science and education in nineteenth-century Portugal (figure 5).

8 On the impossibility of being paid a salary and on the self-sustainability of the colleges, see *Constitutions of the Society of Jesus*, especially parts 4, 5, and 6, which deal with the formation of the Jesuits after the novitiate and with the meaning of their vows. The balance sheets for Campolide, São Fiel, and the novitiate of Barro can be found at the Arquivo Nacional da Torre do Tombo, in Lisbon. See ANTT, *Arquivo das Congregações*, liv. 965–67, and ANTT, *Arquivo das Congregações*, folder 27, document 4.

9 With two excellent sections on mineralogy and zoology, Angeja's museum was one of the most celebrated natural history museums in eighteenth-century Portugal. It became especially renowned for its Egyptian mummy, unique in Portuguese museums. After the republican revolution, the provisional government confiscated the mummy and transferred it to the National Museum of Archaeology. On the history and purchase of this natural history museum, see Maria de Fátima Meneses, "Museus e ensino: Uma análise histórica sobre os museus pedagógicos e escolares em Portugal (1836–1933)" (MSc diss., Universidade Nova de Lisboa, 2003), 194–97; José Leite de Vasconcelos, "Viagem de Pérez Bayer em Portugal, em 1782," *O archeologo portugues* 24 (1920): 108–76, and Grainha, *História do colégio de Campolide*, 105.

1 Laboratories, Cabinets, Museums, and Observatories

After the creation of a natural history museum and a physics cabinet at the Colégio de Campolide in the school year of 1871–72, the Jesuits established a natural history museum and a physics cabinet at the Colégio de São Fiel (*c.*1876–78) and, finally, a physics cabinet at the Colégio de São Francisco (1886), a college founded exclusively for Jesuit scholastics studying philosophy. In 1908, António de Oliveira Pinto (1868–1933), one of the most prominent Jesuit teachers of mathematics and natural sciences during this period, took a further step toward the institutionalization of the teaching and practice of science in Campolide by merging the natural history museum, the physics cabinet, the chemistry laboratory, the scientific library, and the laboratory of natural sciences into a unique institute: the Instituto de Sciencias Naturaes.[10] The foundation of this institute embodied the desire to recreate the daily practices of modern scientific institutions, because its members were expected to contribute directly to its improvement, either by collecting and labeling animals,

FIGURE 5 Jesuit scholastics. Colégio de São Francisco (Setúbal), 1892–93
AB

10 António de Oliveira Pinto: b. January 30, 1868, Covilhã (Portugal); e. August 12, 1882, Barro (Portugal); d. March 17, 1933, Caldas da Saúde (Portugal) (*DHCJ*, 3:3141).

plants, and minerals, or by doing original research work in physics and chemistry. The institute also relied on some student "assistants" who were expected to maintain the collections and the library and contribute to the classification of novel specimens. All former and current students were additionally asked to collect "zoological, botanical, and mineralogical specimens particularly from the colonies" and were thus expected to play an important role in the development of the natural history collections.[11]

By 1908, the collections from the newly founded institute comprised an ethnological collection, which included the renowned Egyptian mummy from Angeja's natural history museum, and important numismatic, heraldic, mineralogical, geological, zoological, and botanical collections. For Oliveira Pinto, the cryptogams, especially the mosses and fungi, and the myxomycetes (slime molds), were among Campolide's finest scientific collections. He was particularly proud of the myxomycetes, which he considered to be "the most comprehensive collection in Europe after the one from the British Museum."[12] This collection, which comprised rare and novel species of myxomycetes, had been established by Camilo Torrend, S.J. (1875–1961) and contained 283 specimens, representing 199 different species.[13] According to Gulielma Lister (1860–1949), a renowned botanist from the British Museum, "the numerous gatherings made by Dr. C. Torrend in Portugal," included in the descriptive catalog of myxomycetes in the herbarium of the British Museum, were "the first examples of this group recorded from that country."[14]

In 1909, the most celebrated acquisition for the zoological section of the museum was a large lion from the Jesuit missionary province of Mozambique. At almost seven feet long (some 2.1 meters), it was the largest specimen ever exhibited in a Portuguese museum. As for the botanical section, the fungi collection was substantially enriched with fifteen hundred new specimens. In that year, the Jesuit botanists also classified and described more than one hundred species for the herbarium of the University of Coimbra, by request of Júlio Augusto Henriques (1838–1928), one of the most eminent contemporary

11 Pinto, "O Instituto de sciencias naturaes," 100–2.

12 Pinto, "O Instituto de sciencias naturaes," 99.

13 Camille (Camilo) Torrend: b. June 21, 1875, Saint-Privat d'Allier (France); e. September 25, 1894, Barro (Portugal); d. June 24, 1961, São Salvador da Bahia (Brazil) (*DHCJ*, 4:3816). For the collection of Campolide, see especially Camilo Torrend, "Les myxomycètes: Étude des espèces connues jusqu'ici," *Brotéria: Botânica* 6 (1907): 5–64; Torrend, "Les myxomycètes: Étude des espèces connues jusqu'ici," *Brotéria: Botânica* 7 (1908): 5–177.

14 Arthur Lister and Gulielma Lister, *A Monograph of the Mycetozoa: A Descriptive Catalogue of the Species in the Herbarium of the British Museum*, 2nd ed. (London: Printed by Order of the Trustees of the British Museum, 1911), 3.

Portuguese botanists. The comprehensive contributions of both teachers and students were crucial in the enlargement of the botanical collections. As in previous years, the Jesuit naturalists organized field trips to collect new specimens of animals and plants. In the 1909–10 school year, for example, they organized a number of short journeys with their students to Lisbon and its outskirts (Lumiar, Queluz, Belas, Alfeite, Vale do Rosal, Sintra) and a longer excursion to Serra de Monchique, the highest mountain range in the Algarve region.[15]

The Jesuits of São Fiel, like their brethren in Campolide, were also enthusiastic collectors,[16] yet it was primarily their role in founding *Brotéria* (1902–2002), the first Jesuit journal exclusively dedicated to science, for which they became internationally renowned. Unlike other well-known Jesuit periodicals, *Brotéria* was not an apologetic magazine with a scientific section, but a strictly scientific journal. Rather than promoting intellectual debates on the compatibility between science and religion, the journal instead sought to publish the works of Portuguese and foreign botanists and zoologists with the aim of contributing to the circulation, identification, description, and classification of novel species of animals and plants. During its centennial existence, *Brotéria* described more than two thousand new species of animals and plants collected in Portugal, Spain, Germany, Angola, Mozambique, Timor, Brazil, and Argentina. Although initially devoted to plant and animal taxonomy, *Brotéria* gradually expanded its scope to the modern subjects of plant physiology, plant breeding, biochemistry, and molecular genetics. The strictly scientific purpose behind its foundation, its unusual longevity, the extensive taxonomical work it published, its international readership, and the role played in the foundation and development of new research fields have all come to characterize *Brotéria* as one of the most significant scientific journals published in twentieth-century Portugal.[17]

The astronomical and geophysical sciences, in a broad sense, represented a major area of research for the restored Society of Jesus.[18] Following the tradition

15 António de Oliveira Pinto, "O Instituto de sciencias naturaes do collegio de Campolide," *O nosso collegio* 6 (1909–10): 143–49.

16 In 1910, their herbarium comprised more than five thousand different species. The herbarium included ninety-three species of Portuguese lichens, 627 species of mosses pertaining both to Portuguese and non-Portuguese floras, 521 species of Portuguese fungi, three thousand diatoms, 855 species and 106 subspecies of Portuguese spermatophytes, and forty-seven species of foreign spermatophytes. See Joaquim da Silva Tavares, "O herbário do Colégio de S. Fiel," *Brotéria: Botânica* 21 (1924): 82–87.

17 For the history of *Brotéria* (1902–2002), see Chapters 6–9.

18 In 1910, the Jesuits directed twenty-four observatories located in various places throughout the world. See Udías, *Searching the Heavens and the Earth*, 2–4.

of establishing observatories in their colleges, the Jesuits in Campolide built a circular tower for astronomical observations in 1886, and a meteorological observatory in São Fiel in 1902.[19] The physics cabinets at Campolide and São Fiel were among the clearest examples of the successful intertwining between theory and practice in Portuguese pre-university institutions. These cabinets comprised three distinctive sections (optics, mechanics and hydrostatics, and electricity) and were equipped with modern instruments that "experimentally illustrated" the most important physics treatises.[20]

The use of experiments when teaching the natural sciences, although a common practice in other European countries from the beginning of the nineteenth century, was somewhat unusual in the Portuguese context. From 1836 to 1895, the Portuguese government promulgated nine curricular reforms with the aim of modernizing secondary education. The reforms also encouraged experimental teaching in the natural sciences and postulated the creation of a botanical garden, a physics cabinet, a chemistry laboratory, and a natural history museum in all public schools.[21] However, implementing the reforms proved challenging, mostly because of financial constraints.[22] During this period, there were clear discrepancies between private and public schools, as far more students attended the private schools than the public ones. To put matters into perspective, between 1868 and 1895, only twenty-one percent of students enrolled in public schools, while the vast majority of students (seventy-nine percent) opted for home-schooling and private colleges.[23] Following the reform introduced by Jaime Moniz (1837–1917) in 1895, the number of enrolments in public schools increased, and a few public schools began offering courses in the natural sciences.

---

19 Carlos Zimmermann, "Observatorio meteorologico do Collegio de São Fiel," *Brotéria* 1 (1902): 185–88, and Grainha, *História do colégio de Campolide*, 96.

20 Pinto, "O Instituto de sciencias naturaes," 100. For a detailed account of the evolution of the usage of these instruments in the nineteenth and early twentieth centuries, see Paolo Brenni, "The Evolution of Teaching Instruments and Their Use between 1800 and 1930," *Science and Education* 21 (2012): 191–226.

21 On the successive attempts to implement the experimental teaching of science in public schools, Inês Gomes, "Os Gabinetes de História Natural dos antigos liceus: Um estudo exploratório a partir dos textos legislativos," in *Actas do Congresso Luso-Brasileiro de História das Ciências*, ed. Carlos Fiolhais, Carlota Simões, and Décio Martins (Coimbra: Imprensa da Universidade de Coimbra, 2011), 1185–202.

22 For a complete and modern analysis of the Portuguese public education system between 1836 and 1975, see Ramos do Ó, *Ensino liceal (1836–1975)*.

23 Maria Cândida Proença, "A reforma de Jaime Moniz: Antecedentes e destino histórico" (PhD diss., New University of Lisbon, 1993), 113.

Like the Jesuit colleges, the Lisbon Military College also distinguished itself in the context of scientific education. Founded in 1806 and devoted to the secondary education of the sons of the military from the court, and to those wishing to pursue a "career of arms," the Lisbon Military College had a prominent status in nineteenth-century Portugal. The collections of the natural history museum, established in 1860, were greatly improved during the following decades by purchases, transfers, and donations made by the Lisbon Polytechnic School, the University of Coimbra, and former alumni, most notably King Carlos (1863–1908, r.1889–1908).[24]

There are also a few Jesuit colleges comparable to Campolide and São Fiel, namely St. Francis Xavier's College, in Liverpool (England), and Colegio Máximo, in Tortosa (Spain). Founded in 1842, St. Francis Xavier College was "the largest Catholic secondary day-school in Britain."[25] In 1875, Superior General Peter Beckx encouraged the reform of the curriculum of the Liverpool college and urged the Jesuits to give more time to the teaching of modern languages, arithmetic, and chemistry. According to Maurice Whitehead, the creation of a chemistry course at the college represented "a significant break with traditional practice."[26] In the following years, the college witnessed the inauguration of a chemistry laboratory (1877) and a physics cabinet (1890), the creation of a "night chemistry class for young men" (1877), and the organization of an independent series of popular scientific lectures (1876–79), most of which were delivered by John Gerard, S.J. (1840–1912).[27] At the Collegio Massimo, the Spanish Jesuits founded an astronomical observatory (1905), a biology laboratory (1906), and a chemistry laboratory (1908), which were subsequently transferred to Barcelona (1916). Because the Collegio Massimo comprised the faculties of philosophy and theology, the creation of these facilities was expected to give the Jesuit scholastics a "solid basis [in the] natural sciences as a part of the philosophical studies."[28]

24 Inês Gomes, "The Natural History Collection at the Lisbon Military College: Tracing the History of a Teaching Collection," *Journal of the History of Collections* 29 (2017): 409–22; doi: 10.1093/jhc/fw036.

25 Maurice Whitehead, "The Jesuit Contribution to Science and Technical Education in Late-Nineteenth-Century Liverpool," *Annals of Science* 43 (1986): 353–68, here 356.

26 Whitehead, "Jesuit Contribution to Science," 357.

27 John Gerard: b. March 30, 1840, Edinburgh (Scotland); e. September 7, 1856, Old Windsor (England); d. December 17, 1912, London (England) (*DHCJ*, 2:1717).

28 Josep Battló and David Altadill, "The Ebre Observatory: Its Path to Ionospheric Research," *Advances in Space Research* 39 (2007): 941–46, here 941.

2 The Scientific Academies: Shaping a Public Image of Science

On the basis of the pedagogical guidelines set out in the *Ratio studiorum* and what had generally been the normal practice in their colleges in the past, the Portuguese Jesuits also promoted the creation of academies in São Fiel and Campolide.[29] Attended by promising students, the academies promoted the teaching and discussion of modern scientific topics during their regular sessions. Once or twice a year, usually before Christmas and the summer holidays, they fostered the organization of solemn sessions, to which all students and their families were invited. In Campolide and São Fiel, the students were the protagonists, being responsible not only for carrying out the experiments but also for describing and explaining their underlying scientific principles. The solemn sessions were theatrical and included screen projections and spectacular experiments with high-frequency electrical discharges, cathode and X-rays,

FIGURE 6 Program of the solemn session held at Campolide on March 16, 1905
APSI

29 On the role played by the academies in the old Society, see Chapter 1.

magnetism, wireless telegraphy, and liquid crystals.[30] The sessions were often presided over by renowned individuals from the political, clerical, and public spheres, with the minister of public education and the royal princes being among the most illustrious chairmen of the solemn sessions held in Lisbon in the years preceding the republican revolution. Because these festive events, initiated in the 1870s, publicly displayed the Jesuits' modern scientific knowledge to a broad audience, they played a significant role in their efforts to recover their educational and scientific credibility.

One of the most significant sessions held at Campolide, which took place on March 16, 1905, was presided over and dedicated to Luís Filipe (1887–1908) and Manuel de Bragança (1889–1932, r.1908–10) (figure 6).[31] At this session, the royal princes became honorary members of the scientific academy of Campolide and witnessed the presentation and experimental demonstration of three different theses. By putting together experiments involving high-frequency electrical discharges, liquid carbon dioxide and nitrogen, and wireless telegraphy, the Jesuits tried to impress their audience, and especially the royal princes, with their use of theater to convey their knowledge of modern science. This theatricality was already made clear in the beautifully conceived program, which included engravings of the instruments that were to be used during the solemn session. With these public sessions, the Portuguese Jesuits were clearly trying to replicate the European practice of organizing scientific spectacles while also seeking to evoke their scientific and educational history and the achievements of Jesuit polymath Athanasius Kircher (*c.*1602–80).[32]

30 On the solemn sessions of the scientific academies held at Campolide and São Fiel until 1910, see Francisco Malta Romeiras, *Ciência, prestígio e devoção: Os jesuítas e a ciência em Portugal* (*séculos XIX e XX*) (Cascais: Lucerna, 2015), 91–106.

31 After the assassination of his elder brother and father, in 1908, Manuel (1889–1932, r. 1908–10) became the last king of Portugal.

32 On the history of public scientific demonstrations and their importance for the popularization of science, see Agustí Nieto-Galan, *Los públicos de la ciencia: Expertos y profanos a través de la historia* (Madrid: Fundacíon Jorge Juan, Marcial Pons Historia, 2011), 81–121; Bernadette Bensaude-Vincent and Anne Rasmussen, eds., *Science and Spectacle in the European Enlightenment* (Aldershot: Ashgate, 2008); and Aileen Fyfe and Bernard Lightman, eds., *Science in the Marketplace: Nineteenth-Century Sites and Experiences* (Chicago: University of Chicago Press, 2007). Known to Renaissance and early modern scholars as "the last man who knew everything," Athanasius Kircher founded an impressive museum in the Collegio Romano. Visited by popes, cardinals, princes, and intellectuals, his museum was one of the first of its kind in early modern Europe, and one of the most celebrated institutions in Rome. At the museum, visitors were amazed by the natural history collections and spectacular machines. Because it brilliantly combined science and spectacle, Kircher's museum was, in the words of Mark A. Waddell, a "theater of the unseen." On Kircher and the Roman College Museum, see Paula Findlen, "Scientific Spectacle in

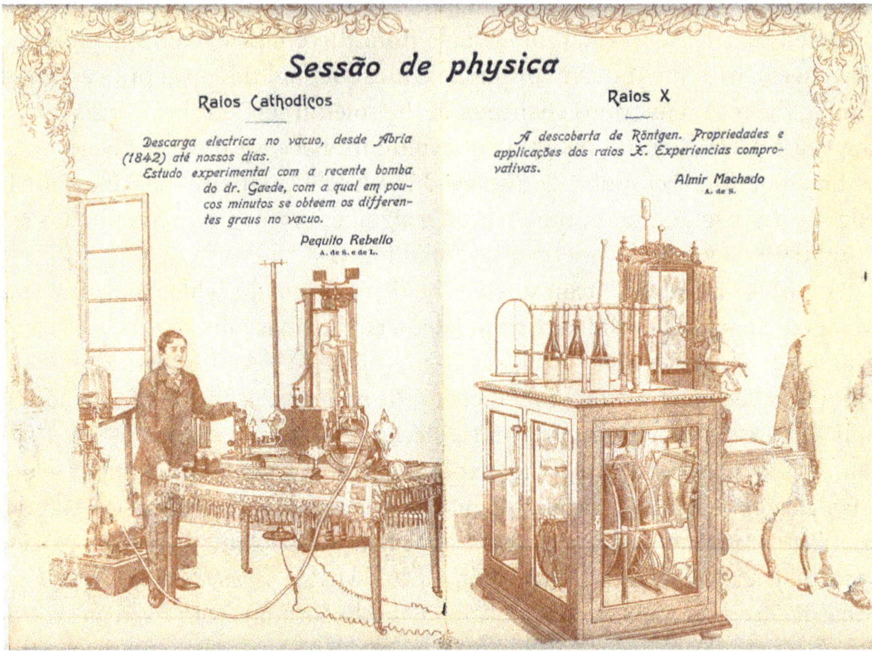

FIGURE 7 Program of the solemn session held at Campolide on May 23, 1906
FMR

The following session of the scientific academy of Campolide was presided over by Abel Pereira de Andrade (1866–1958), minister of education (figure 7). The session took place on May 27, 1906 and was dedicated to the study of crystals.[33] The first act of the perfectly choreographed session sought to explain "crystal systems," "elementary notions on the polarization of light," and birefringence, and it was planned to be an "introduction to what will be exposed in the second [act]."[34] During the second act, the audience was invited to observe sixteen-year-old João Maria Berquó d'Aguiar (1889–1954) re-enact the "very recent" experiments on the birefringence of some liquids, performed by Otto Lehmann (1855–1922) between 1904 and 1906.

Baroque Rome: Athanasius Kircher and the Roman College Museum," in Feingold, *Jesuit Science and the Republic of Letters*, 225–48, Paula Findlen, ed. *Athanasius Kircher: The Last Man Who Knew Everything* (London: Routledge, 2004), and Mark A. Waddell, "A Theater of the Unseen: Athanasius Kircher's *Museum in Rome*," in *World-Building and the Early Modern Imagination*, ed. Allison B. Kavey (New York: Palgrave Macmillan, 2010), 67–90.

33 Colégio de Campolide, *Á sua celeste padroeira* (Lisbon: n.p., 1906); Colégio de Campolide, "Sessão solemne de physica," *O nosso collegio* 3 (1906–7): 17–20.

34 Colégio de Campolide, "Sessão solemne de physica," 17–18.

For the execution of experiments with liquid crystals in 1906, the Jesuits had purchased an instrument that had been "expressly conceived for the Colégio de Campolide" by Zeiss, the international manufacturer. Purchasing new scientific instruments for research and public demonstrations was one of the most distinctive features of the Jesuits' academies in the Portuguese educational context. In the solemn session of 1907, for instance, the acquisition of "the recent pump of Dr. Gaede" and X-ray equipment was crucial for planning new demonstrations.[35] The pump built by Wolfgang Gaede (1878–1945) created a vacuum in just four minutes, thus simplifying the experiments with high-frequency electrical discharges, while the X-ray apparatus was used to radiograph a cat in a box.

The Campolide and São Fiel scientific academies also hosted sessions dedicated to other subjects, such as astronomy (São Fiel, December 1905), volcanology (São Fiel, May 1907), animal cytology and histology (Campolide, March 1908), and seismology (São Fiel, March 1909), but most of the sessions were dedicated to physics. It is likely that this was not entirely unplanned, because experimental physics delivered the best and well-proven combination between science and spectacle.

3 The Solar Eclipses of 1900 and 1905

As well as organizing public sessions, the students and teachers of the scientific academies engaged in other periodic activities, such as journeys to collect new specimens of animals and plants and field trips to observe eclipses. The Jesuits began observing eclipses in the 1890s at the Colégio de São Francisco, in Setúbal, and in 1900 they organized their first expedition to observe this popular astronomical event (figure 8). On May 28, 1900, some naturalists from São Fiel were in Benespera and Capinha, two small parishes not far from the college premises, to observe the total solar eclipse.[36] During the days before the eclipse, several professional astronomers had already reached Portugal

35 Colégio de Campolide, "Sessão anual de sciências da academia campolidense," *O nosso collegio* 4 (1907–8): 29–31. This vacuum pump was the first instrument designed by physicist Wolfgang Gaede (1878–1945). It was presented in 1906 at a conference in Merano, Italy. During the following years, Gaede would continuously improve his vacuum pump. On Gaede, see Manfred Dunkel, "Wolfgang Gaede: An Appreciation of His Life on the Occasion of the 50th Anniversary of the Invention of the Diffusion Pump," *Vacuum* 13 (1963): 501–3.

36 Colégio de São Fiel, *Eclipse do Sol de 28 de Maio de 1900: Observações dos professores do collegio de São Fiel* (Lisbon: La Bécarre, 1900).

FIGURE 8 Jesuit scholastics observing an eclipse in Setúbal, *c.* 1890
AB

and Spain to observe this phenomenon, which was only completely visible in certain regions of the Iberian Peninsula. By the turn of the century, these expeditions were undertaken quite frequently because they provided a perfect opportunity for the spectrophotometric analysis of the atmosphere and for the observation of the solar corona. There were also a number of amateur expeditions, such as the one from São Fiel. Frederico Oom (1864–1930), the head of Lisbon's astronomical observatory, encouraged these amateur observations by raising public awareness of their importance and establishing a network of correspondence with both amateur and professional astronomers.[37]

After the publication of the *Eclipse do Sol de 28 de Maio de 1900* (The solar eclipse of May 28, 1900), a booklet containing a summary of their observations, the Jesuits continued to correspond with Oom, whose suggestions played an important role in the creation of a meteorological observatory in São Fiel. In November 1901, two months before its official opening, Carlos Zimmermann, S.J. (1871–1950) asked Oom to confirm the list of scientific instruments already

37 Luís Miguel Carolino and Ana Simões, "The Eclipse, the Astronomer and His Audience: Frederico Oom and the Total Solar Eclipse of 28 May 1900 in Portugal," *Annals of Science* 69 (2011): 215–38.

available at São Fiel.[38] Because Zimmermann wanted to establish a reliable observatory, one capable of integrating the national network of meteorological stations, Oom's suggestions were extremely valuable and contributed to the inclusion of São Fiel's data in the annual reports of the national meteorological observatory from 1902 onward.

On August 30, 1905, the Jesuits of São Fiel and Campolide organized a collective expedition to observe a total solar eclipse in Spain; and on December 8, 1905, the eponymous feast of the Colégio de Campolide, they published a new booklet containing the results of their observations.[39] This popular work, which was dedicated to the royal princes, did not set out to communicate new astronomic discoveries but instead sought to summarize the Jesuits' observations and "enrich the archive of empirical knowledge, always significant when considering a phenomenon so restricted in time and space."[40] The amateur astronomers had observed the eclipse from three different cities: Tortosa, Palencia, and Burgos. The committee led by Oliveira Pinto had traveled from Campolide to Tortosa, and the one from São Fiel, guided by Joaquim da Silva Tavares, S.J. (1866–1931) and Valério Cordeiro, S.J. (1877–1940), to Palencia.[41] The group heading to Burgos included one naturalist from each college and two students from the scientific academy of Campolide: José Pequito Rebelo (1892–1983) and Simeão Pinto de Mesquita (1889–1989). As a meeting point for astronomers from all over Europe, Burgos was the most suitable location for the Jesuit committee to interact with amateur and professional astronomers. Comfortably installed in a Jesuit college, Pinto de Mesquita and Pequito Rebelo were responsible for photographing and illustrating the eclipse; Pequito Rebelo's drawing, which was later included in the booklet, was highly

38 Karl (Carlos) Zimmermann: b. May 28, 1871, Ehingen (Germany); e. September 7, 1890, Lyon (France); d. October 21, 1950, Salvador (Brazil) (Francisco Malta Romeiras and Henrique Leitão, "Jesuítas e ciência em Portugal II: Carlos Zimmermann S.J. e o ensino da microscopia vegetal," *Brotéria* 174 [2012]: 113–25; Doriedson Ferreira Gomes et al., "Father Zimmermann (1871–1950): The First Brazilian Diatomist," *Diatom Research* 27 [2012]: 177–88). Carlos Zimmermann to Frederico Oom, November 12, 1901 (OAL, MS C469).

39 Luís Gonzaga Cabral, *O eclipse total do Sol no dia 30 de Agosto de 1905: Observações feitas pelas comissões das academias scientificas dos collegios de S. Fiel e Campolide* (Lisbon: La Bécarre, 1905). On this Jesuit expedition, see also Joaquim da Silva Tavares, "Eclipse total do sol em 30 de Agosto de 1905," *Brotéria* 5 (1906): 254–60.

40 Cabral, *O eclipse total do Sol*, 8.

41 Joaquim da Silva Tavares: b. August 17, 1866, Casais de S. Bento, Vila de Rei (Portugal); e. November 13, 1880, Barro (Portugal); d. September 2, 1931, Paris (France) (*DHCJ*, 4:3706). Valério Cordeiro: b. August 17, 1877; e. September 8, 1898, Barro (Portugal); d.1940 (*Mendizábal*, "Elenchus scriptorum qui, in restitutam Societatem Iesu cooptati, eam deseruerunt," 606).

praised by Oom and some foreign astronomers who were staying in Burgos at that time.[42]

4 Masters of Theory and Practice

The Jesuits' teaching and research skills and their willingness to keep up with modern scientific practices played a crucial part in the success of their colleges (figure 9). However, during this period, some of their opponents accused them of lacking the necessary scientific credentials, especially because they did not meet the standard criteria for being high school teachers. After an official visit to São Fiel in 1880, Joaquim de Sousa Refóios (1853–1905) concluded that it was directed by Jesuits and accused them of harming the population of Louriçal do Campo and its outskirts. Attesting to the existence of a modern

FIGURE 9 António de Oliveira Pinto, S.J. (1868–1933) in Setúbal, 1891–92
APSI

42 Cabral, *O eclipse total do Sol*, 41. See also Francisco Malta Romeiras and Henrique Leitão, "Jesuítas e ciência em Portugal III: As expedições científicas e as observações dos eclipses solares de 1900 e 1905," *Brotéria* 174 (2012): 227–37.

physics cabinet and a chemistry laboratory, the criticism was directed at the hygienic conditions and the teaching of history and philosophy, which he deemed highly conservative and reactionary. Updating the popular accusation of supporting tyrannicide, Sousa Refóios's report also asserted that the Jesuits believed that patricide was not a sin if committed in the service of God. Sousa Refóios recognized that most families sent their children to São Fiel because they wanted them to receive an excellent scientific and humanistic education, but he added that they did so "hoping that, when leaving the college, it would be easy to remove the vestiges of Jesuit education."[43] Thirty years later, Pedro Ferrão (dates unknown) voiced similar accusations against the Jesuits of São Fiel, alleging that they were ambitious, deceitful, and secretive and claiming they were inept teachers.[44]

From 1880 to 1910, the Jesuits' opponents often claimed that they should be prevented from teaching due to their lack of state-certified training. The most detailed response to these accusations came from physician António Mendes Lages (1838–1916), who claimed that the Jesuits' teaching abilities and their profound knowledge of the subjects they taught were more important than an official diploma.[45] The future Jesuit added that, since state training was not legally enforced, the assessment of teaching skills should be left to the rectors of educational institutions.[46]

As in the old Society, the Jesuits professed their first vows after a two-year novitiate. Following a three-year philosophy course, which also included mathematics and natural philosophy, each scholastic would either become a missionary or a teacher for a number of years. The time it took to complete this period of training was somewhat variable, and it could take several years

43 Refóios, *O collegio de São Fiel*, XII.

44 Pedro Ferrão, *A educação jesuitica: O Collegio de S. Fiel; Subsidios para a historia contemporanea dos jesuítas* (Lisbon: Guimarães & Ca, 1910), 81–83. His book was mostly based on Refóios's report.

45 António Mendes Lages: b. January 2, 1838, Loriga (Portugal); e. November 12, 1908, Barro (Portugal); d. January 11, 1916, Murcia (Spain) (*Mendizábal* 13,789). Following the passing of his wife and an intense spiritual experience during an eight-day retreat based on the *Spiritual Exercises*, Mendes Lages asked to be admitted to the Society of Jesus. He was seventy years old when he became a novice. He was ordained a priest in exile on May 8, 1911.

46 António Mendes Lages, *O Sr. Marianno de Carvalho e o collegio de S. Fiel* (Lisbon: Typographia da Cruz do Operário, 1888), 23. The memoirs of two former students of São Fiel, the Nobel laureate Egas Moniz and Luís Cabral de Moncada (1888–1974), similarly emphasized the quality of Jesuit education, especially in the experimental teaching of the natural sciences. See, e.g., Moniz, *A nossa casa*; Luís Cabral de Moncada, *Memórias ao longo de uma vida* (Lisbon: Editorial Verbo, 1992).

before they enrolled in theology studies. As the Society of Jesus did not offer a theology course in Portugal, the Jesuits were always sent abroad to study theology for a period of three to five years. After a process that lasted a minimum of ten years, they were finally ordained priests and their future tasks were then determined by the provincial superior. Typically, those who had distinguished themselves as skillful teachers during their training returned to the colleges. Given this prolonged formation, it is clear that the Jesuits' scientific training, which only took place as part of the philosophy course, is insufficient to explain the success of the teaching of the natural sciences at Campolide and São Fiel. To understand this issue fully, it is necessary to consider the life and work of Oliveira Pinto, one of the most prominent Jesuit teachers of mathematics and natural sciences during this period.

Oliveira Pinto was born in Covilhã, in the interior of Portugal, in 1868, and he became a novice after turning fourteen years old (figure 9). After the novitiate, he taught mathematics in São Francisco (1892–93) and physics, chemistry, and natural history in São Francisco (1891–92) and São Fiel (1893–95). From 1895 to 1897, he studied theology in Õna, before being ordained priest in Vals-près-le-Puy (Haute-Loire) in the following year. From 1901 to 1910, he taught mathematics, physics, chemistry, and natural history at Campolide, where he was also responsible for the Instituto de Sciencias Naturaes. A member of several national and international societies, he earned a prestigious place in the history of physics in Portugal by performing some of the first experiments with telegraphy (1902) and the first recorded experiments with radioactivity (1910), both at Campolide.[47]

Oliveira Pinto's interest in radioactivity dated back to the First International Congress of Radiology and Electricity, held in Liège in 1905.[48] Five years later, he attended the Second International Congress of Radiology and Electricity (Brussels, 1910) and presented the results of his experiments on the radioactivity of Portuguese mineral waters.[49] Despite their interest in radioactivity, the

47 Francisco Malta Romeiras, "Os colégios dos jesuítas e o ensino da física em Portugal (1858–1910)," *Gazeta de física* 40 (2017): 2–9; Francisco Malta Romeiras and Henrique Leitão, "Jesuítas e ciência em Portugal I: António Oliveira Pinto S.J. e as primeiras experiências com radioactividade em Portugal," *Brotéria* 174 (2012): 9–20.

48 António de Oliveira Pinto, "Primeiro congresso internacional de radiologia e ionização," *Brotéria* 5 (1906): 129–34.

49 António de Oliveira Pinto, *Primeira contribuição para o estudo da radioactividade das aguas mineraes de Portugal* (Porto: Typographia Occidental, 1910); Pinto, "Première contribution à l'étude de la radioactivité des eaux minérales du Portugal," in *II Congrès international de radiologie et d'électricité* (Brussels: Imprimerie Médicale et Scientifique L. Severeyns, 1911), 3–8.

lack of appropriate experimental training and the difficulties encountered in acquiring radioactive sources prevented the physicists from the University of Coimbra working with radioactivity until 1915.[50] Although he lacked a formal university scientific education, Oliveira Pinto became acquainted with radioactivity in 1905, and in 1910 Pierre Curie (1859–1906) and Marie Skłodowska-Curie (1867–1934) welcomed him to their laboratory. He worked there for a short period of time in which he became acquainted with the most recent methods and instruments for measuring radioactivity.[51] Oliveira Pinto had the necessary expertise and access to the coveted radioactive source, in this case a solution of radium bromide, to carry out his experiments. The radium bromide solution, essential for the calibration of the electroscopes, was supplied by Ellen Gleditsch (1879–1968).[52]

When the Society of Jesus was restored in 1858, Portuguese Jesuits were very much aware of the longevity of the eighteenth-century accusations of obscurantism. As part of their efforts to counter these claims, and while ultimately aspiring to the "greater glory of God," they established observatories, laboratories, and museums in their colleges. By putting the teaching and practice of science at the top of their educational priorities and by projecting their public image as modern scientists and pedagogues, they gradually regained their credibility with the royal family, the highest ranks of the nobility and bourgeoisie, and the scientific community. In addition, the restoration of their credibility represented a fundamental step in the approval of new apostolic missions by the crown, namely the foundation of colleges and missions in China, India, Mozambique, and Timor. The colleges of Campolide and São Fiel thus played a major role in the restoration of the Society of Jesus and in the diffusion of Ignatian spirituality, through the restoration of the Marian congregations and through the organization of Spiritual Exercises and other silent retreats. With the objective of providing an education grounded on the "alliance between religion and science," these colleges took a further step toward the formal scientific training of a new elite, which included not only future scientists but also military officers, priests, bishops, politicians, and celebrated artists. As with the reform of the scientific curriculum of St. Francis Xavier's College in Liverpool, it is possible that the Jesuit curia in Rome urged the Portuguese Jesuits

50 António José Leonardo, Décio Martins, and Carlos Fiolhais, "A física na Universidade de Coimbra de 1900 a 1960," *Gazeta de física* 34 (2011): 9–15.

51 Acácio Casimiro, *P. António da Costa e Oliveira Pinto* (Porto: Tipografia Pôrto Médico, 1933), 4. Between December 1909 and January 1910, Oliveira Pinto traveled to Belgium, the Netherlands, and France and became acquainted with the most recent methods for measuring radioactivity.

52 Romeiras and Leitão, "António Oliveira Pinto," 15–16.

to dedicate themselves to the teaching of scientific courses in their colleges. However, the extant sources suggest that the selection of the courses was decided locally. It is worth remembering that because scholastics only received a scientific education during the philosophy course, the engagement of Jesuits in the teaching and practice of the natural sciences reflected their active involvement in amateur practices in physics, astronomy, zoology, and botany.

The promotion of amateur expeditions to observe eclipses and to collect new specimens of animals and plants, the systematic collection of meteorological data, their attendance at international conferences, and engagement in original research work all suggest that the Jesuits' informal training, and their participation in national and international scientific networks, proved to be more relevant for their training as modern scientists and pedagogues than any official diploma or examination. Thus, by teaching subjects directly related to their own scientific practices, they were able to intertwine the teaching and practice of science in their colleges. Furthermore, the direct involvement of students in duties of diverse complexity and the desire to recreate the modern practices of scientific institutions were probably the most distinctive features of these colleges, not only in the Portuguese context but also in the wider Jesuit educational framework.

CHAPTER 5

The Republican Exile and the Confiscation of the Natural History Collections

Fifty years after a reticent return, the implicit acceptance of the Jesuits' presence by the royal family and the government, the success of their colleges, and the increasing national and international prestige of their scientific works seemed to represent a timely conjuncture for the Jesuits living and working in Portugal. Between 1858 and 1910, the colleges thrived, with the Jesuits being responsible for the education of more than four thousand students in Portugal, East Africa, India, Macao, and Timor.[1] However, the order was in a precarious situation, especially due to significant anti-Jesuit opposition within the larger mass of nineteenth-century anticlericalism. Influential writers, poets, and politicians frequently insisted that there was a clear conflict between liberty, progress, and science on the one hand, and clericalism on the other.[2]

The usual accusations of obscurantism gained momentum from the 1880s onward, and in 1910 the Jesuits were banned from the Portuguese Empire for the third time.[3] The new republican government subsequently closed their colleges and seized their possessions. Although the confiscation of their books, instruments, and collections generated widespread opposition both in Portugal and elsewhere, the protests of Portuguese and foreign naturalists were mostly ignored, with a few notable exceptions.

1 Rodrigues, *A formação intellectual do jesuíta*, 596. According to Francisco Rodrigues, in 1910 the Jesuits taught 807 students in continental Portugal and 3,365 students in the Portuguese Empire.

2 On nineteenth-century religious history in Portugal, see Manuel Clemente and António Matos Ferreira, eds., *Religião e secularização* (Lisbon: Círculo de Leitores, 2002); Clemente, *Igreja e sociedade portuguesa: Do liberalismo à República* (Porto: Assírio & Alvim, 2012); Luís Machado de Abreu and José Eduardo Franco, eds., *Ordens e congregações religiosas no contexto da I República* (Lisbon: Gradiva, 2010); and Machado de Abreu, *Ensaios anticlericais* (Lisbon: Roma Editora, 2004).

3 For a brief summary of the history of the Jesuits in Portugal, including both the history of the Portuguese assistancy and the history of the restored Society of Jesus, see António Lopes and António Araújo, "Jesuítas," in *Dicionário histórico das ordens, institutos religiosos e outras formas de vida consagrada católica em Portugal*, ed José Eduardo Franco (Lisbon: Gradiva, 2010), 195–206.

 | DOI:10.1163/9789004382367_006

1 The Rise of Anti-Jesuitism

Anti-Jesuitism occupied a highly significant place at the core of the more general anticlericalism of the nineteenth century, and the opposition to the Jesuits was publicly displayed and widely diffused through plays, books, treatises, and journals that portrayed them as sinful, decadent, ambitious, deceitful, reactionary, and mediocre.[4] One of the most common themes put forward was the argument that the victory of the Enlightenment demanded that religion be vanquished. To late nineteenth-century anticlerical poets, historians, and politicians, the Jesuits were the main culprits in the propagation of "non-thought."[5] The central message of these works was that "whoever believes [in God] does not think, and whoever thinks does not believe."[6]

Hence accusations of obscurantism and of promoting idiocy were still widespread. Renowned poet Guerra Junqueiro (1850–1923), for instance, followed Pombal's line of reasoning in accusing the Jesuits of kidnapping children and enrolling them in a "dark seminary" that transformed them into "stupid night-birds."[7] Borges Grainha, alumni and outspoken critic, repeatedly insisted on the poor quality of Jesuit schools and the backwardness of their teaching. For him, their undeclared objective had always been to control the nobility and the bourgeoisie through education and to control the people by instilling the fear of God in them.[8] This argument was also dear to Teófilo Braga (1843–1924), future interim president of the Portuguese republic (in office 1915) and one of the most active popularizers of positivism. In his work on the history of Christian legends, Braga defined Jesuitism as the third and final phase of the history of Christianity. He considered the dawn of Christianity to be a spontaneous and popular movement emotionally and mythically developed in response to

4 Araújo, *Jesuítas e antijesuítas*, 10.

5 The most influential anti-Jesuit writers in this period were Francisco Adolfo Coelho (1847–1919), Antero de Quental (1842–91), Abílio Guerra Junqueiro (1850–1923), Miguel Bombarda (1851–1910), Sampaio Bruno (1857–1915), Teófilo Braga (1843–1924), Manuel Borges Grainha (1862–1925), Joaquim de Sousa Refóios (1853–1905), and José Ramos Preto (1871–1949).

6 "Quem crê não pensa, e quem pensa não crê": Manuel Joaquim de Carvalho Júnior, *Nem Deus, nem diabo: Solução da philosophia positiva* (Lisbon: Typographia Elzeviriana, 1884), 7.

7 Abílio Guerra Junqueiro, "Como se faz um monstro," in *A velhice do padre eterno* (Porto: Livraria Minerva, 1885), 65–69.

8 Before publishing *História do colégio de Campolide da Companhia de Jesus* (Coimbra: Imprensa da Universidade de Coimbra, 1913) (reprint *Histoire du collège de Campolide et de la résidence des jésuites à Lisbonne* [Lisbon: A Editora Limitada, 1914]), Borges Grainha wrote several anti-Jesuit books, including *Os jesuitas e as congregações religiosas em Portugal nos ultimos trinta annos* (Porto: Empreza Litteraria e Typographica, 1891) and *O Portugal jesuita* (Lisbon: Typographia e Sterotypia Moderna, 1893).

a fabricated divinity, and Catholicism as a restructuring phase corresponding to the political and cultural reorganization of Europe. Jesuitism, on the other hand, was defined as a morally relaxed period that sought to reinstate papal theocracy and popular piety through the diffusion of increasingly elaborate Christian myths.[9]

In the renowned Casino Lectures, held in Lisbon in 1871, Antero de Quental (1842–91) argued that tercentennial Iberian decadence was threefold (moral, political, and economic) and identified the Catholic Reformation, and its effective implementation by the Jesuits, as the single most important cause for the moral decadence of Portugal and Spain in the seventeenth, eighteenth, and nineteenth centuries.[10] Similarly, in the final lecture, Francisco Adolfo Coelho (1847–1919) retrieved the arguments set out in *Compêndio histórico* and held the Jesuits responsible for the decline of the University of Coimbra. His main thesis was that the Jesuits had been the leading force behind the educational backwardness of Portugal from the sixteenth century onward.[11] Twenty years later, this argument was also promoted by Braga in his voluminous *História da Universidade de Coimbra* (History of the University of Coimbra [1892–1902]).[12] With almost three thousand pages, this positivist history of the University of Coimbra represented the highpoint of the Pombaline narrative of obscurantism. Likewise, in his two seminal works *História da civilização ibérica* (History of the Iberian civilization [1879]) and *História de Portugal* (History of Portugal [1880]), Joaquim Pedro de Oliveira Martins (1845–94), one of the most respected historians of his generation, accused the Jesuits of having converted Christian ethics into casuistry and portrayed them as ambitious, decadent, and illiterate.[13]

9 Teófilo Braga, *As lendas christãs* (Porto: Livraria Internacional de Ernesto Chardron, 1892) [reprint, Lisbon: Edições Vercial, 2012]. See also *Origens poéticas do christianismo* (Porto: Livraria Universal de Magalhães & Moniz, 1880) and *Soluções positivas da politica portugueza: Obras políticas* (Porto: Livraria Chardron, de Mello & Irmão, 1912).

10 Antero de Quental, *A decadência dos povos peninsulares* (Porto: Typographia Comercial, 1871) (reprint Carlos Reis, ed., *As conferências do casino* [Lisbon: Alfa, 1990], 95–128). He repeated the thesis defended by Alexandre Herculano (1810–77) in his *História da origem e estabelecimento da Inquisição em Portugal* (Lisbon: Imprensa Nacional, 1854–59) (reprint John C. Branner, trans., *History of the Origin and Establishment of the Inquisition in Portugal* [New York: Ktav Publishing House, 1972]).

11 Francisco Adolfo Coelho, *A questão do ensino* (Porto: n.p., 1872), 33 (reprint in Reis, *As conferências do casino*, 143–77).

12 Teófilo Braga, *História da Universidade de Coimbra* (Lisbon: Typographia da Academia Real das Ciencias, 1892–1902). For the period concerning the Jesuits in Coimbra, see vols. 2 (1555–1700) and 3 (1700–1800).

13 Joaquim Pedro Oliveira Martins, *História da civilização ibérica* (Lisbon: Bertrand, 1879) (reprint [Lisbon: Guimarães Editores, 2007]), and Martins, *História de Portugal* (Lisbon: Bertrand, 1880) (reprint, Lisbon: Guimarães Editores, 2010).

In 1901, civil engineer Tomás Lino de Assumpção (1844–1902) published the *História geral da Companhia de Jesus* (General history of the Society of Jesus).[14] At more than six hundred pages, his *magnum opus* was one of the most extensive and influential anti-Jesuit accounts of this period. As the title suggests, the work provided an overview of the Society of Jesus's history, from the conversion of Ignatius of Loyola to the nineteenth-century controversies surrounding the so-called illegal restoration. It consisted of 114 short and, often disconnected, chapters, including commentaries on the *Constitutions*, the *Monita secreta*, and the *Spiritual Exercises*—"an obscure, sinister, damned, and destructive night"—and brief descriptions of European and overseas episodes.[15] In the passages discussing the Portuguese case, *História geral da Companhia de Jesus* referred to *Dedução cronológica* as its main primary source, a work Lino de Assumpção deemed "a true monument of erudition and an example of honorable courage."[16] When concluding the succinct biography of the Jesuit António Vieira (1608–97), Lino de Assumpção conceded his erudition but claimed that "if he had not worn that cassock he would have been one of the greatest [literary splendors] of his time."[17] Throughout his diachronic account, Lino de Assumpção mixed fictional dialogues with authentic citations in order to reinforce his portrayal of the Jesuits as "infernal magnets attracting disgrace and ruin."[18]

The most significant quarrel in which the Jesuits were involved during this period was the controversy that opposed physician Miguel Bombarda (1851–1910) and Manuel Fernandes Santana, S.J. (1864–1910).[19] In 1897, Bombarda delivered a series of lectures in Lisbon inspired by the neurological discoveries of Santiago Ramón y Cajal (1852–1934) in which he argued that

14 Tomás Lino de Assumpção, ed., *História geral da Companhia de Jesus, desde a fundação até aos nossos dias* (Lisbon: Empreza da História de Portugal, 1901) (reprint *História geral dos jesuítas* [Rio de Janeiro: Moraes Editores, 1982]).

15 Assumpção, *História geral da Companhia de Jesus*, 69.

16 Assumpção, *História geral da Companhia de Jesus*, 378.

17 António Vieira: b. February 6, 1608, Lisbon (Portugal); e. May 5, 1623, Bahia (Brazil); d. July 18, 1697, Bahia (Brazil) (*DHCJ*, 4:3948–51). Assumpção, *História geral da Companhia de Jesus*, 371.

18 Assumpção, *História geral da Companhia de Jesus*, 480.

19 Manuel Fernandes Santana: b. May 14, 1864, Funchal, Madeira (Portugal); e. October 15, 1880, Barro (Portugal); d. May 2, 1910, Lisbon (Portugal) (*DHCJ*, 4:3499). On the history of this controversy, see João Lobo Antunes, "Ciência e fé," *Brotéria* 152 (2001): 331–48, and Vítor Neto, "Miguel Bombarda e Manuel Fernandes Santana: Um confronto de ideias," *Revista portuguesa de história* 2 (2003): 303–11.

conscience and free will were mere delusions.[20] In response, Fernandes Santana published a number of articles in the Catholic periodical *Correio nacional* (National courier) criticizing Bombarda's stance, Ernst Haeckel's (1834–1919) monism, and several other positivist, materialist, and mechanical philosophers and biologists.[21] The controversy escalated with the publication of Bombarda's *A sciencia e o jesuitismo: Réplica a um padre sábio* (Science and Jesuitism: response to a learned father [1900]).[22] More than a simple rebuttal, *A sciencia e o jesuitismo* soon became one of the most influential anticlerical accounts of the day. By including references to recent (and not so recent) anti-Jesuit works, Bombarda aimed to present the Catholic Church and the Society of Jesus as rigid and obscurantist institutions.[23] As well as referring to the usual charges of hypocrisy, immorality, and ambition, he also added a new accusation: invested with scientific authority as a renowned psychiatrist, Bombarda concluded his book by arguing that Jesuits were, by nature, mentally insane—as far as he was concerned, no one could become a member of the Society of Jesus without being mentally ill because "there are minds predisposed to that evil, as there are minds set for ordinary madness."[24]

From 1882 to 1910, the narrative of obscurantism was also echoed in the public speeches of a number of parliamentarians who reasserted in both houses of Parliament that the Society's expulsion and the secularization of education had been Pombal's greatest achievement in the modernization of Portugal.[25] In 1882, during the preparations for the centennial commemoration of Pombal's death and the construction of a monumental bronze statue, some deputies even went so far as to claim that Pombal should be given the moniker "liberator of the kingdom" for having expelled the Jesuits in the eighteenth century.

---

20 His lectures at the Sociedade Médica de Lisboa were compiled and published in 1898. See Miguel Bombarda, *A consciência e o livre-arbítrio* (Lisbon: Parceria António Maria Pereira-Livraria Editora, 1898).

21 Santana's articles were compiled and published in a two-volume work: Manuel Fernandes Santana, *Questões de biologia: O materialismo em face da sciencia; A propósito da consciência e livre arbítrio do Sr. Prof. Miguel Bombarda*, 2 vols. (Lisbon: Typographia da Casa Catholica, 1899–1900).

22 Miguel Bombarda, *A sciencia e o jesuitismo*: *Réplica a um padre sábio* (Lisbon: António Maria Pereira, Livraria Editora, 1900).

23 He included references to *Compêndio histórico* (1771); Grainha, *Os jesuítas* (1891); Grainha, *O Portugal jesuíta* (1893); Refóios, *O collegio de São Fiel* (1883); and Paul Bert, *La morale des jésuites* (Paris: G. Charpentier, 1883).

24 Bombarda, *A sciencia e o jesuitismo*, 186.

25 Portugal's parliamentary system was constituted by two houses, the lower Câmara dos Senhores Deputados and the higher Câmara dos Pares do Reino, inspired by the British House of Commons and House of Lords, respectively.

Implicitly, and sometimes explicitly, these parliamentarians agreed that the expulsion of the Jesuits had played a major part in Portugal's progress and development. Praise for Pombal's actions could be stretched to almost ridiculous proportions, as was the case when a deputy suggested the following parallel:

> Just as they pray on the church of Saint Ignatius of Loyola on July 31, let us pray to Saint Sebastião José de Carvalho e Melo on the altar of the country and on the temple of freedom on May 8; let us canonize that figure and let us state, at least here, with heads held high, that the centennial celebration of the marquis of Pombal also asserts that we worship him for being the grand statesman who expelled the Jesuits, because that expulsion was incontestably a step on the road to freedom.[26]

The contradictions in these accusations were manifest, as several anti-Jesuit politicians chose the colleges of the Society of Jesus for the education of their children. The duplicity of their position on Jesuit education was explicitly condemned by parliamentarian Tomás Ribeiro (1831–1901), who accused Portuguese politicians of living a life of "constant lies" because they did the opposite of what they claimed to stand for in public.[27]

Nevertheless, during the final years of the monarchist regime, demands for the Jesuits' expulsion frequently featured in parliamentary speeches. In 1908 and 1909, for instance, Afonso Costa (1871–1937), arguably the leading figure during the first years of the new republic, presented a bill to ban the Jesuits and close their colleges.[28] After being rejected twice, his bill was eventually promulgated on October 8, 1910 (i.e., three days after the republican revolution), thus becoming his first decree as minister of justice of the new republic (in office 1910–11).

2 The Republican Expulsion

The old regime disintegrated rapidly following the assassination of King Carlos and Prince Luís Filipe de Bragança on February 1, 1908. During the following

26 Câmara dos Senhores Deputados, "Sessão de 17 de Abril de 1882," in *Diário da Câmara dos Senhores Deputados* (Lisbon: Imprensa Nacional, 1882), 1127–49, here 1145.

27 Câmara dos Pares do Reino, "Sessão de 15 de Junho de 1891," in *Diário da Câmara dos Dignos Pares do Reino* (Lisbon: Imprensa Nacional, 1891), 1–7, here 2.

28 Câmara dos Senhores Deputados, "Sessão de 27 de Julho de 1908," in *Diário da Câmara dos Senhores Deputados* (Lisbon: Imprensa Nacional, 1908), 1–33, here 4–5; Câmara dos Senhores Deputados, "Sessão de 28 de Julho de 1909," in *Diário da Câmara dos Senhores Deputados* (Lisbon: Imprensa Nacional, 1909), 1–13, here 7.

two years, young King Manuel tried to reconcile monarchists and republicans, but his efforts soon proved ineffective. With the support of the masonic Grande Oriente Lusitano and the para-masonic Carbonária, republicans multiplied their allies throughout the country and gradually infiltrated the military. When, on October 3, 1910, masonic sergeants began the revolution with a few hundred soldiers in Lisbon, no one in Portugal or abroad tried to defend the weak monarchist regime, and within two days the republic was officially proclaimed.[29] Republicans associated the monarchy's decline with the alleged obscurantism of the Jesuits and made their expulsion a priority.[30] On October 8, Costa revoked the controversial decree of Hintze Ribeiro and restored two monarchist laws, the Pombaline decree that had expelled the Jesuits in 1759, and the liberal decree that had banned all religious orders in 1834.[31] Besides expelling the Jesuits and all religious orders and confiscating their property, the government also banished Cardinal José Almeida Neto (1841–1920), former patriarch of Lisbon (in office 1883–1907), as he was a member of the Franciscan order.[32]

Following the promulgation of Hintze Ribeiro's decree in 1901, the influence of the religious orders had increased significantly, and within a decade they were responsible for 164 institutions throughout the country. The steady growth of the Apostleship of Prayer during this period was perhaps one of the clearest expressions of the Jesuits' apostolic achievements after the restoration, as in 1909 this lay association had around fifteen hundred branches and a claimed membership of more than two million individuals, about half the population.[33] For republicans, the expulsion of the orders was mandatory, not only because the "country was a dependency of Jesuitism" but also because "political reaction went hand in hand with religious reaction."[34] As well as expelling the religious orders and confiscating their property, the new government also issued several laws to control traditional religious practices: the republicans abolished all Catholic feasts, banned the wearing of cassocks, forbade religious oaths, closed the Faculty of Theology, prohibited the teaching

29 Rui Ramos, "O fracasso do reformismo liberal (1890–1910)," in Ramos, Vasconcelos e Sousa, Monteiro, *História de Portugal*, 549–76, here 573–76.

30 Francisco Malta Romeiras and Henrique Leitão, "Jesuítas e ciência em Portugal V: Os Colégios de Campolide e de São Fiel e a implantação da República," *Brotéria* 174 (2012): 425–40.

31 *Diário do governo* 4 (October 10, 1910): 17–18.

32 C. [Carlton] H. Hayes and E. [Edward] M. Sait, "Record of Political Events," *Political Science Quarterly* 25 (1910): 727–60, here 757.

33 Richard Alan Hodgson Robinson, "The Religious Question and the Catholic Revival in Portugal, 1900–30," *Journal of Contemporary History* 12 (1977): 345–62, here 348.

34 Robinson, "Religious Question," 351. For the original quotation, see Eurico de Seabra, *A egreja, as congregações e a república*, 2nd ed. (Lisbon: Clássica, 1914), 292.

of Catholic doctrine in primary schools, allowed divorce, and made civil registration of births, marriages, and deaths mandatory.[35] All this culminated in a final decree that postulated the separation of the state from the church. Promulgated on April 20, 1911, the decree nationalized all church property; restricted religious processions, funerals, and prayer after sunset; reduced the number of seminaries; and, most importantly, submitted all parishes to state control.[36] As Costa supposedly declared in a masonic meeting, the goal was to "completely expunge" Catholicism "within two generations."[37]

During the initial years following the revolution, the new regime consequently viewed the eradication of Catholic influence in Portugal as its main priority, something that did not go unnoticed in the international press: "Preparations for the general expulsion of the religious congregations are absorbing much of the attention of the new Republican Government, which yesterday decreed their exile by virtue of a law passed as long ago as 1834."[38] Even those who defended the pretensions of republicans expressed their doubts on the severity of the measures being taken against the religious orders:

> We may regret the bloodshed and violence with which the new Portuguese government has been ushered into existence; we may even question the wisdom of the wholesale expulsion of the religious orders, but few will contend that the new republic is not at least as capable of fulfilling its international obligations as was the old, bankrupt, corrupt, and reactionary regime. We hold no brief for Portugal, but we believe that the beginning of a new and brighter era has dawned upon that unhappy land.[39]

---

35 Robinson, "Religious Question," 352.

36 On the decree that postulated the separation between the state and church, see especially João Seabra, *O estado e a igreja em Portugal no início do século XX: A Lei da Separação de 1911* (Cascais: Princípia, 2009); Luís Salgado Matos, *A separação do estado e da igreja* (Lisbon: Dom Quixote, 2011); and Sérgio Ribeiro Pinto, *Separação religiosa como modernidade: Decreto-Lei de 20 de Abril de 1911 e modelos alternativos* (Lisbon: Centro de Estudos de História Religiosa–Universidade Católica Portuguesa, 2011).

37 Rui Ramos, "A república antes da guerra," in Ramos, Vasconcelos e Sousa, Monteiro, *História de Portugal*, 577–603, here 586. This statement was broadcasted by the republican press and was refuted by Afonso Costa a few years later. The masonic meeting was allegedly held on March 26, 1911.

38 "Republic Sending Nuns from Portugal," *New York Times* (October 10, 1910): 4.

39 James Brown Scott, "The Revolution in Portugal: The Recognition of New Governments," *American Journal of International Law* 5 (1911): 178–81, here 181.

In the revolution's immediate aftermath, republicans reached out to the international press, looking for recognition for the newly formed regime. The editorial team of the *New York Times* reassured the paper's readers that the "outline of the programme of the victorious republicans" as conveyed by Costa "justifies strong hope for the stability of the newly formed republic" and that the "outlook, at present, is gratifying." Although the newspaper clearly supported the new regime, the wholesale expulsion of the religious orders was seen as a "somewhat too violent measure,"[40] and in the following days, some readers wrote letters in response to the republican propaganda. On October 9, Merwin-Marie Snell (1863–1921) warned that the new regime's objective was to "destroy the last vestiges of political, religious and pedagogical liberty, and to bring about a far closer union of Church and State." In his letter to the editor of the *New York Times*, Snell compared the "disregardful" expulsion of the religious orders to the hypothetical expulsion of "all married people, or all the unmarried, or all widows and widowers, or all carpenters and plumbers, or all Methodists and Baptists," concluding that a "nation that closes congregational schools is as hostile to liberty of education as one would be that closed all schools."[41]

During the turbulent month of October 1910, the international press sometimes tried to appear impartial, but the republican version of events was usually presented as the more accurate and truthful. When referring to the decree of October 8, for example, the *New York Times* referred to the assault on the Jesuit provincial curia, in the Rua do Quelhas, and presented both versions of the story:

> The hoisting of the British flag over the Jesuit Monastery in the Rua do Quelhas saved the occupants from the attacking Republican troops last night. One priest was killed and the others arrested. The attack began during the night after Republican troops had been fired upon from the windows of the monastery, or, according to some of to-day's papers, when the Jesuits threw bombs upon the heads of the parading pupils of the Polytechnic School, killing three of the youths. The troops hurried to the scene and fired a volley at the windows of the building. The Jesuits responded with more bombs.[42]

40 "The Situation in Portugal," *New York Times* (October 8, 1910): 10.

41 Merwin-Marie Snell, "The Revolution in Portugal," *New York Times* (October 9, 1910): 12.

42 "Religious Orders Exiled," *New York Times* (October 9, 1910): 2.

According to the *New York Times*, the "attacks upon the convents and particularly the assault upon the Jesuit monastery in the Rua do Quelhas" had caused "great excitement."[43] To the accusation that the Jesuits had started the altercation in the Rua do Quelhas, the Holy See appealed to the "common sense of the people of any country as to the improbability" of the Jesuits and other ecclesiastics throwing "bombs and other explosives without provocation" and insisted that the "mob must have attacked the religious houses, the inmates defending themselves."[44] Quoting an English Catholic residing in Lisbon for many years, *The Sacred Heart Review* assured its readers that "this bogey of Jesuitry in Portugal is a trumped up affair," and that "the stories about underground tunnels and bomb-throwing monks" would be "withdrawn as soon as they have done their inflammatory work."[45] Regardless of who was responsible for the altercations in Lisbon, Costa insisted that the congregations were all illegal, adding that their dissolution will "occur without trouble" and that the "confiscation of property will follow in due course."[46]

The attack against the Rua do Quelhas took place a few days after the bombardment of the Colégio de Campolide, which had been invaded by the military and a rebellious crowd on October 5 in a fruitless search for guns, explosives, and secret tunnels. The search for secret tunnels was motivated by a popular myth claiming that the queen mother Maria (1847–1911) used a "secret underground passage" to travel by coach between her palace and the Colégio de Campolide.[47] The *London Daily News* reported that the military declared it had found a "long tunnel no less than two miles in length," and that during the course of the investigations, one soldier and a civilian were seriously wounded and were found lying "on the ground in a pool of blood." Although the Jesuits were initially blamed for the incident, the English newspaper reported that "on calm consideration" it seemed that "the affair of the shooting might have been due to a mistake by one of the sentinels."[48]

43 "Republic Sending Nuns from Portugal," *New York Times* (October 10, 1910): 4.

44 "Vatican Defends Jesuits," *New York Times* (October 10, 1910): 4.

45 "The Republic in Portugal," *Sacred Heart Review* 44 (October 29, 1910): 8–9, here 9.

46 "Republic Sending Nuns from Portugal," *New York Times* (October 10, 1910): 4.

47 Robinson, "Religious Question," 351.

48 "New Portugal," *London Daily News* (October 13, 1910): 1–2, here 1. For other coetaneous accounts of the republican revolution and expulsion of the religious congregations in American and English newspapers, see also "Portugal Expels Monks and Nuns," *Los Angeles Herald* 33 (October 11, 1910): 16; "The Orders in Portugal," *New York Times* (October 12, 1910): 8; "The Orders in Portugal," *New York Times* (October 15, 1910): 10; "Republican Plans: Jesuits to Be Expelled," *Globe* (October 8, 1910): 7; and "Revolution within Revolution: Republicans against the Church," *Illustrated London News* (October 15, 1910): 3.

The Jesuit journal *America* conveyed an account of the republican expulsion as seen by scholastic Leon Meyer (1898–1942).[49] In his letter to a Spanish scholastic in Tortosa, Meyer reported that at one o'clock in the morning of October 4, the Jesuits of Campolide "were aroused by the discharge of artillery and musketry."[50] As the "rattle of musketry was continuous," some Jesuits dressed "themselves in citizen's clothes" and went to the train station nearby:

> When we got into the street that leads to the station we found that it was filled with people who, on seeing us, began to whisper amongst themselves, and here and there I heard the words, "They are disguised." We had taken only a few steps when the children began to shout, "Kill them, kill them!" following up their shouts with a shower of stones, and some older persons hidden amongst the trees were throwing such large ones that it would surely have been the end of us had we been struck in the head.[51]

Following the revolution, the military imprisoned 134 Jesuits for about a month in the penitentiaries of Limoeiro, in the center of Lisbon, and Caxias, in the outskirts of Cascais, one of the most popular summer holiday destinations for the royal family and aristocracy. During their captivity, the Jesuits were interviewed by Costa himself, who insistently accused them of kidnaping children to enroll them in the novitiate, thus distressing their families, as well as starving the novices for several consecutive days and concealing explosives in their colleges.[52] In the course of the interviews, the minister of justice allegedly declared to the captives that he would "go much harder with the Jesuits, because they had made vigorous warfare upon the Republicans."[53] A few months later, Provincial Luís Gonzaga Cabral (1866–1939, in office 1908–10) condemned the Jesuits' imprisonment.[54] In his words, the Jesuits had been "clapped into gaols and dungeons as notorious criminals" and "exposed to barbarous sufferings," including the "cruel humiliation" of going to an anthropometric station to be

49 Leon Meyer: b. June 2, 1878, Baden (Germany); e. September 27, 1898; d. November 13, 1942, Baturité, Ceará (Brazil) (*Mendizábal* 22,444).

50 "How the Portuguese Jesuits Were Driven Out," *America* 4 (1910): 131–32, here 131.

51 "How the Portuguese Jesuits Were Driven Out," 131.

52 On the description of the Jesuits' captivity and expulsion, see especially Luís Gonzaga de Azevedo, *Proscritos: Notícias circunstanciadas do que passaram os religiosos da Companhia de Jesus na revolução de Portugal de 1910*, 2 vols. (Valladolid: Florencio de Lara, 1911), 1:262–326.

53 "How the Portuguese Jesuits Were Driven Out," 132.

54 Luís Gonzaga Cabral: b. October 1, 1866, Foz do Douro, Porto (Portugal); e. October 9, 1882, Barro (Portugal); d. January 28, 1939, Salvador da Baía (Brazil) (*DHCJ*, 1:591–92).

"described, photographed, and measured in every detail, down to the joints of their fingers." To assess their religious monomania, the provisional government measured and took photographs of all captives without distinction, including "venerable elders," "distinguished men of science," "religious venerated for their virtue," and "youths still almost boys."[55] These photos, which were widely published in the republican press, aimed to convey the modernity and intellectual superiority of a republican scientific elite; yet, when seen from the perspective of today, they represent the most vivid testimonies of the farcical absurdity that was the republican expulsion (figure 10).[56] Whereas the conservative and obscurantist clerics under examination included "distinguished men of science" acknowledged for their educational and scientific work, the modern and enlightened physicians used and believed in phrenology, an ill-reputed pseudoscience.

FIGURE 10 "Os jesuítas em Portugal." Adapted from *Illustração portugueza* 246 (1910): 582–88 HML

55 Luís Gonzaga Cabral, *Expulsion of the Jesuits from Portugal: Protest by the Provincial of the Society of Jesus in Portugal* (London: Burns & Oates, 1911), 4–5.

56 The weekly *Illustração portugueza* published several pieces on the expulsion of the religious orders, with a special emphasis on the expulsion of the Jesuits. See especially the following articles: "As buscas nos conventos," *Illustração portugueza* 243 (October 17, 1910): 507–8; "O Ministro da Justiça e as Congregações Religiosas: Visita aos hospícios do Telhal e Idanha e à escola Penha Longa," *Illustração portugueza* 244 (October 24, 1910): 522–23; "Em busca dos subterraneos dos conventos," *Illustração portugueza* 244 (October 24, 1910): 537; "Fecham-se os conventos," *Illustração portugueza* 245 (October 31, 1910): 549; "Os jesuítas em Portugal," *Illustração portugueza* 246 (November 7, 1910): 582–88; "A expulsão dos jesuítas," *Illustração portugueza* 247 (October 14, 1910): 608–11; "O embarque das irmãnzinhas dos pobres," *Illustração portugueza* 258 (January 30, 1911): 143; "O espolio da Companhia de Jesus: A quinta do Valle do Rosal," *Illustração portugueza* 263 (March 6, 1910): 297.

3 The Natural History Collections: Diplomacy and Scientific Networks in Action

The bombardment of the Colégio de Campolide, at the height of the revolution, put an abrupt end to a timely conjuncture for Jesuit science and education.[57] Following the bombardment, the Jesuits' scientific instruments, books, and collections were destroyed or stolen during an assault that involved the military and a rebellious crowd. And what was not lost on that day was later confiscated by the new republic.[58] Most of the scientific instruments, books, and collections from São Fiel escaped unharmed, largely because the revolution only reached São Fiel a few days later.

After the expulsion of the religious orders, the provisional government nominated two commissions to catalog every object pertaining to the Jesuit colleges.[59] Headed by local Jesuit opponent José Ramos Preto (1871–1948), the commission of São Fiel suggested the relocation of the scientific instruments and collections to the University of Coimbra, the nearest public high school (the Liceu de Castelo de Branco), the local hospital, the municipal library, and the city hall. The commission sent the instruments and the physics, geology, zoology, and botany collections to the Liceu, which despite all legislative efforts still lacked a natural history museum and a physics cabinet, whereas the radiography equipment was transferred to the Hospital da Misericórdia "to be used in the treatment of poor patients." The commission also determined that the meteorological devices should be sent to the city hall in order to establish an observatory, while the science books ought to be included in the collections of the municipal library.[60] Given its importance for the study of botany and zoology, the herbarium of São Fiel and the collections of moths and butterflies belonging to Cândido de Azevedo Mendes, S.J. (1847–1943) were sent to the University of Coimbra.[61]

---

57 On the bombardment and invasion of the Colégio de Campolide, see Azevedo, *Proscritos*, 1:32–39.

58 Azevedo, *Proscritos*, 1:176.

59 Araújo, *Jesuítas e antijesuítas*, 56.

60 Maria Adelaide Neto Salvado, "O colégio de S. Fiel: Centro difusor da ciência," in *Fé, ciência, cultura:* Brotéria; *100 anos*, ed. Hermínio Rico and José Eduardo Franco (Lisbon: Gradiva, 2003), 211–31, here 218–19. For a detailed list of São Fiel's instruments and collections before the republican expulsion, see also Maria Adelaide Neto Salvado, *O colégio de S. Fiel: Centro difusor da ciência no interior da Beira* (Castelo Branco: Semedo-Soc. Tipográfica, 2001).

61 Cândido de Azevedo Mendes: b. January 17, 1874, Torres Novas (Portugal); e. September 7, 1888, Barro (Portugal); d. December 16, 1943, Ceará (Brazil) (*DHCJ*, 3:2618).

Azevedo Mendes attempted to retrieve his collections while exiled in Salamanca, Spain. Together with the diplomatic efforts to retrieve his moths and butterflies, Azevedo Mendes also published two booklets refuting the official republican account of the Jesuits' expulsion, focusing particularly on contradicting the version conveyed by lawyer Ramos Preto.[62] As well as denying the typical and generic accusations of wealth, avarice, and influence, Azevedo Mendes vehemently rebutted the allegations of educational backwardness. For the Jesuit, Ramos Preto's assertion that São Fiel was not superior to the Liceu, except for the existence of a physics cabinet, a chemistry laboratory, and a zoology museum, was a deliberate misrepresentation. As a local jurist and, perhaps most importantly, as a subscriber to the Jesuit scientific journal *Brotéria*, Ramos Preto was certainly aware of the Jesuits' contributions to zoology and botany, as well as their efforts to establish a reliable meteorological observatory in São Fiel. For those reasons, Azevedo Mendes failed to understand why the Jesuits' successive attempts to retrieve their books, instruments, and collections had been unsuccessful.[63]

Nevertheless, despite the loss of his possessions, Azevedo Mendes continued to collect, describe, and classify new specimens of moths and butterflies during his twenty-one-year exile in Spain. In 1932, however, history repeated itself. Expelled from the Spanish territories together with his fellow Jesuits, Azevedo Mendes lost his collections once again. One of the differences between the 1910 banishment and the 1932 expulsion resided in the final destination of Azevedo Mendes's collections: in 1910, his collections of moths and butterflies were confiscated by the Portuguese provisional government and were later transferred to the University of Coimbra (where they still remain), whereas in 1932 the government of Manuel Azaña (1880–1940, in office 1931–33, 1936) eventually ordered his collections to be returned.[64]

Besides the individual efforts to rescue their possessions, the exiled naturalists also submitted a joint petition to the provisional government asking for the return of their scientific instruments books, journals, and manuscripts, as well

62 Cândido de Azevedo Mendes, *A* Brotéria *no exílio* (Madrid: Imprenta Ibérica, 1913), and Mendes, *O collegio de S.Fiel: Resposta ao advogado Sr. José Ramos Preto* (Madrid: López del Horno, 1911).

63 José Ramos Preto, *Relatório sobre o extincto collegio de São Fiel da Companhia de Jesus* (Lisbon: Typographia Mauricio, 1911), 26; Mendes, *O collegio de S.Fiel*, 66.

64 The moths and butterflies that Azevedo Mendes collected between 1911 and 1932 can be seen at the Jesuit college Instituto Nun'Alvres (Caldas da Saúde, Santo-Tirso). Afonso Luisier, "R.P. Cândido Azevedo Mendes (S.J.)," *Brotéria: Ciências naturais* 13 (1944): 44–45.

as the zoological, botanical, geological, and mineralogical collections.[65] The petition was widely diffused by the Jesuits and their acquaintances and generated a commotion both nationally and internationally.[66] When American botanist Theodore Cockerell published a plea in favor of the exiled naturalists in *Science*, he explicitly mentioned the significance of the entomological and botanical collections, the natural history library, and the microscopy laboratory that had been confiscated by the republican regime. He was particularly concerned with the fate of the collections of São Fiel because the commission appointed by the government "did not include a single naturalist; instead it consisted of a veterinarian, a physician, a professor and two lawyers" and was "presided over by a particular enemy of the college." Cockerell warned his readers that the collections could be lost for science because "in many cases the specimens are not labelled, and in others they are marked with numbers, abbreviations, etc., intelligible only to their original readers."[67]

For the Jesuits, appealing to the common sense of a broader community of naturalists seemed to be the best strategy to get their collections back. It was also a way of responding to Costa, who supposedly claimed that even "if your collections are lost to you, they are not lost to science."[68] Cockerell conceded that there might be "another side to this question," but added that "granting the accusations of their enemies, that they are Jesuits, and (I suppose) opposed to a republican form of government, there is still no justification for the action taken in depriving them of their scientific materials." He consequently proposed to "appoint and pay the expenses of a man who would enquire into all the facts" to determine if the Jesuits' case was well grounded. If a "carefully considered report" supported their claim, then "organized protests from the scientific bodies of different countries might be appropriate."[69]

Portuguese scientists also expressed their repugnance at the confiscation of the Jesuits' collections. Public denunciations came from the most famous Portuguese chemist, António Ferreira da Silva (1853–1923), founder and president of the Portuguese Chemistry Society, renowned physician Marck Athias (1875–1946), as well as from agronomists Luís de Castro (1868–1928) and José

65 The petition was signed by some of the most prominent Jesuit scientists working in Portugal in 1910, including Afonso Luisier (1872–1957), António de Oliveira Pinto (1868–1933), Camilo Torrend (1875–1961), Cândido de Azevedo Mendes (1874–1943), Carlos Zimmermann (1871–1950), and Joaquim da Silva Tavares (1866–1931).

66 Luís de Castro, "Protesto de homens de sciencias," *Boletim da Associação central da agricultura portuguesa* 1 (1912): 165–67.

67 Cockerell, "Exiled Naturalists of Portugal," 714.

68 Cockerell, "Exiled Naturalists of Portugal," 714.

69 Cockerell, "Exiled Naturalists of Portugal," 714–15.

Veríssimo de Almeida (1834–1915).[70] While Ferreira da Silva and Castro were close to the Jesuits and, most likely, monarchists, Veríssimo de Almeida was a staunch republican and a freemason. As Silva Tavares put it: "Despite being a free thinker and almost sectarian in matters of religion," Veríssimo de Almeida publicly supported the Jesuits' claims.[71]

Institutionally, the Jesuits owed much of their support to the Portuguese Society of Natural Sciences. Established in 1907 at the initiative of physicians Athias and Augusto Celestino da Costa (1884–1956), the society included seven Jesuits among its twenty-two founding members: Silva Tavares, Azevedo Mendes, Zimmermann, and Manuel Rebimbas (1873–1944), professors of natural sciences of the Colégio de São Fiel; and Afonso Luisier (1872–1957), Torrend, and Oliveira Pinto, professors of natural sciences of the Colégio de Campolide.[72] During the exile, Tavares, Zimmermann, and Rebimbas acknowledged the encouragement they had received from the society and its members in the "defense of our case."[73]

Notwithstanding the formal and informal support the Jesuits received from the scientific community, the provisional government ignored most of the protests. Eventually, however, some collections were partially recovered thanks to the intervention of foreign diplomats. Luisier retrieved part of his collection of mosses in April 1913, for example, following the intervention of the Swiss consul, whereas Torrend recovered his complete collection of fungi with the support of the French ambassador. This benign attitude toward foreign Jesuits reveals the inner and outer tensions that the new regime was experiencing. Striving to be internationally recognized as lawful rulers, and fearing

70 Castro, "Protesto de homens de sciencias," 165–67; António Ferreira da Silva, "Director e redatores da *Brotéria*: Um appello aos homens de boa vontade," *Revista de chimica pura e applicada* 6 (1910): 362–63; Silva, "Os redactores da *Brotéria* exilados: As suas collecções scientificas confiscadas e perdidas," *Revista de chimica pura e applicada* 7 (1911): 229–31; Silva, "*Brotéria*," *Revista de chimica pura e applicada* 7 (1911): 377; Silva, "Testemunho em favor da *Brotéria*," *Brotéria:Vulgarização científica* 16 (1918): 190–91; Silva, "*Brotéria* e os seus naturalistas," *Revista de chimica pura e applicada* (1918): 75–79; Joaquim da Silva Tavares, "José Veríssimo de Almeida," *Brotéria: Botânica* 13 (1915): 57–60.

71 Tavares, "José Veríssimo de Almeida," 57.

72 Joaquim da Silva Tavares, "A Sociedade Portuguesa de Sciencias Naturais," *Brotéria: Vulgarização Científica* 6 (1907): 127–34. Manuel Rebimbas Tavares: b. January 27, 1873, Pardelhas, Aveiro (Portugal); e. September 18, 1894, Barro (Portugal); d. December 26, 1944, Braga (Portugal) (*Mendizábal* 23,295). Alphonse (Afonso) Luisier: b. February 6, 1872, Fregnoley, Valais (Switzerland); e. October 2, 1891, Barro (Portugal); d. November 4, 1957, Caldas da Saúde (Portugal) (*DHCJ*, 4:2440–41).

73 APSI, Joaquim da Silva Tavares, Carlos Zimmermann and Manuel Rebimbas to President of the Portuguese Society of Natural Sciences, n.d. (after 1910).

diplomatic reprisals from France and Switzerland, republicans only sacrificed their political (and religious) agenda in extreme circumstances.[74]

Zimmermann, who had been exiled in Brazil since March 1911, also tried to recover his collections by means of diplomacy. In October 1911, however, the German consul informed him that he could not proceed without proof of his citizenship. After contracting a lawyer in Germany to act on his behalf and get the coveted documents, the Jesuit was notified by the government that the efforts to recover his collections could cause "a serious conflict" between the two nations.[75] Zimmermann subsequently abandoned his efforts to retrieve his collections via diplomatic means, instead resorting to his network of correspondents, and especially his friend José da Silva e Castro (1842–1928), who played a pivotal role in the return of the collections.[76]

Like Zimmermann, Silva e Castro had been invested in the identification and classification of diatoms, one of the largest taxa of microscopic algae, since the 1890s. He exchanged diatoms with Zimmermann and frequently resorted to the Jesuit and his scientific correspondents, including French botanists Joannes Albert Tempère (1847–1926) and Maurice Peragallo (1853–1931), Zimmermann's former professor in Paris, to corroborate his taxonomical research.[77] Along with the letters and trade of specimens, Zimmermann also provided Silva e Castro with better solutions, including a mounting media

---

74 For comparison, see also Filipa Lowndes Vicente, "Travelling Objects: The Story of Two Natural History Collections in the Nineteenth Century," *Portuguese Studies* 19 (2003): 19–37.

75 Jorge de Brito e Abreu (henceforth JBA), Carlos Zimmermann to José da Silva e Castro (October 17, 1911). Extant in a private archive, this letter was also included in Francisco Malta Romeiras, "Constituição e percurso das colecções científicas dos jesuítas exilados pela 1ª República: O caso de Carlos Zimmermann S.J. (1871–1950)," *Archivum historicum Societatis Iesu* 168 (2015): 287–327, here 320–21. On Zimmermann, see also Romeiras and Leitão, "Jesuítas e ciência em Portugal II: Carlos Zimmermann S.J.," and José Eduardo Franco, "Os directores da *Brotéria*: Carlos Zimmermann," in Rico and Franco, *Fé, ciência, cultura*, 149–50.

76 For a biography, see Rolanda Maria Albuquerque de Matos, *José da Silva e Castro: O naturalista e a sua obra, 1842–1928* (Lisbon: Instituto Português de Malacologia, 2009).

77 JBA, Carlos Zimmermann to José da Silva e Castro (April 25, 1910). Published in Romeiras, "Constituição e percurso das colecções científicas," 313. On the importance of networks of scientific correspondence for the development of botany and zoology in the nineteenth and twentieth centuries, see especially Eugene Cittadino, "Botany," in *The Cambridge History of Science*, vol. 6, *Modern Biological and Earth Sciences*, ed. Peter J. Bowler and John V. Pickstone (Cambridge: Cambridge University Press, 2009), 225–42, and Mario A. Di Gregorio, "Zoology," in Bowler and Pickstone, *Cambridge History of Science*, 6:204–24.

especially conceived by Tempère for observing diatoms.[78] Indebted to the exiled Jesuit, Silva e Castro agreed to act as his representative and in November 1911 retrieved Zimmermann's microscopic preparations and manuscripts, as well as his photographs and sketches of diatoms.[79]

The successful return of the collections to Silva e Castro also owed much to the efforts of naturalist António Machado (1883–1969). Son to the minister of foreign affairs, and future president (in office 1915–19 and 1925–29), Bernardino Machado (1851–1944), and member of the commission responsible for cataloging the assets of Campolide, Machado was particularly well positioned to negotiate the return of the collections. Simply referred to in Jesuit sources as a "friend who is also a friend of the new regime," Machado cataloged Zimmermann's possessions and kept them in his house before delivering them personally to Silva e Castro, because it was not safe to keep them in Campolide nor to send them by train.[80] It is also likely that the commission itself was favorable to the return of Zimmermann's collections, because unlike the commission of São Fiel, it included distinguished personalities from the political, scientific, and public spheres, such as historian and politician Anselmo Braamcamp Freire (1849–1921), anthropologist José Leite de Vasconcelos (1858–1941), and, most importantly, botanists António Xavier Pereira Coutinho (1851–1939) and Veríssimo de Almeida, and zoologist Antero de Seabra (1897–1983).[81] Following the return of the collections, Machado continued to correspond with the Jesuits and, encouraged by Luisier, started to publish his botanical and zoological research in *Brotéria*.[82]

Fifty years after their de facto restoration, the Portuguese Jesuits were recognized by national elites and, most importantly, by their peers, as competent

78 JBA, Carlos Zimmermann to José da Silva e Castro (May 17, 1910). Published in Romeiras, "Constituição e percurso das colecções científicas," 314–15.

79 JBA, Carlos Zimmermann to José da Silva e Castro (November 27, 1911). Letter included in Romeiras, "Constituição e percurso das colecções científicas," 321–22.

80 Mendes, *A Brotéria no exílio*, 26–27.

81 JBA, José da Silva e Castro to [António Machado?] (after September 1911). See Romeiras, "Constituição e percurso das colecções científicas," 325–26.

82 António Machado, "Notas de bryologia portuguesa: 1; Plantas novas para Portugal," *Brotéria: Botânica* 15 (1917): 8–11; Machado, "Notas de bryologia portuguesa: 2; Duas excursões às Serras da Estrela e do Gerês," *Brotéria: Botânica* 15 (1917): 49–63; Machado, "Notas de zoologia," *Brotéria: Ciências naturais* 11 (1942): 97–105; Machado, "Problemas da sistemática (As grandes divisões ou tipos zoológicos)," *Brotéria: Ciências naturais* 14 (1945): 62–71; Machado, "Revisão de algumas hipóteses filogenéticas," *Brotéria: Ciências naturais* 14 (1945): 149–55; Machado, "Contribuição para os Miriápodes de Portugal," *Brotéria: Ciências naturais* 15 (1946): 5–37; Machado, "Miriápodes de Portugal," *Brotéria: Ciências naturais* 21 (1952): 65–170.

pedagogues and scientists. Their books, instruments, and collections were perhaps the most vivid testimony of their educational and scientific legacy in early twentieth-century Portugal, which helps explain why they resorted to all possible means to retrieve their possessions after the 1910 expulsion. Whereas Luisier's and Torrend's collections were returned following the interventions of other countries, Zimmermann was only able to retrieve his collections with the support of an informal network constituted by himself and naturalists Silva e Castro and Machado. Unlike his foreign brethren, Silva Tavares had to wait until 1927 before his collections were returned. Although they had been proscribed since 1910, the Jesuits started to return discreetly in 1922, and with the fall of the republic in 1926, the political situation became more favorable to the Society of Jesus. In 1927, probably as a token of good faith, the new military regime (1926–33) returned the botanical and entomological collections to Silva Tavares as well as a microtome essential for his cytological studies.[83]

The Jesuits' expulsion, the closure of their colleges, and the new government's unwillingness to return their scientific collections, instruments, and books can be interpreted as a predictable consequence of a struggle for cultural hegemony between the Jesuits, as intellectual representatives of the old monarchist regime, and the republicans, who sought to broadcast their own values and principles through the creation of a new educational order.[84] In a period when a variety of political and religious groups were struggling for educational hegemony throughout Europe, the republicans viewed the expulsion of the Jesuits and all religious orders as a necessary first step toward achieving social and political stability in the newly formed regime.

83 António Lopes, *Roteiro histórico dos jesuítas em Lisboa* (Braga: Livraria Apostolado da Imprensa–Editorial Apostolado da Oração, 1985), 143–46; Joaquim da Silva Tavares, "A *Brotéria* nos primeiros 25 anos da sua existência," in *Brotéria: Número especial no XXV aniversário da sua fundação* (Caminha: Brotéria, 1927), 19–37, here 37.

84 Gramscian concepts such as hegemony, counter-hegemony, and intellectuals have been reconsidered by historians of science in the past two decades. See Agustí Nieto-Galan, "Antonio Gramsci Revisited: Historians of Science, Intellectuals, and the Struggle for Hegemony," *History of Science* 49, no. 4 (2011): 453–78. For a book using this historiographical framework, see Oliver Hochadel and Agustí Nieto-Galan, eds., *Barcelona: An Urban History of Science and Modernity, 1888–1929* (London: Routledge, 2016). On the struggle between different religious groups for cultural and educational hegemony in the nineteenth and twentieth centuries, see also Barry J. Hake, Tom Steele, and Alessando Tiana-Ferrer, eds., *Masters, Missionaries and Militants: Studies of Social Movements and Popular Adult Education 1890–1939* (Leeds: Leeds Studies in Continuing Education, 1996); Barry J. Hake and Tom Steele, eds., *Intellectuals, Activists and Reformers: Studies of Cultural, Social and Educational Reform Movements in Europe, 1890–1930* (Leeds: Leeds Studies in Continuing Education, 1998); John Rudolph, *Scientists in the Classroom: The Cold War Reconstruction of American Science Education* (New York: Palgrave, 2002).

During their exile in Spain, Belgium, and Brazil, the Portuguese Jesuits established new missions and colleges, namely the Colégio Antônio Vieira, in São Salvador da Baía, Brazil, and the Instituto Nun'Alvres.[85] The latter institute, the proclaimed heir of Campolide, was dedicated to Nuno Álvares Pereira (1360–1431), a celebrated fourteenth-century war hero who became a Carmelite friar after his wife's passing.[86] Dedicating their college in the diaspora to Manuel II's forefather was certainly not fortuitous. First established in Dieleghem (1912–14), on the outskirts of Brussels, the institute was consecutively transferred to Los Placeres (1914–16) and La Guardia (1916–32), both in Spain, and finally to Santo-Tirso (1932–), in Portugal, where it still functions today.

During the exile, the Jesuits continued to classify and describe animals and plants in the pages of *Brotéria*, which was subsequently awarded a gold medal in two international exhibitions in Brazil (Baía, 1913, and Rio de Janeiro, 1922). When they returned to Portugal after two decades in exile, the Jesuits' contributions to botany and zoology were publicly acknowledged by the scientific community and they were granted honorary doctorates and bestowed with the highest state honors. Faithful to recent history and their scientific tradition, the Portuguese Jesuits continued to publish *Brotéria* in the following decades.

---

85 On the history of the Instituto Nun'Alvres, see José Carvalhais, *80 anos na educação (1912–1992): Instituto Nun'Alvres* (Caldas da Saúde: Instituto Nun'Alvres, 1992). On the history of the Colégio Antônio Vieira, see Waldir Freitas Oliveira and Edilece Souza Couto, *Colégio Antônio Vieira (1911–2011): Vidas e histórias de uma missão jesuíta* (Salvador da Baía: EDUFBA, 2011).

86 Although it was uncommon to dedicate a college to a Carmelite friar, especially one that was not beatified, the choice of Álvares Pereira was probably motivated by political factors. Besides his pivotal role in the foundation of the Avis dynasty (1385–1580), he was also tightly connected with the ruling dynasty in 1910, the house of Bragança (1640–1910), because his daughter Beatriz (1380–1415) and his son-in-law Afonso (1377–1461) were the first dukes of Bragança. The subject of popular veneration for centuries, Nuno Álvares Pereira was beatified by Benedict XV (1854–1922, r.1914–22) on January 23, 1918 and canonized by Benedict XVI (1927–, r.2005–13) on April 26, 2009.

CHAPTER 6

The Journal *Brotéria*, the Book of Nature, and the Greater Glory of God

During the Republic of Letters, the increasing amount of academic correspondence, the creation of well-known salons and academies, and the foundation of the first scientific journals played crucial roles in the circulation of knowledge, especially in Europe and North America.[1] The French Jesuits were aware of the importance of participating in this metaphorical and learned nation and followed the scientific tradition of the previous centuries by launching the Society's first learned periodical. Founded in Trévoux with the support of Louis-Auguste de Bourbon (1620–1726), duke of Maine and legitimized son to King Louis XIV (1638–1715, r.1643–1715), the journal was officially entitled *Mémoires pour l'histoire des sciences et des beaux arts* (Memoirs for the history of sciences and fine arts [1701–67]). Its foundation at the beginning of the eighteenth century was associated with two central religious affairs in this period: the controversies surrounding the expansion of Jansenism and the revocation of the Edict of Nantes (1598) in 1685, which led to the expatriation of French Calvinists. With the publication of a periodical such as the *Journal de Trévoux*, the French Jesuits had found an efficient vehicle to put forth their points of view and participate in those religious quarrels.[2]

For over six decades, the *Journal de Tréxoux* played a highly significant role in the dissemination of modern scientific research to a learned, yet broad, audience.[3] Despite being known especially for its role in the defense of Cartesianism in France, as well as its articles on physics and mathematics, it was in fact a very eclectic periodical. The wide array of subjects it covered—including theology, philosophy, history, law, economics, natural history, astronomy, fine arts, medicine, and agriculture—and the high-quality of its

1 Among the most renowned initiatives of the Republic of Letters are the foundation of the Académie Française (1635), the Royal Society (1660), and the Académie des Sciences (1666), and the publication of two learned journals: the *Journal des Sçavans* (first printed in January 1665) and the *Philosophical Transactions* (first printed in March 1665).

2 Jean Laponce, "The Jesuits and Science in Eighteenth-Century France: An Analysis of Scientific Writings in the *Journal of Trévoux*" (MSc. diss., University of British Columbia, 1990), 23–24.

3 Marie-Hélène Froeschlé-Chopard and Michel Froeschlé, "'Sciences et arts' dans les *Mémoires de Trévoux* (1701–1762)," *Revue d'histoire moderne et contemporaine* 48 (2001): 30–49.

 | DOI:10.1163/9789004382367_007

scholarship distinguished the *Journal de Trévoux*, together with the *Journal des Sçavans*, as one of the most important learned journals published during the French Enlightenment.[4]

Following the success of the *Journal de Trévoux*, in 1750 Italian Jesuit Francesco Antonio Zaccaria (1714–95) founded an encyclopedic journal entitled *Storia letteraria d'Italia* (Literary history of Italy [1750–59]).[5] Each volume contained a section exclusively dedicated to science with articles commenting on modern books and scientific topics in numerous fields, including mathematics, physics, chemistry, medicine, and astronomy. Together with summary descriptions of scientific novelties, which constituted the core of the scientific section, the journal sometimes published articles containing original research, especially astronomical observations.[6] Yet *Storia letteraria d'Italia* would have a short and controversial life. As Zaccaria and his collaborators failed to rebut their most outspoken critics, Superior General Ignazio Visconti (1682–1755, in office 1751–55) intervened and instituted pre-publication censorship. When the controversies continued to linger, Superior General Lorenzo Ricci (1703–75, in office 1758–73) requested its suspension "for the better interest of the Order."[7] Despite the efforts of Francesco III, duke of Modena (1698–1780), to secure its subsistence, publication of the *Storia letteraria d'Italia* was finally suspended in 1759.

In the nineteenth century, the Jesuits revived their determination to edit learned periodicals directed at a broad audience. *La civiltà cattolica* (Catholic civilization), which was founded in Naples by Carlo Maria Curci, S.J. (1809–91) in April 1850, was the first journal published by the Jesuits after the restoration.[8] According to its inaugural editorial, the main aim of the journal was to provide a careful and logical explanation of Catholic doctrine and moral principles.[9] Following the foundation of a house of writers in 1866, *La civiltà cattolica* fell

---

4 For a thematic index of the articles published in the *Journal de Trévoux*, see Carlos Sommervogel, *Table méthodique des* Mémoires de Trévoux *(1701–1775)* (Paris: Imprimerie de E. Donneaud, 1864–65).

5 Francesco Antonio Zaccaria: b. March 27, 1714, Venice (Italy); e. August 18, 1731, Vienna (Austria); d. August 10, 1795, Rome (Italy) (*DHCJ*, 4:4063–64).

6 Brendan Dooley, "The *Storia letteraria d'Italia* and the Rehabilitation of Jesuit Science," in Feingold, *Jesuit Science and the Republic of Letters*, 433–73. During its short existence, *Storia letteraria d'Italia* published the observations of the solar eclipses of January 8 and June 19, 1750, the lunar eclipse of June 8, 1751, and the transit of Mercury on May 5, 1753.

7 Dooley, "*Storia letteraria d'Italia*," 459.

8 Carlo Maria Curci: b. September 4, 1809, Naples (Italy); e. September 13, 1826, Naples (Italy); d. June 8, 1891, Florence (Italy) (*DHCJ*, 1:1025–26).

9 "Il giornalismo moderno ed il nostro programma," *La civiltà cattolica* 1 (1850): 5–24.

under the jurisdiction of the Roman curia and of Pius IX, who urged the Jesuits to dedicate themselves to refuting heterodox doctrines circulating in the press. In the period that followed, the authors of *La civiltà cattolica* would participate in the main political, philosophical, theological, and scientific quarrels of the day, thus establishing the journal as one of the most important cultural periodicals published in the Italian Peninsula.[10]

In 1856, the Jesuits founded the cultural magazine *Études religieuses, philosophiques, historiques et littéraires* (Religious, philosophical, historical, and literary studies) in Paris. As a periodical aimed at a wide audience, *Études*, as it came to be known, published articles on subjects such as architecture, archeology, history, geography, demography, education, sociology, law, politics, philosophy, and theology. It also discussed controversial topics of particular interest to the defense of the Catholic faith, such as freemasonry, positivism, Protestantism, and Jansenism.[11] Following the previous examples of the *Journal de Trévoux* and of the *Storia letteraria d'Italia*, *Études* also published popular scientific articles on botany, physiology, physics, chemistry, astronomy, meteorology, medicine, and hygiene, and covered polemical issues in the history of science such as Darwinism and the Galileo affair.[12] However, the period in which the French Jesuits most actively engaged in popularizing scientific research only corresponds to the first quarter of the history of *Études* (1856–80), as the journal's focus on scientific themes gradually diminished in the following decades.

The success of *La civiltà cattolica* and *Études* inspired the foundation of other Jesuit cultural magazines such as *The Month* (England, 1864), *Stimmen der Zeit* (Voices of the times [Germany, 1865]), *Razón y fé* (Reason and faith

10 On the history of *La civiltà cattolica* from 1850 to 1999, see Giuseppe de Rosa, La civiltà cattolica: *150 anni al servizio della chiesa, 1850–1999* (Rome: La Civiltà Cattolica, 1999).

11 For the reconstruction of the main subjects covered by the journal *Études* from 1856 to 1924, see the following indexes: *Études religieuses, philosophiques, historiques et littéraires publiée par des pères de la Compagnie de Jésus: Table générale des vingt-cinq premières années (1856–1880)* (Paris: Victor Retaux & Fils, Éditeurs, c.1880); *Études publiées par des pères de la Compagnie de Jésus: Table générale des années 1880 a 1900* (Paris: Victor Retaux, Libraire-Éditeur, c.1900); *Études: Tables des années 1911–1924* (Paris: Bureau des Études, 1926).

12 From 1856 and 1880, the following articles criticized Darwinism and commented on the Galileo affair: V. Becker, "Un duel darwiniste (Haeckel et His)," *Études* 21 (1877): 672–88; J. de Bonniot, "Darwinisme et logique," *Études* 16 (1872): 423–43; I. Carbonnelle, "Le darwinisme, théorie et objections," *Études* 14 (1869): 472–82; E. Desjardins, "M. Mézières et le procès de Galilée," *Études* 21 (1877): 273–81; A. de Gabriac, "Galilée devant la science, la religion et la littérature," *Études* 12 (1867): 528–47.

[Spain, 1901]), and *America* (United States, 1909), to name but a few.[13] With the publication of *Razón y fé*, as the title itself suggested, the Spanish Jesuits aspired to participate in the controversies related to the so-called conflict between science and religion and thus contribute to the discussion and acceptance of the compatibilist thesis. In its opening issue, they justified the publication of the new journal by arguing that the "public tribune of the press" was the best stage to dispute the "clamor of false science, untruthful philosophy, and corrupting literature" that had been undermining Catholic doctrine in recent decades.[14] For the editors, the foundation of *Razón y fé* was firmly rooted in the ideological and political struggles of Pius VII (1742–1823, r.1800–23), Leo XII (1760–1829, r.1823–29), Gregory XVI (1765–1846, r.1831–46), and Pius IX against modernism, with the journal serving as a practical means to foster the development of a Catholic culture in nineteenth-century Europe. In one editorial contesting the views of anticlerical critics, Italian Jesuits recalled the encyclical *Immortale Dei* (1885), in which Leo XIII (1810–1903, r.1878–1903) openly declared the church's support of science: "Whatsoever spreads the range of knowledge will always be willingly and even joyfully welcomed by the Church. She will always encourage and promote, as she does in other branches of knowledge, all study occupied with the investigation of nature."[15] While describing the publication of the journal as part of a broader apostolic mission, the editorial's authors also resorted to the warlike metaphors attributed to Ignatius of Loyola in presenting themselves as the last soldiers of an army striving for the greater glory of God.[16]

The journal sought to publish scientific articles that would appeal both to specialists and non-specialists. During the initial years of its publication, *Razón y fé* documented some of the Society's scientific activities in Spain, with articles describing the astronomical and meteorological observatory in Granada, for instance, or the philosophical voyage of a Jesuit naturalist traveling through the region of Galicia.[17] In addition to presenting the Jesuits as defenders of

13 On the history of *America*, see especially Michael F. Lombardo, *Founding Father: John J. Wynne, S.J. and the Inculturation of American Catholicism in the Progressive Era* (Leiden: Brill, 2017), 231–94.

14 "À los lectores," *Razón y fé* 1 (1901): 1–5, here 2.

15 Leo XIII, *Immortale Dei* (Rome: November 1, 1855). English translation available from http://w2.vatican.va/content/leo-xiii/en/encyclicals/documents/hf_l-xiii_enc_01111885_immortale-dei.html (accessed March 3, 2019).

16 "À los lectores," 4–5.

17 On the observatory of Granada, see Juan Granero, "El observatorio astronómico, geodinámico y meteorológico de Granada," *Razón y fé* 3 (1902): 222–25; Granero, "Observatorio de Granada," *Razón y fé* 3 (1902): 512–20; Ramón Martínez, "Observatorio de Granada: Sección meteorológica," *Razón y fé* 4 (1902): 478–90; Juan Granero, "Observatorio de Granada: Sección astronómica," *Razón y fé* 5 (1903): 339–47. Apart from the textual descriptions,

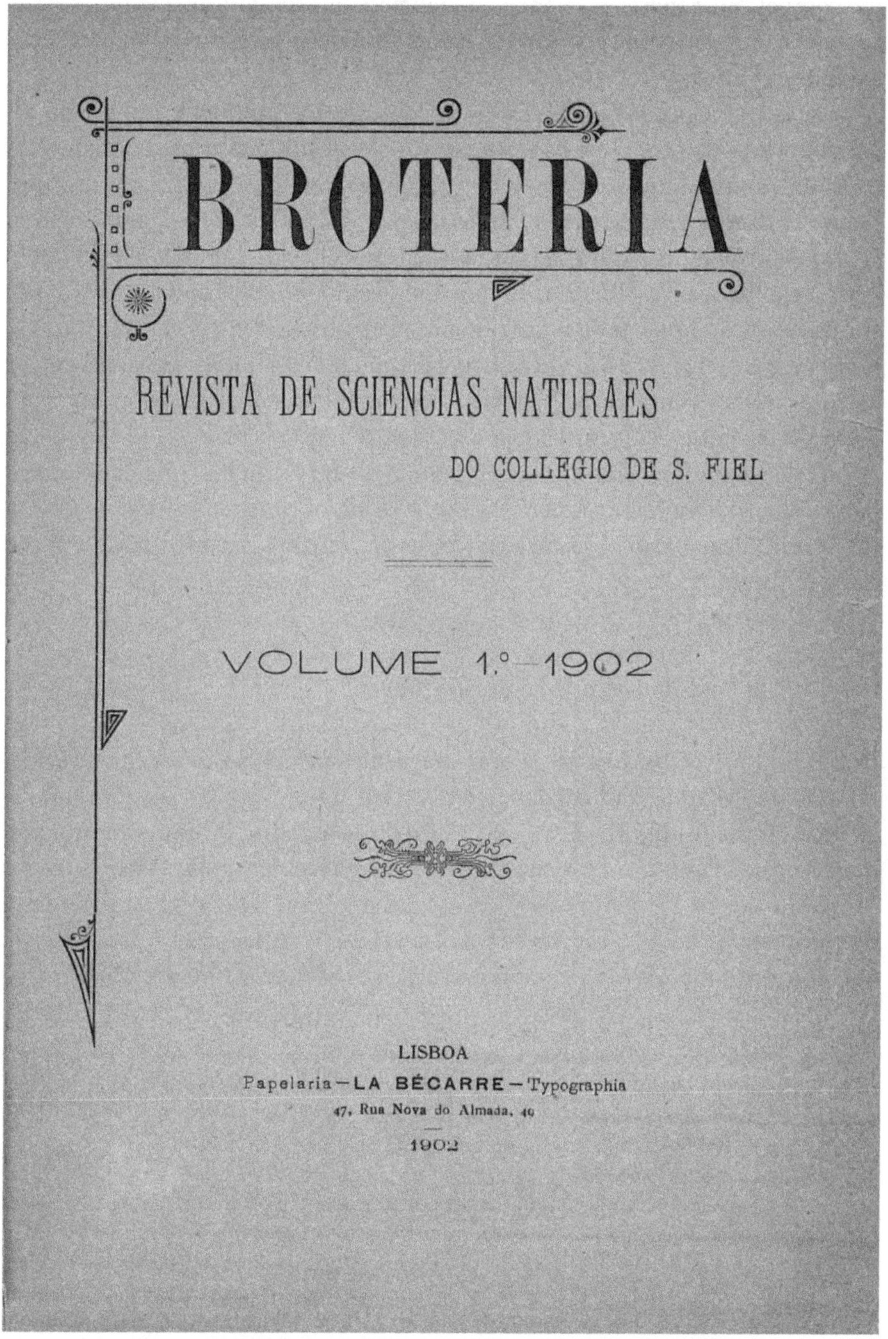

BROTERIA

REVISTA DE SCIENCIAS NATURAES

DO COLLEGIO DE S. FIEL

VOLUME 1.º—1902

LISBOA

Papelaria—LA BÉCARRE—Typographia

47, Rua Nova do Almada, 49

1902

FIGURE 11 First issue of *Brotéria*, October 1902
AB

the compatibilist thesis of science and religion, *Razón y fé* aimed to portray the Spanish Jesuits as active and modern practitioners of astronomy, meteorology, and natural history.

Despite being published in very different historical, political, and geographical contexts, all of the Society's journals served a common purpose in allowing the Jesuits to participate in the major philosophical, theological, political, and scientific debates of their time. It was against this background, and inspired by their companions, that the Portuguese Jesuits decided to publish a learned periodical themselves, though, unlike other Jesuit periodicals, *Brotéria* was not conceived as an apologetic magazine but rather as a strictly scientific journal. From 1902 to 2002, *Brotéria* published around 1,300 scientific articles covering a variety of fields, including botany, zoology, plant breeding, biochemistry, and molecular genetics.[18] Its initial focus on identifying, describing, and classifying novel species collected in Portugal, Spain, Austria, Germany, Angola, Mozambique, Timor, Brazil, and Argentina played a significant part in the evolution of the botanical and zoological sciences, especially in the early decades of the twentieth century.

1 The Foundation of a Scientific Journal

In 1902, a group of teachers from the Colégio de São Fiel began to publish *Brotéria*, the first Jesuit periodical devoted exclusively to science.[19] The journal was created by the Jesuits Silva Tavares, Azevedo Mendes, and Zimmermann and borrowed its name from the prominent Portuguese naturalist Félix de Avelar Brotero (1744–1828). As its title suggested, the journal sought to contribute to the progress of botany and zoology through the identification, classification, and description of new species of animals and plants. While the initial focus

these articles were also accompanied by photographs and illustrations of the scientific instruments. On the journey through Galicia, see Baltasar Merino, "Viajes de herborización por Galicia," *Razón y fé* 1 (1901): 95–98; *Razón y fé* 2 (1902): 82–89; *Razón y fé* 2 (1902): 367–73; *Razón y fé* 4 (1902): 82–93; *Razón y fé* 5 (1903): 348–60.

18 See Table 7 in the appendix.

19 The complete title of this journal in the opening issue was *Brotéria: Revista de sciencias naturaes do Collegio de S. Fiel*. However, it became known simply as *Brotéria*. On the scientific history of *Brotéria*, see Francisco Malta Romeiras, "The Journal *Brotéria* (1902–2002): Jesuit Science in the 20th Century," *HoST: Journal for the History of Science and Technology* 6 (2012): 100–9; and Romeiras and Henrique Leitão, "One Century of Science: the Jesuit Journal *Brotéria* (1902–2002)," in *Exploring Jesuit Distinctiveness: Interdisciplinary Perspectives on Ways of Proceeding within the Society of Jesus*, ed. Robert Maryks (Leiden: Brill, 2016), 235–58.

on botany and zoology corresponded to the founders' own scientific interests, it also reflected the growing popularity of these scientific fields for members of the Jesuit order as a whole. As Steven J. Harris has noted, the natural sciences were among the main scientific fields in which the Jesuits were engaged in the eighteenth century. In contrast to the situation in the sixteenth and seventeenth centuries, when Aristotelian philosophy, mathematical sciences, astronomy, and natural philosophy were clearly predominant, in the eighteenth century natural history accounted for fifteen percent of all scholarly publications produced by Jesuits worldwide.[20] Hence it was not surprising to find the natural sciences, together with astronomy and the geophysical sciences, resurge as topics of special interest after the official papal restoration of the Society in 1814.

To justify their scientific program and gain the necessary approval to publish *Brotéria*, Silva Tavares, Azevedo Mendes, and Zimmermann recalled in the first editorial the famous metaphor of the "Two Books," which postulated that true knowledge of God could be obtained either by reading the Book of Revelation (i.e., the scriptures) or by studying the Book of Nature.[21] Framed in a classic natural theology program, the publication of *Brotéria* was therefore intended to promote the advancement of science as well as contribute to a deeper understanding of the divine mysteries:

> We rejoice in the idea that our contributions, as insignificant as they might be, can disseminate the taste for the natural sciences in our country. The natural world is a vast book, which has many pages to be opened. The name of our grand Creator is written on each of them. When opening these pages, what greater satisfaction could anyone have than unveiling the greatness of God, which is stamped equally on the immensity of the world and on the myriad of tiny animals and plants, whose existence only the microscope can uncover?[22]

---

20 Harris, "Jesuit Ideology and Jesuit Science," 146–67. See also Asúa, *Science in the Vanished Arcadia*, 29.

21 For a detailed historical account of the metaphor of the "Two Books," see Olaf Pedersen, *The Two Books: Historical Notes on Some Interactions between Natural Science and Theology* (Vatican: Vatican Observatory Foundation, 2007). For an overview of the relations between science and religion in the nineteenth and twentieth centuries, see especially James Moore, "Religion and Science," in Bowler and Pickstone, *Cambridge History of Science*, 6:541–62, and Jonathan Topham, "Natural Theology and the Sciences," in *The Cambridge Companion to Science and Religion*, ed. Peter Harrison (Cambridge: Cambridge University Press, 2010), 59–79.

22 "Duas palavras de introdução," *Brotéria* 1 (1902): v–vii, here v.

However, despite the apparently apostolic intention behind its creation, *Brotéria* was mainly conceived as a scientific journal that would focus on the publication of original botanical and zoological research, not only by Jesuits but also by lay scientists. Rather than promoting intellectual debates on the compatibility of science and faith, the Portuguese Jesuits sought to present their own scientific accomplishments, and thus their direct participation in the progress of the botanical and zoological sciences. Their undeclared objective, however, was to project a public image as active and competent scientists, something that could not be achieved by producing an apologetic journal. Thus, in much the same way as the success of the colleges of Campolide and São Fiel helped to project an image of the Jesuits as skillful teachers, *Brotéria* aimed to present them as competent naturalists.

In parallel with the colleges, the decision to start publishing a scientific journal was also made in response to the adverse cultural and political milieu in which the restoration of the Society had occurred.[23] The Portuguese Jesuits not only had to face contemporary adversaries but also a historical legacy that was seriously tainting their image. The need to contradict the Pombaline accusations of obscurantism, together with the increasing importance of science in the late nineteenth century, consequently played a pivotal role in the foundation of *Brotéria* in early twentieth-century Portugal. Its creation can broadly be interpreted as the highpoint of a set of initiatives that fostered the teaching and practice of science in the Jesuit colleges of Campolide and São Fiel, including the creation of modern chemistry and physics laboratories, the curation of natural history collections, the establishment of astronomical and meteorological observatories, and the foundation of scientific academies.[24]

In the Portuguese context, *Brotéria* was also unique, for no religious order nor a boarding school had directly supported a strictly scientific journal. In the eighteenth and early nineteenth centuries, the publication of scientific periodicals was almost exclusively an affair of learned societies and academies.[25]

---

23 For a broader discussion of the context in which the 1759 expulsion and the 1858 restoration took place, see especially Chapters 2 and 3.

24 On the Jesuits' contributions to science and science education in the Colégio de Campolide and the Colégio de São Fiel, see Chapter 4.

25 On scientific periodicals published in eighteenth- and nineteenth-century Portugal, see especially Fernando Egídio Reis, ed., *Felicidade, utilidade e instrução: A divulgação científica no* Jornal enciclopédico dedicado à Rainha, *1779; 1788–1793; 1806* (Porto: Porto Editora, 2006); Reis, "Scientific Dissemination in Portuguese Encyclopaedic Periodicals, 1779–1820," *History of Science* 45 (2007): 83–118; Ana Duarte Rodrigues, *Horticultura para todos* (Lisbon: Biblioteca Nacional de Portugal–Junta de Freguesia de Alvalade, 2017); Rodrigues and Ana Simões, "Horticulture in Portugal 1850–1900: The Role of Science and Public Utility in Shaping Knowledge," *Annals of Science* 74 (2017): 192–213.

The publishing activities of the Royal Academy of Sciences, the country's most important scientific academy, were particularly prolific. Beginning in the late 1780s, the academy published a number of scientific journals covering a variety of subjects, from agriculture to the mathematical and physical sciences, the most relevant being the *Memórias da Academia real das sciencias de Lisboa* (Memoirs of the Royal Academy of Sciences of Lisbon [1797–1856]) and the *Jornal de sciencias mathematicas, physicas e naturais* (Journal of mathematical, physical, and natural sciences [1866–1927]). With the foundation of *O Instituto: Revista científica e literária* (The Institute: scientific and literary journal [1852–1981]), the University of Coimbra also began publishing scientific journals.[26] Following the creation of this eclectic journal, the university began to support the publication of more specialized bulletins, including the *Boletim da sociedade Broteriana* (1880–1970),[27] which was edited by the renowned botanist from Coimbra, Júlio Henriques. The *Boletim* shared with *Brotéria* a clear taxonomic program, as well as its patron, the eighteenth-century naturalist Brotero.

The extensive taxonomical work published by *Brotéria* soon caught the attention of Portuguese and foreign naturalists. For Henriques, *Brotéria* had the potential to make a noteworthy contribution to Portuguese science. Remarking on the first issue of the journal of São Fiel, the dean of botany at the University of Coimbra highlighted the works of Torrend and Zimmermann.[28] While the former had produced an extensive catalog of fungi collected in the coastal city of Setúbal, the latter had written a practical essay on the importance of microscopy for the teaching of botany.[29] In his review of the first issue of *Brotéria*, Veríssimo de Almeida also praised the work Torrend had undertaken in mycological studies, an undervalued branch of botany in twentieth-century Portugal.[30] The significance of *Brotéria* as a taxonomical journal is clear not only from the praise the journal received from some of Portugal's most prominent naturalists but also by the inclusion of the species it described in the annual

26 António José Leonardo, "O Instituto de Coimbra e a evolução das ciências físico-químicas em Portugal de 1852 a 1952" (PhD diss., Universidade de Coimbra, 2011).

27 Despite its importance for the history of botany in nineteenth- and twentieth-century Portugal, there are no detailed studies on the *Boletim da sociedade Broteriana*. Nevertheless, the vast majority of its issues can easily be found online.

28 Joaquim da Silva Tavares, "A *Brotéria* e as revistas scientificas extrangeiras," *Brotéria* 2 (1903): 234–35, here 234.

29 Camilo Torrend, "Contribuições para o estudo dos fungos da região setubalense," *Brotéria* 1 (1902): 93–150; Carlos Zimmermann, "Microscopia vegetal," *Brotéria* 1 (1902): 49–75.

30 "A *Brotéria* e as revistas scientificas," *Brotéria* 3 (1904): 366–67, here 366. See also: José Veríssimo de Almeida, "Bibliografia: Primeira contribuição para o estudo dos fungos da região setubalense," *Revista agronómica* 1 (1903): 368–69.

catalogs of renowned specialized journals such as the *American Naturalist*, the *Journal of Mycology*, and the *Bulletin of the Torrey Botanical Club*, the first American journal devoted to botany.[31]

The geographical extension of *Brotéria*'s influence was remarkable for a journal published in such a small country in Mediterranean Europe. In addition to selling *Brotéria* to their readers, the Jesuits exchanged copies of the journal with scientific journals published all over the globe. This editorial practice, still common today, allowed the Portuguese Jesuits to diffuse their works and facilitated their access to other modern scientific journals. In Europe alone, *Brotéria* could easily be found in the headquarters of major scientific journals published in Madrid, Paris, Rome, Brussels, Berlin, and Stockholm.[32] Although the practice of exchanging the journal with societies elsewhere was the main factor in the worldwide circulation of *Brotéria*, learned societies and academies also played their part by subscribing to the journal. Among the most illustrious institutional subscribers to *Brotéria* was the Royal Society of London, which purchased the first issue of the Jesuit journal as early as 1903.[33] In that same year, the Linnean Society of New South Wales, Australia, also included *Brotéria* among the long list of scientific periodicals recently donated (or exchanged).[34] By integrating myriad scientific periodicals and monographs in their collections and, most importantly, by making them available to their fellows, learned institutions aimed to support the growth of an all-encompassing Republic of Letters.

As well as exchanging their journal with more than one hundred national and international scientific magazines in Europe, North America, and South America, the Portuguese Jesuits also exchanged specimens with naturalists worldwide.[35] This had always been a fairly common practice among naturalists,

31 On the national and international reception of *Brotéria*, see especially Francisco Malta Romeiras and Henrique Leitão, "Jesuítas e ciência em Portugal IV: A revista *Brotéria: Sciencias naturaes* e a sua recepção nacional e internacional," *Brotéria* 174 (2012): 323–33.

32 "Revistas scientificas que trocam com a Brotéria," *Brotéria* 3 (1904): 352–55. The geographical distribution of the interchanges in 1904 was as follows. In Europe: Portugal (twenty-four), Germany (eleven), France (eleven), Italy (eleven), Spain (ten), Austria and Hungary (seven), Switzerland (three), Belgium (three), Russia (two), England (two), Netherlands (two), Denmark (one), and Sweden (on). In the Americas: United States (nine), Mexico (three), Canada (two), Brazil (two), Chile (two), and Argentina (one).

33 "Accessions to Serials," *Year-Book of the Royal Society of London* (London: Harrison and Sons, 1904), 122.

34 "Donations and Exchanges," *Proceedings of the Linnean Society of New South Wales* 28 (1903): 946–69, here 964.

35 On the importance of networks of correspondence for the development of botany and zoology in the nineteenth century, see Di Gregorio, "Zoology," 205–24, and Citadino, "Botany," 225–42.

yet the exchange of animals and plants increased significantly from the eighteenth century onward.[36] Both the scale of their missions and networks of correspondence before the papal suppression of 1773 inspired the early twentieth-century Portuguese naturalists in the Society of Jesus. They sought, as a result, to reestablish that global network of scientific correspondence. By reaching out to naturalists working all around the globe, Silva Tavares and his Jesuit companions diffused their research to a global audience, built long-distance networks of expertise, and brought what seemed to be a minor taxonomy journal into the transnational network of scientific periodicals.

2 Silva Tavares and the Advent of Taxonomy

Under the direction of Silva Tavares between 1902 and 1931, *Brotéria* became a center for the accumulation, systematization, and circulation of knowledge, especially in the fields of botany and zoology. During this period, most of the articles published in *Brotéria* were written by Silva Tavares himself, Azevedo Mendes, Zimmermann, Luisier, and Torrend. As their fields of research did not overlap, the issues of *Brotéria* were particularly varied. Articles describing new species of moths and butterflies, for instance, appeared side by side with others identifying new species of microscopic algae.

Besides making *Brotéria* an appealing journal to both botanists and zoologists, the wide coverage of taxa also contributed to the development of increasingly specialized networks of correspondence. Generally, the Portuguese correspondents participated in these networks by collecting and sending the specimens to the Jesuit naturalists, whereas the foreign correspondents assisted in the resolution of ambiguities over the identification and classification of those specimens. Following some changes to the editorial structure in 1907, the articles on botany and zoology were published in two separate series—*Botânica* and *Zoologia*[37]—both of which continued to cover a variety of taxa (figure 12). The botanical articles included the classification of new species of algae, fungi, mosses, lichens, and phanerogams (i.e., plants that produce seeds), while the zoological articles focused on the identification of new species of flies, bees, wasps, beetles, spiders, moths, butterflies, and net-winged

36 On the importance of establishing long-distance networks for the development of Jesuit science before the suppression, see especially Harris, "Confession-Building, Long-Distance Networks and the Organization of Jesuit Science," 287–318.

37 The series *Botânica*, *Zoologia*, and *Vulgarização científica* were published alternately. The series *Vulgarização científica* was published in January, March, May, July, September, and November; the series *Zoologia* in February, June, and October; and the series *Botânica* in April, August, and December. On the history of *Vulgarização científica*, see Chapter 7.

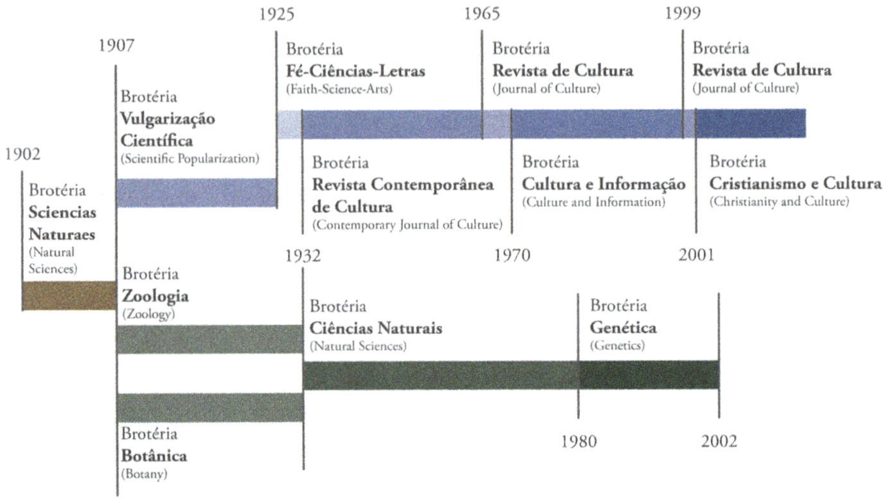

FIGURE 12 Editorial evolution of the journal *Brotéria*. Adapted from Hermínio Rico and José Eduardo Franco, eds., *Fé, ciência, cultura:* Brotéria; *100* anos (Lisbon: Gradiva, 2003), 101

insects. This interest in entomology was shared by two of *Brotérias*'s founders, Silva Tavares and Azevedo Mendes.

Silva Tavares was born in 1866 in Casais de São Bento, a small village in the interior of Portugal, and entered the Jesuit novitiate at the age of fourteen. After the profession of his first vows in 1882, he left for São Francisco, where he studied humanities and philosophy for six years (figure 13). Before engaging in theological studies, he taught physics, chemistry, and natural history for five consecutive years at Campolide (1889–90) and São Fiel (1890–94). Like all of his Portuguese companions in the Society of Jesus, Silva Tavares studied theology abroad. Between 1894 and 1897, he studied theology in Uclés, Spain, and was ordained a priest on June 20, 1897. He later spent a year in Vals-près-le-Puy, France (1897–98), studying dogmatic theology, and another in Vienna, Austria (1898–99). When he returned to Portugal, he soon resumed his science courses, initially teaching physics, chemistry, and mechanics at São Francisco (1899–1900), and then physics, chemistry, and natural history at São Fiel (1900–7), where he was also responsible for the herbarium, the curation of the natural history museum, and the publication of *Brotéria*.[38]

38 The most significant dates and scientific contributions of Silva Tavares can be found in Serafim Leite, "J.S. Tavares," *Brotéria: Fé, ciências, letras* 13 (1931): 273–97; Afonso Luisier, "In memoriam: Le R.P.J. da Silva Tavares, S.J.," *Brotéria: Ciências naturais* 1 (1932): 9–34;

FIGURE 13 Joaquim da Silva Tavares, S.J. (1866–1931)
AB

Beginning in the 1900s, Silva Tavares became involved in the study of *cecidia*, abnormal plant growths caused by parasites commonly known as plant galls. For three decades, Silva Tavares described hundreds of specimens of *zoocecidia* (plant galls caused by insects) collected in Portugal, Spain, Austria, Brazil, Argentina, and Mozambique. In his first research article, he produced a thorough catalog of Portuguese *zoocecidia*. As well as listing 234 species of insects, he also published a description of twenty-four new species of insects and identified seventy new host plants.[39] Although he was only an amateur naturalist, his early contributions to taxonomy were soon acknowledged by the scientific community. When he was appointed corresponding fellow of the Royal Academy of Sciences in 1903, the most exclusive and prestigious scientific society in Portugal, his peers justified the nomination by referring to the originality of his 1900 catalog of *zoocecidia*.

As *Brotéria*'s editor-in-chief, Silva Tavares was particularly aware of the importance of publishing his works exclusively in the journal, as doing so would enable his research to reach a more international readership, comprising both Jesuits and lay naturalists alike. Yet it would also contribute to the consolidation of *Brotéria* as a noteworthy taxonomy publication. Given the reception of his early taxonomical works, Silva Tavares went on to publish a series of articles on plant galls in the first issue of the journal, including an addenda to the 1900 catalog, where he identified fifteen new species of *cecidia*.[40] Inspired by the work of the French naturalist priest Jean-Jacques Kieffer (1857–1925), who had published a detailed and valuable catalog of European plant galls in 1901, Silva Tavares started to work on a synopsis of Portuguese species.[41] A network of correspondence was critical to ensure the success of this endeavor, and as was generally the case, Portuguese correspondents assisted Silva Tavares by assembling specimens while foreign correspondents clarified some of the ambiguities regarding their classification. Gonçalo Sampaio (1865–1937) and Augusto Nobre (1865–1946), from the Polytechnic Academy of Porto,

---

Domingos Maurício, "O fundador da *Brotéria* na Academia," *Brotéria: Revista contemporânea de cultura* 23 (1936): 449–55. For more succinct biographies, see especially *DHCJ*, 4:3706, and *Grande enciclopédia portuguesa e brasileira* (Lisbon: Verbo, 1936–60), 30:811–3; José Eduardo Franco, "Os directores da *Brotéria*: Joaquim da Silva Tavares," in Rico and Franco, *Fé, ciência, cultura*, 143–46.

39 Joaquim da Silva Tavares, "As zoocecidias portuguesas: Enumeração das espécies até agora encontradas em Portugal e descrição de dezanove ainda não estudadas," *Anais de ciências naturais* 7 (1900): 17–106.

40 Joaquim da Silva Tavares, "As zoocecidias portuguesas: Addenda," *Brotéria* 1 (1902): 5–48.

41 Jean-Jacques Kieffer, "Synopsis de zoocécidies d'Europe," *Annales de la Société entomologique de France* 70 (1901): 233–580; Joaquim da Silva Tavares, "Synopse das zoocecidias portuguesas," *Brotéria* 4 (1905): v–xii; 1–123.

as well as the Jesuit naturalist Luisier, collected specimens of plant galls and sent them to the editor-in-chief of *Brotéria*. In turn, French Jesuits Joseph de Joannis (1854–1932) and León de Joannis (1843–1919), as well as Kieffer himself, aided the Portuguese Jesuit in the classification of several species.[42]

To supplement his original compendium, and probably to draw attention to its significance, Silva Tavares included fourteen images depicting 240 different plant galls. Rather than following the customary practice of including drawings, he decided to use photographs, which were able to display small objects "with greater fidelity."[43] As in the first issue of *Brotéria*, he stated that his work was inspired by a desire to understand God's mysteries; in the same way as the "immensity of heavens," "the tiniest insect," and the "simplest cell" displayed the "glory of the Creator," so too did the "most beautiful and gracious" *cecidia*, the "admirable instinct" of insects, and the "intimate relation between a zoological specimen and the shape of a plant gall."[44]

In the following years, Silva Tavares's work was able to reach ever more distant regions of the world as his network of correspondents gradually expanded. Thanks to amateur botanist Carlos Azevedo de Meneses (1863–1928), he was able to describe seventy-two types of *zoocecidia* extant in Madeira, including twelve new species.[45] For his study of plant galls collected in Mozambique and Brazil, the missionary network of the Society of Jesus was crucial, as Jesuit missionaries often collected specimens of animals and plants overseas before sending them to the premises of *Brotéria* for identification and description by their Jesuit companions. For the description of fifty-four types of plant galls collected in Mozambique, for instance, including eleven new species, Silva Tavares benefited from the work of Jesuit missionary and collector Luís Lopes (1867–1954).[46] Similarly, the study of forty types of Brazilian *cecidia* was

42 For a complete list of contributors, see Tavares, "Synopse das zoocecidias portuguesas," ix–xi. Joseph de Joannis: b. June 6, 1854, La Meignanne (France); e. August 21, 1871, Angers (France); d. August 26, 1932, Paris (France) (*DHCJ*, 3:2150–1). León de Joannis: b. October 23, 1843, Angers (France); e. October 25, 1862; d. July 30, 1919, Vannes (France) (*Mendizábal* 15,144).

43 Tavares, "Synopse das zoocecidias portuguesas," ix.

44 Tavares, "Synopse das zoocecidias portuguesas," xi.

45 Joaquim da Silva Tavares, "Zoocecidias da ilha da Madeira," *Brotéria* 2 (1903): 179–86; *Brotéria* 4 (1905): 221–27; *Brotéria: Zoologia* 12 (1914): 193–97.

46 Joaquim da Silva Tavares, "Contributio prima ad cognitionem cecidologiae regionis Zambeziase," *Brotéria: Zoologia* 7 (1908): 133–74. Luís Lopes: b. February 13, 1867, Louriçal do Campo (Portugal); e. September 20, 1882, Barro (Portugal); d. May 5, 1954, Salvador da Baía (Brazil) (*Mendizábal* 26,414).

only possible due to the joint efforts of Jesuits Johannes Rick (1869–1946) and Johann Bruggman (1863–1922).[47]

Following the republican revolution, Silva Tavares went into exile. Since the revolution did not reach Louriçal do Campo as early as Lisbon, Silva Tavares, who was by then rector of São Fiel (in office 1908–10), managed to escape unharmed, reaching Salamanca, Spain, on October 13, 1910. Less than a month later, he went to Cádiz and embarked to Buenos Aires, Argentina, arriving on November 25, 1910. Like his companions, Silva Tavares also tried to retrieve his collections by negotiating with the provisional government. Apparently, António José de Almeida (1866–1929), minister of interior and future president (in office 1919–23), supported his pretentions. Nevertheless, António José de Almeida preferred to avoid inner tensions within the recently proclaimed regime and thus handed the final decision to his colleague, the progressive minister of justice Afonso Costa, who, faithful to his republican principles, consistently disregarded the overarching support of the Portuguese scientific community and refused to return Silva Tavares's *cecidia*.[48]

During his exile, Silva Tavares resumed the publication of *Brotéria* and focused on studying *cecidia* collected in Argentina and Brazil (1910–14) and Spain (1914–28) (figure 14). As a result of his short stay in Argentina, he identified seventy-three local species of *cecidia*.[49] The period in which he worked in Brazil was particularly fruitful, as he was able to collect and describe firsthand eleven genera and thirty species of plant galls.[50] During his stay in São Paulo (1911–13) and Salvador da Baía (1913–15), Silva Tavares saw his influence in the field of cecidology and his network of correspondents increase considerably. After he returned to Portugal, his Brazilian confrères frequently sent him specimens, thus allowing him to work from a distance. Thanks to naturalists Francisco Dias da Rocha (1869–1960), Hermann von Ihering (1850–1930), and Gustavo

47 Joaquim da Silva Tavares, "Contributio prima ad cognitionem cecidologiae regionis Braziliae," *Brotéria: Zoologia* 8 (1909): 5–29. Johannes Rick: b. January 19, 1869, Hatlerdorf (Austria); e. August 1, 1887, Bleijenbeek (Netherlands); d. May 6, 1946, São Salvador (Brazil) (*DHCJ*, 4:3358–59). Johann Bruggman: b. August 30, 1863, Magdenau (Germany); e. September 30, 1884; d. November 9, 1922, São Sebastião do Sul (Brazil) (*Mendizábal* 15,975).

48 In 1927, shortly after the fall of the republic, Silva Tavares retrieved his collection of *cecídia*: Joaquim da Silva Tavares, "A *Brotéria* nos primeiros 25 anos da sua existência," in *Brotéria: Número especial no XXV aniversário da sua fundação* (Caminha: Brotéria, 1927), 19–37, here 37.

49 Joaquim da Silva Tavares, "Cécidologie argentine," *Brotéria: Zoologia* 13 (1915): 88–128.

50 Joaquim da Silva Tavares, "Cecidologia brasileira," *Brotéria: Zoologia* 16 (1918): 21–28; *Brotéria: Zoologia* 18 (1920): 82–125; *Brotéria: Zoologia* 19 (1921): 76–112; *Brotéria: Zoologia* 20 (1922): 5–48.

Peckolt (1861–1923), Silva Tavares complemented his study of Brazilian *cecidia*, describing one new genus and five new species of plant galls.[51] While in Pontevedra and La Guardia, Spain, he worked extensively on a catalog of plant parasites found in the Iberian Peninsula. Between 1916 and 1924, he published a series of articles on gall midges (*Cecidomyiidae*) and gall wasps (*Cynipidae*), identifying, for the first time, eight genera, thirty-three species, ten subspecies, and one new variety of flea.[52]

Following the fall of the republic on May 28, 1926, the Jesuits were once again welcome to return to Portugal. Once back in his homeland (1928–31), Silva Tavares was invited to work in the prestigious Institute Rocha Cabral, and he also managed to secure state funding for the publication of *Brotéria: Botânica* and *Brotéria: Zoologia* in 1930. By then, the prestige of the Jesuit naturalists, mirrored in the scientific series of *Brotéria*, was undisputed. As such, the decision of the recently founded government agency Junta de Educação Nacional (1929–36) to support the publication of *Brotéria* is unlikely to have met with any resistance. Given that the new military rulers wanted to further distance themselves from the republican legacy, the funding of *Brotéria* and the retrieval of Silva Tavares's possessions was an important political issue.

In Lisbon, the Jesuit entomologist soon resumed his study of plant parasites. Between 1927 and 1931, he published a series of articles describing six new genera and six new species of Iberian gall wasps.[53] And in 1930, after thirty years of research, he decided to publish a copiously illustrated and all-encompassing catalog of Iberian gall wasps.[54]

Acknowledged for his contributions to the study of gall midges, gall wasps, and plant galls, Silva Tavares's scientific reputation preceded him.[55] He was a

51 Joaquim da Silva Tavares, "Nova contribuição para o conhecimento da cecidologia brasileira," *Brotéria: Zoologia* 22 (1925): 6–55.

52 Joaquim da Silva Tavares, "Espécies novas de cynípides e ceccidomyas da Península Ibérica, e descrição de algumas já conhecidas," *Brotéria: Zoologia* 14 (1916): 65–136; *Brotéria: Zoologia* 16 (1918): 130–42; *Brotéria: Zoologia* 17 (1919): 5–101; *Brotéria: Zoologia* 18 (1920): 43–69; *Brotéria: Zoologia* 20 (1922): 97–155; *Brotéria: Zoologia* 21 (1924): 5–48.

53 Joaquim da Silva Tavares, "Os cynípides da Península Ibérica," *Brotéria: Zoologia* 24 (1927): 16–78; *Brotéria: Zoologia* 25 (1928): 11–152; *Brotéria: Zoologia* 26 (1930): 25–53; *Brotéria: Zoologia* 27 (1931): 5–100.

54 Joaquim da Silva Tavares, *Cynipidae peninsulae Ibericae*, 2 vols. (Lisbon: Brotéria, 1931). The first volume compiled the articles published until 1928 and included five plates and seventy-one figures. Although incomplete, the second volume comprised the articles published between 1930 and 1931 and included four plates and forty-eight figures.

55 For modern reappraisals of his taxonomic work, see especially Marcela Skuhravá et al., "Gall Midges (Diptera: Cecidomyiidae) of the Iberian Península," *Boletín de la Asociación española de entomología* 20 (1996): 41–61, and Skuhravá et al., "Gall Midges (Diptera: Cecidomyiidae) of the Iberian Península 2: Zoogeographical Analysis of the Gall Midge

member of numerous Portuguese, Spanish, French, and Brazilian specialized societies, as well as generic learned academies, including the prestigious Accademia Pontificia dei Nuovi Lincei. Shortly after returning to Portugal in 1928, Silva Tavares was granted a very significant distinction, the election as fellow for life by the Lisbon Academy of Sciences. According to a Jesuit alumnus and future Nobel laureate, Egas Moniz, the decision had been a unanimous one because Silva Tavares's "noteworthy contributions to natural history" honored the centennial institution.[56] As with the invitation to work at a prestigious laboratory, and the financial support from a state agency, this election also represented an official apology for the 1910 expulsion.

In July 1931, three years after his formal return, the Jesuit entomologist followed the advice of his physician and traveled once again to Royat, France, to cure his chronic fatigue. On his way back to Portugal, Silva Tavares stopped in Paris to attend the Colonial Exhibition and write an article for *Brotéria*. However, his health quickly deteriorated, and he passed away on September 2, 1931. His most important legacy was arguably the consolidation of *Brotéria* as an international journal of taxonomy.

3 Tailored Networks

Under Silva Tavares's guidance, *Brotéria* expanded its interests to include virtually every field of interest for Jesuit naturalists. Publishing an inclusive taxonomy journal resulted in the proliferation and growth of increasingly specialized networks of correspondents. For the resolution of ambiguities in the classification of plant galls, Silva Tavares, for instance, resorted to his tailored network, which included the aforementioned Kieffer, Rick, Jesuit brothers Joannis, as well as French entomologist Ernest Marie Louis de Bedel (1849–1922), and Italian naturalists Alfredo Corti (1880–1973) and Mario Bezzi (1868–1927).

Azevedo Mendes, Torrend, Luisier, and Zimmermann also built their own tailored networks. These networks included collectors, amateur naturalists, and professional botanists and zoologists, and though there were some occasional overlaps, they were mostly independent, because the principle that

Fauna," *Boletín de la Asociación española de entomología* 30 (2006): 93–159. Dedicated to the memory "of the great Portuguese entomologist Prof. Joaquim da Silva Tavares," the 1996 article acknowledged the role played by Silva Tavares in the foundation of cecidology in the Iberian Peninsula.

56 APSI, Egas Moniz to Silva Tavares (March 6, 1928).

FIGURE 14 The editorial team of Brotéria in São Salvador da Baía, 1913. Seated: Manuel Rebimbas Tavares (1873–1944), António de Oliveira Pinto (1868–1933), António Menezes (1869–1919), Carlos Zimmermann (1871–1950). Standing: Joaquim da Silva Tavares (1866–1931), Atanásio Silvano (1878–1947), Joseph Foulquier (1887–1962), António Coutinho (1858–1917)
APSI

guided their somewhat extemporaneous growth was the interest in a common taxon. Besides their role in the circulation and classification of natural specimens, these networks could also be valuable in different situations. In the case of Zimmermann, for example, the smooth functioning of a network centered on the study of microscopic algae played a pivotal role in the return of his collections after the 1910 expulsion. Most importantly, these tailored networks definitely contributed to the characterization of *Brotéria* as a center of calculation, in Latourian terms.[57] Botanical and zoological knowledge was accumulated, systematized, and circulated through *Brotéria*. The accumulation of specimens, the systematization of their taxonomic description, and the subsequent publication of their classification owed much to these networks.

Each Jesuit scholar developed a network specific to his research interests, as was the case with Zimmermann, whose background led him to correspond with

57 Bruno Latour, *Science in Action: How to Follow Scientists and Engineers through Society* (Cambridge, MA: Harvard University Press, 1987), 215–57.

botanists worldwide. Born in Ehingen, Germany, on May 28, 1871, Zimmermann entered the Jesuit novitiate in Lyon at the age of nineteen but soon came to Portugal to study humanities at the novitiate of Barro (1890–92). Following three years of philosophy and natural sciences at São Francisco (1892–95), he went to São Fiel (1895–1902), where he taught a variety of courses, including sciences (physics, chemistry, natural history), humanities (Latin, German, history, geography), and fine arts (drawing, music). After an unusually long regency, he went abroad to study theology, first in Canterbury, England (1902–5), and then in the college of Saint Ignatius in Galway, Ireland (1905–6). Back in Portugal, he stayed in the novitiate of Barro for his tertianship (1906–7) and then moved to Porto to establish a natural history museum. After the closure of the Jesuit college of Porto in 1909, Zimmermann returned to São Fiel, where he had found his calling for botany and microscopy.[58] When only a scholastic, he had assembled a herbarium (1897–98), established a meteorological observatory (1902), and founded *Brotéria* (1902). By 1910, the herbarium comprised over five thousand different species, including three thousand species of diatoms, Zimmermann's most precious microscopic algae.[59] Alongside his desire to observe and study these algae, Zimmermann displayed a genuine interest in promoting microscopy as a method to teach botany to high-school students.[60] Before the expulsion, the German Jesuit studied in Paris with French botanist Maurice Peragallo, author of an impressive work on the classification of diatoms.[61] His network of correspondents also included French naturalist Joannes Albert Tempère, enthusiastic collector of diatoms, as well as a manufacturer of mounting media tailored for observing these algae through microscope lenses. In Portugal, Zimmermann represented an international hub for botanists working with diatoms, including his friend Silva e Castro.[62]

During his exile in Brazil from 1911, Zimmermann taught at the Colégio Antônio Vieira, where he also directed the library and the natural history museum. Founder of the Portuguese Society of Natural Sciences and fellow of the Royal

58 Romeiras and Leitão, "Jesuítas e ciência em Portugal II: Carlos Zimmermann S.J.," 113–25; Gomes et al., "Father Zimmermann," 177–88; Franco, "Os directores da *Brotéria*," in Rico and Franco, *Fé, ciência, cultura*, 149–50.

59 Tavares, "O herbário do Colégio de S. Fiel," 82–87.

60 Carlos Zimmermann, "Microscopia vegetal," *Brotéria* 1 (1902): 49–75; *Brotéria* 2 (1903): 5–40; *Brotéria* 4 (1905): 137–59; *Brotéria* 5 (1906): 229–44.

61 Although there are no precise dates, Zimmermann mentioned to José da Silva e Castro that Peragallo had been his professor in Paris. See: JBA, Carlos Zimmermann to José da Silva e Castro (April 25, 1910). Published in Romeiras, "Constituição e percurso das colecções científicas," 313.

62 Romeiras, "Constituição e percurso das colecções científicas," 287–327.

Microscopical Society of London, he studied thousands of diatoms collected in Portugal, Madeira, Porto Santo, Mozambique, and Brazil, naming eighty new species.[63] In Portugal alone, he identified twenty-seven families, seventy-three genera, and around four hundred species of diatoms, 340 of which firsthand, while in Madeira he registered two hundred different species.[64] As in the case of Silva Tavares, the study of African specimens was only possible due to the efforts of Jesuit missionaries in the field. Collected "a few years ago" in the Nhamadzi, Cathunda, and Quelimane (or Bons Sinais) Rivers by missionaries Luís Lopes and Manuel da Fonseca (1871–1962), these specimens traveled from Mozambique to São Fiel, and from there to the Colégio Antônio Vieira, Brazil, where Zimmermann studied them.[65] From 1913 onward, he focused on studying Brazilian diatoms.[66] To undertake this "laborious and fastidious" venture, he began by exhorting Brazilian naturalists to send him new material. Although he considered the collection of diatoms the "simplest" task, he promised to send "detailed instructions" to those who needed them.[67] Despite this request, by 1913 his network of correspondents was already considerable: when he published his first contribution to the study of Brazilian diatoms, the network included twenty-five different collectors working all around Brazil.[68] These correspondents allowed him to study twenty families, forty genera, and 113 species of diatoms. Until 1919, he continued to study diatoms collected by

63 Gomes et al., "Father Zimmermann," 177–88.

64 Carlos Zimmermann, "Catálogo das diatomaceas portuguezas," *Brotéria* 5 (1906): 245–51; *Brotéria: Botânica* 8 (1909): 89–103; *Brotéria: Botânica* 9 (1910): 95–102; *Brotéria: Botânica* 12 (1914): 115–24; Zimmermann, "Contribuição para o conhecimento das diatomaceas da provincia de Moçambique," *Brotéria: Botânica* 12 (1914): 155–62; Zimmermann, "Beitrag zur Kenntnis der Diatomaceen Flora der Inseln Madeira und Porto Santo," *Brotéria: Botânica* 8 (1909): 114–27; *Brotéria: Botânica* 9 (1910): 103–20.

65 Zimmermann, "Contribuição para o conhecimento das diatomaceas da provincia de Moçambique," 155.

66 Carlos Zimmermann, "Contribuições para o estudo das diatomaceas dos Estados Unidos do Brazil," *Brotéria: Botânica* 11 (1913): 149–64; *Brotéria: Botânica* 12 (1914): 5–12; *Brotéria: Botânica* 13 (1915): 37–56; 65–71; 124–46; *Brotéria: Botânica* 14 (1916): 85–103; 130–57; *Brotéria: Botânica* 15 (1917): 30–45; *Brotéria: Botânica* 16 (1918): 8–24; 113–22; *Brotéria: Botânica* 17 (1919): 5–16.

67 Zimmermann, "Contribuições para o estudo das diatomaceas dos Estados Unidos do Brazil," 149. Manuel da Fonseca: b. October 18, 1871, Soudos, Torres Novas (Portugal); e. September 7, 1888, Barro (Portugal); d. February 10, 1962, Recife (Brazil) (*Mendizábal* 29,093).

68 The complete list of collectors is as follows: Bailey, Cleve, Deby, Ehrenberg, Eulenstein, Firth, Griffin, Grove, Gruendler, Grunow, Hardman, Jelinek, Kitton, Marcacci, Martius, Mosén, Neijahr, Novara, Perry, Puigarri, Era, Schmidt, Van Heurck, Warming, and Weissflog.

himself or by his correspondents in the states of Baía, Minas Gerais, Rio de Janeiro, São Paulo, and Rio Grande do Sul. His activity was especially prolific, as he was able to produce an all-encompassing catalog of Brazilian diatoms listing around eight hundred microscopical algae. In 1919, he also compiled all the articles he had written for *Brotéria* in a manuscript entitled *Diatomées*,[69] which unfortunately remained unpublished, probably because he left the Society of Jesus shortly after. In 1921, the Holy See dismissed him from the clerical state. With the exception of his residence and date of death, little is known about his remaining years as a layman. After leaving the Society, he lived in Salvador da Baía and passed away on October 21, 1950. It is unclear whether he continued to teach botany and study diatoms or how he made a living after 1921.

While still a Jesuit, Zimmermann had worked alongside Azevedo Mendes, who had created his own network of scientific researchers. Born in Torres Novas on January 17, 1874, Azevedo Mendes entered the Society of Jesus at the age of fourteen (figure 15). He studied humanities at the novitiate of Barro (1888–90) and São Francisco (1890–93), and philosophy and natural sciences at São Fiel (1893–96). Between 1896 and 1902, he taught mathematics, physics, chemistry, and natural history at São Fiel, where he also collaborated with Silva Tavares and Zimmermann in the foundation of *Brotéria*. Following the customary practice, he studied theology abroad, at the Gregorian University (1902–6), in Rome, being ordained a priest in 1905. After his tertianship (Barro, 1906–7), he returned to São Fiel to teach mathematics, physics, chemistry, and the natural sciences.

Fellow of the Portuguese Society of Natural Sciences and the Italian Society of Entomology, Azevedo Mendes was mostly interested in studying lepidopterans (i.e., species of moths and butterflies). Between 1902 and 1913, he published a series of articles in *Brotéria* identifying more than eight hundred species of lepidopterans collected near São Fiel.[70] In order to do so, he relied on four Jesuit correspondents: Joseph and León de Joannis, José da Cruz Tavares (1847–1916), and Luís Alves Correia (1862–1935).[71] Whereas Cruz Tavares and Alves Correia aided Azevedo Mendes by collecting diurnal and nocturnal butterflies, the French naturalists contributed to his research by describing

69 Gomes et al., "Father Zimmermann," 181.

70 Cândido de Azevedo Mendes, "Lepidopteros de S. Fiel (Portugal)," *Brotéria* 1 (1902): 151–71; *Brotéria* 2 (1903): 41–80; *Brotéria* 3 (1904): 223–54; *Brotéria* 4 (1905): 166–77; *Brotéria: Zoologia* 10 (1912): 161–82; *Brotéria: Zoologia* 11 (1913): 15–44.

71 José da Cruz Tavares: b. December 9, 1847, Covilhã (Portugal); e. April 20, 1861, Barro (Portugal); d. September 6, 1916, Pontevedra (Spain) (*Grande enciclopédia portuguesa e brasileira*, 30:813–14). Luís Alves Correia: b. February 1, 1862, Torres Novas (Portugal); e. October 8, 1876, Barro (Portugal); d. November 16, 1935, Caldas da Saúde, Santo Tirso (Portugal) (*Mendizábal* 19,866).

FIGURE 15 The brothers Azevedo Mendes: João (1883–1940), Cândido (1874–1943), and Manuel (1880–1960)
APSI

some new species. The circulation of specimens between Azevedo Mendes and Joseph Joannis also led to the identification of a new genus of lepidopterans in 1904. As a homage to the Portuguese Jesuit, Joannis named it *Mendesia*.[72] Five years after the description of *Mendesia echiella*, Azevedo Mendes returned the compliment by baptizing a novel species of this genus as *Mendesia joannisiella*.[73] The circle of tributes was thus complete.

After compiling his catalog of the moths and butterflies of São Fiel, Azevedo Mendes turned to the study of lepidopterans collected in the premises of other Jesuit colleges and residences in Portugal, namely in Campolide, Torres Vedras, Guimarães, and Val do Rosal. The Portuguese entomologist continued to send moths and butterflies to León and Joseph de Joannis, who willingly added Portuguese specimens to their significant collection. The moths and butterflies collected by Azevedo Mendes in São Fiel and Torres Vedras, which are currently housed in the Muséum National d'Histoire Naturelle, in Paris, continue to be used for taxonomic purposes, allowing for the description of new species of lepidopterans, namely *Symnoca serrata* Gozmány (1985), *Pleurota gallicella* Huemer & Luquet (1995), and *Homoeosoma soaltheirellum* Roesler (1965).[74]

As was also the case with Silva Tavares and Zimmermann, Azevedo Mendes used the Portuguese missionary network to study African entomological fauna. The correspondence with three Jesuit missionaries in Angola and Mozambique, including his younger brother João de Azevedo Mendes, S.J. (1883–1940), allowed Cândido to describe 180 species of lepidopterans during his exile in Salamanca (figure 15).[75]

Azevedo Mendes gradually reduced his taxonomical work, presumably for lack of spare time, as from 1913 onward he became particularly engaged in governance. After administering the Portuguese residence in Tuy, Spain (1913–17), and the writers' residence in Pontevedra, Spain (1917–19), Azevedo Mendes served the Society of Jesus as provincial of Portugal for eleven years (1919–24 and 1927–33). In 1934, at the age of sixty, he was sent to Baía, Brazil, to direct the

72 Joseph de Joannis, "Note sur un microlépidoptère nouveau de Portugal," *Bulletin de la Société entomologique de France* 71 (1902): 230–32.

73 Cândido de Azevedo Mendes, "Altera nova Mendesia ex Lusitania," *Brotéria: Zoologia* 8 (1909): 65–67.

74 Martin Corley, "The Lepidoptera Collections of Deceased Portuguese Entomologists," *Entomologist's Gazette* 59 (2008): 145–71; Corley, "The Lepidoptera Collections of Deceased Portuguese Entomologists: II," *Entomologist's Gazette* 66 (2015): 25–49.

75 Cândido de Azevedo Mendes, "Lepidoptera Africana I: Ex Zambezia Lusitana," *Brotéria: Zoologia* 10 (1912): 183–91; Mendes, "Lepidoptera africana II: Ex Angola lusitana," *Brotéria: Zoologia* 10 (1912): 191–93. João de Azevedo Mendes: b. April 27, 1883, Torres Novas (Portugal); e. September 7, 1889, Barro (Portugal); d. August 10, 1940, Barcelos (Portugal) (*Mendizábal* 21,652).

FIGURE 16 Camilo Torrend, S.J. (1875–1961)
APSI

mission of northern Brazil, becoming its first vice-provincial four years later (1938–42). Following the vice-provincialate, he served as master of novices. His office, however, was rather brief, as on December 16, 1943, one year after assuming his new role, Azevedo Mendes passed away. By then, he was living in Baturité, in the Brazilian state of Ceará.

Like Zimmermann and Azevedo Mendes, Torrend built a network of correspondence around a specific taxon. He was born in Saint-Privat d'Allier, France, and joined the Society of Jesus on June 21, 1875 after completing his studies at the apostolic school of Montciel, France (figure 16). Because Torrend wanted to become a missionary like his elder brother Jules Torrend, S.J. (1861–1936), he joined the Society in Portugal.[76] After three years in the novitiate of Barro (1894–97), he studied philosophy and natural sciences at São Fiel (1897–98) and São Francisco (1900–2). Between 1904 and 1908, he studied theology in Dublin, where he was ordained a priest on July 29, 1907.[77]

76 Jules Torrend: b. October 4, 1861, Saint-Privat d'Allier (France); e. January 13, 1879, Arlon (Luxembourg); d. March 11, 1936, Kabwe (Zambia) (*DHCJ*, 4:3817–18).

77 See especially Pinto, "O Instituto de Sciencias Naturaes," 99, and Lister, *Monograph of the Mycetozoa*, 3.

For fifty years, Torrend described hundreds of species of fungi collected in Portugal, Madeira, Mozambique, Timor, and Brazil in the pages of *Brotéria*. In his first work, he produced a comprehensive catalog of fungi listing 512 different species.[78] Although Torrend had focused only on the region of Setúbal, Júlio Henriques, the dean of botany from the University of Coimbra, as well as agronomist Veríssimo de Almeida, deemed the catalog one of the most valuable contributions to Portuguese mycology.[79]

While in Dublin, Torrend published an account focusing on the mycological flora of Mozambique,[80] a work in which, once again, the smooth functioning of the Jesuit missionary network played a crucial role. The thirty-six species Torrend listed had been collected by his older brother Jules and Aloysius (Luís) Gonzaga Dialer, S.J. (1866–1943), missionaries in Miruru, in the district of Zumbo, Mozambique. With the assistance of Father Giacomo Bresadola (1847–1929), illustrious naturalist and founder of the Société mycologique de France, Torrend described two new species of fungi: *Psalliota dialerii* and *Fomes dialerii*.[81] Torrend then assembled a copiously illustrated catalog of slime molds. Throughout 230 pages, the French Jesuit listed the 271 species of myxomycetes known to date.[82] He later focused on studying the mushrooms of Madeira. Thanks to two local naturalists, Fathers Jaime de Gouveia Barreto (1887–1963) and Manuel da Silveira (dates unknown), and to the aforementioned Carlos Azevedo de Meneses, Torrend was able to identity 338 varieties of mushrooms, including thirty-one new species.[83]

When the republican revolution began in October 1910, Torrend was completing his tertianship at the novitiate of Barro. Although he was subsequently arrested together with eighty-nine of his companions, he was soon released, possibly due to the intervention of a French official. Minister of Justice Afonso Costa offered him a professorship at the University of Coimbra, but Torrend declined the offer and instead embarked to London on October 18. After the expulsion, he never returned to Portugal.

78 Camilo Torrend, "Contribuições para o estudo dos fungos da região setubalense," *Brotéria* 1 (1902): 93–150; *Brotéria* 2 (1903): 123–48; *Brotéria* 4 (1905): 207–11.

79 "A *Brotéria* e as revistas scientificas," 366; Almeida, "Bibliografia: Primeira contribuição para o estudo dos fungos da região setubalense," 368–69.

80 Camilo Torrend, "Primeira contribuição para o estudo da flora mycologica da provincia de Moçambique," *Brotéria* 4 (1905): 212–21.

81 Aloysius (Luís) Gonzaga Dialer: b. June 1, 1866, Innsbruck (Austria); e. January 23, 1890; d. January 7, 1943, Recife (Brazil) (*Mendizábal* 22,521).

82 Camilo Torrend, "Les myxomycètes: Étude des espèces connues jusqu'ici," *Brotéria: Botânica* 6 (1907): 5–64; *Brotéria: Botânica* 7 (1908): 5–177.

83 Camilo Torrend, "Contributions pour l'étude des champignons de l'île de Madère," *Brotéria: Botânica* 8 (1909): 128–44; *Brotéria: Botânica* 10 (1912): 29–49; *Brotéria: Botânica* 11 (1913): 165–81.

Between 1912 and 1914, Torrend taught at the Instituto Nun'Alvres, Dieleghem, Belgium. With the outbreak of the First World War, the Jesuit college was relocated to Los Placeres, Spain, and Torrend left for Salvador da Baía, Brazil, where he dedicated himself to further research. For almost forty years, he taught apologetics and biology at the Colégio Padre Antônio Vieira (1914–53). Besides teaching at a Jesuit school, he was also invited to teach botany at several other institutions, namely the diocesan seminary, the Catholic School of Philosophy, and the Agronomic School of Baía (1932–43).

After arriving in Brazil, Torrend focused almost exclusively on the Brazilian mycological flora, identifying 115 different varieties of Brazilian fungi, including one new species, between 1920 and 1935.[84] Although *Brotéria* was his preferred outlet for publishing his taxonomical works, Torrend also wrote for several other scientific periodicals, including the *Boletim da Sociedade portuguesa de ciências naturais* (Bulletin of the Portuguese Society of Natural Sciences), the *Bulletin de la Société mycologique de France* (Bulletin of the Micologycal Society of France), *Mycological Notes*, and *The Irish Naturalist*. A fellow of numerous academies, including the Portuguese Society of Natural Sciences, the Royal Academy of Sciences, the Real Società Agraria di Torino, and the Mycological Society of America, Torrend was an esteemed botanist in Brazil and a trusted advisor on botanical and geographical research, being appointed multiple times to spearhead scientific expeditions throughout the country.[85] In 1954, when he established the Institute of Mycology at the University of Pernambuco, Augusto Chaves Batista (1916–67) paid tribute to his former professor by dedicating the herbarium to Torrend. The Herbário Pe. Camille Torrend, which contains eighty-eight thousand records of fungi and forty-six thousand specimens, is one of the largest mycological herbarium in South America.[86] In 1957, the University of Recife recognized Torrend's work by granting him an honorary doctorate.[87] He passed away four years later, on June 24, 1961, after living in Salvador da Baía for forty-seven years.

Brotéria's foundation by a small group of teachers in a Jesuit college belies its distinctive history as a scientific journal. Unlike other contemporary Jesuit journals, *Brotéria* was not a cultural periodical; nor was it a religious or

84 Camilo Torrend, "Les Polyporacées du Brésil," *Brotéria: Botânica* 18 (1920): 23–43; 121–42; *Brotéria: Botânica* 20 (1922): 107–12; *Brotéria: Botânica* 22 (1924): 12–42; *Brotéria: Botânica* 24 (1926): 5–19; *Brotéria: Ciências naturais* 31 (1935): 108–20.

85 *DHCJ*, 4:3816.

86 Leandro Agra, Andrea Bezerra, and David Barbosa, "URM Mycological Herbarium: Revision of the Myxomycetes Collection," *Brazilian Journal of Botany* 37 (2014): 299–313.

87 Everett Smith Beneke, "Honors, Degrees, Promotions, Invitational Lectures," *Newsletter of the Mycological Society of America* 8 (1957): 28–31, here 30.

apologetic magazine with a scientific section. *Brotéria* was instead a strictly and programmatically scientific journal. Under the direction of Silva Tavares, *Brotéria* established itself as a noteworthy journal of taxonomy. Between 1902 and 1931, the functioning and extemporaneous growth of tailored networks of correspondence allowed for the circulation of thousands of specimens between collectors, amateur naturalists, and professional botanists, and zoologists. As in the sixteenth and seventeenth centuries, the communication between Jesuits proved crucial for the circulation of knowledge and for the identification of new species of animals and plants around the globe.[88] Because they acknowledged the Jesuits' individual contributions to taxonomy and the significance of their journal among the multitude of specialized scientific botanical and zoological periodicals, lay naturalists also corresponded regularly with the Jesuits and published their works in *Brotéria*. Thus, besides including works by Portuguese, Spanish, French, Swiss, German, and Austrian Jesuits, *Brotéria* also published works by Portuguese, Spanish, Italian, Brazilian, and Swedish botanists and zoologists.[89]

However, the encouraging responses to the Portuguese Jesuits' botanical and zoological discoveries were not sufficient to cover the expenses of publishing a specialized journal. Accordingly, shortly after its foundation, Silva Tavares had to devise a new strategy to secure funding for his scientific journal. In 1907, as part of this strategy, he decided to publish a popular scientific magazine: *Brotéria: Vulgarização científica* (figure 12). Written entirely in Portuguese, this magazine was specifically designed to make a profit and secure funding for the Jesuits' taxonomical work.

88 On the importance of networks of travel for the classification of new specimens of animals and plants, see especially Steven J. Harris, "Networks of Travel, Correspondence, and Exchange," in *The Cambridge History of Science*, vol. 3, *Early Modern Science*, ed. Katharine Park and Lorraine Daston (Cambridge: Cambridge University Press, 2006), 341–62.

89 The most important authors in *Brotéria* between 1902 and 1932 can be subdivided according to two mutually non-exclusive criteria: nationality and belonging to the Society of Jesus. According to these criteria, the most relevant authors in this period are: (a) Jesuit botanists living in Portugal—Luisier, Torrend, Zimmermann, and Silva Tavares; (b) foreign Jesuit botanists—Baltasar Merino (1845–1917), Longino Navás (1858–1938), Rick, Theissen, and Jaime Pujiula (1869–1958); (c) Portuguese botanists—José d'Ascenção Guimarães (1862–1922), Gonçalo Sampaio, and Júlio Henriques; (d) foreign botanists—Erwin Baur (1875–1933), Romualdo González Fragoso (1862–1928), Carlos Pau y Español (1857–1937), and Emili Huguet i Serratacó (1871–1951), also known as Huguet del Villar; (e) Jesuit zoologists living in Portugal—Azevedo Mendes and Silva Tavares; (f) foreign Jesuit zoologists—Navás, Pelegrín Franganillo-Balboa (1873–1955), Jean-Jacques Kieffer; (g) Portuguese zoologists—José Maximiano Corrêa de Barros and Carlos França (1877–1926); and (h) foreign zoologists—Alfredo Corti, Charles Brues (1879–1955), Mario Bezzi, Paul Choffat (1849–1919), and Per Olof Christopher Aurivillius (1843–1928).

CHAPTER 7

The Journal *Brotéria: Vulgarização científica* and the Popularization of Science, Technology, and Medicine

Between 1902 and 1906, *Brotéria* was directed at a broad academic audience. In order to ensure that the journal reached as wide a readership as possible, the Portuguese Jesuits exchanged their journal with more than a hundred national and international scientific magazines. But doing so also led to significant financial problems. As the journal specialized in identifying and classifying new species of animals and plants, its primary readership was made up of professional botanists and zoologists.; because these scientists directly collaborated with some of the journals that exchanged with *Brotéria*, they could easily access the journal and, more importantly, they could do so for free.

At the same time, the number of paying non-academic subscribers was largely insignificant, which meant that the revenue from subscriptions was insufficient to cover the costs of publication. In 1907, to compensate for the lack of individual subscriptions, Silva Tavares restructured *Brotéria* by dividing it into three separate periodicals: the two scientific series already mentioned, *Botânica* and *Zoologia*, and a popular scientific periodical, *Vulgarização científica* (figure 17). The latter journal, which was written entirely in the vernacular, was directed at "less instructed people" and was designed to make a profit.[1] The journal published more than 450 articles on chemistry, physics, agriculture, commerce and industry, medicine and hygiene, seismology, geography, and archaeology between 1907 and 1924, when it ceased publication, and it played an important role in disseminating knowledge of these subjects to a non-specialist Portuguese readership.

---

1 "Prólogo," *Brotéria: Vulgarização científica* 6 (1907): 5–9, here 7. Unlike the articles published in *Vulgarização científica*, the articles in the *Zoologia* and *Botânica* series were written in a multitude of languages, including Latin, Portuguese, German, English, French, and Spanish. The Jesuits used a variety of strategies to make a profit and cover the expenses of the scientific series. In 1919, *Vulgarização científica* began to include a plethora of advertisements in its initial pages for banks, hotels, typographies, factories, drugstores, bookstores, and jewelry stores. Beginning in the late 1920s, the support of benefactors seems to have been crucial as well. For forty consecutive years (1927–67), the Jesuits printed a list of benefactors on the back cover of *Brotéria*, which included Jesuit alumni and a few notable personalities from the clerical and aristocratic spheres.

 | DOI:10.1163/9789004382367_008

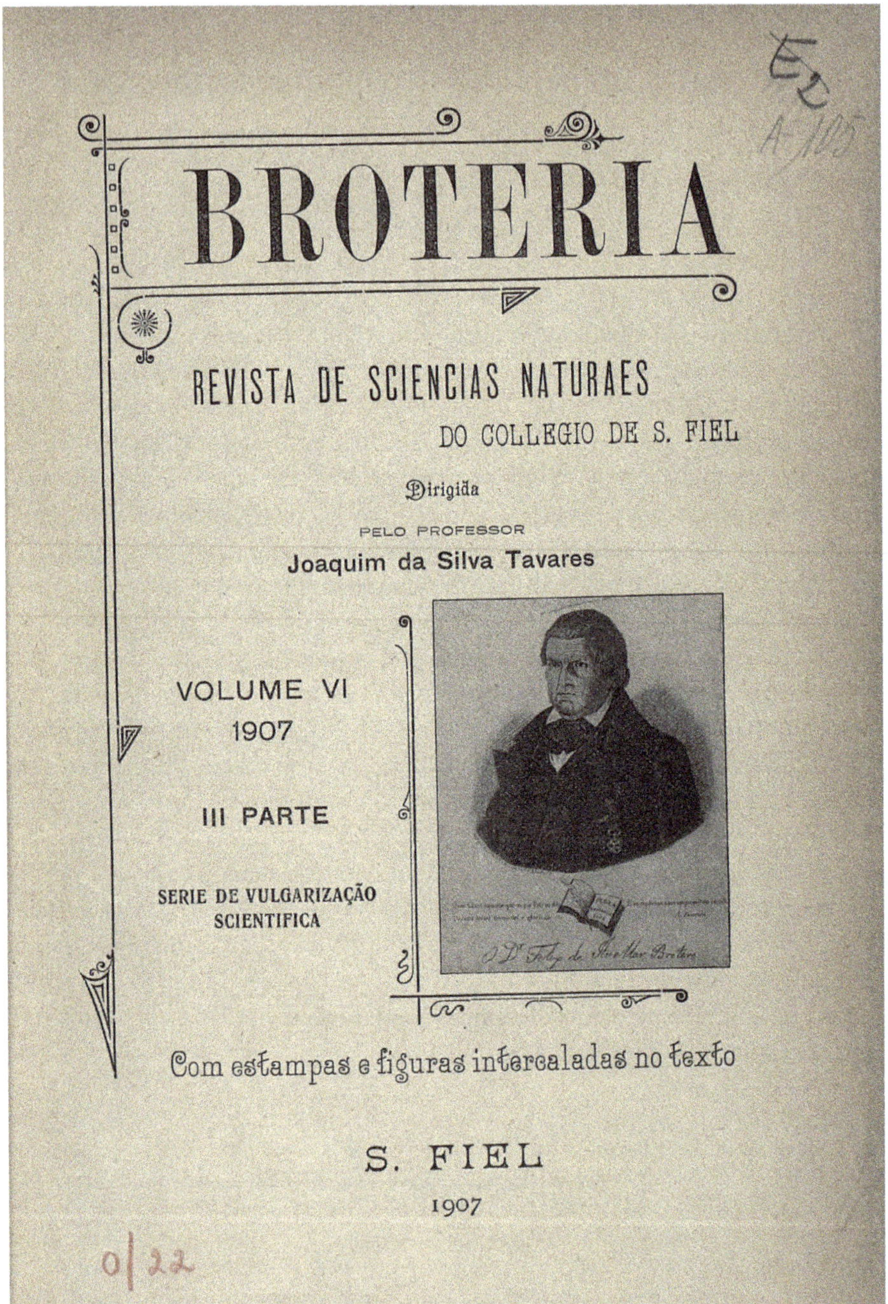

BROTERIA

REVISTA DE SCIENCIAS NATURAES

DO COLLEGIO DE S. FIEL

Dirigida

PELO PROFESSOR

Joaquim da Silva Tavares

VOLUME VI

1907

III PARTE

SERIE DE VULGARIZAÇÃO SCIENTIFICA

Com estampas e figuras intercaladas no texto

S. FIEL

1907

FIGURE 17 First issue of *Brotéria: Vulgarização científica*, 1907
AB

As well as publishing longer essays, *Vulgarização científica* also published shorter articles, including more than 350 book reviews and over a thousand brief news reports referring not only to scientific novelties in agriculture, chemistry, physics, botany, zoology, and medicine but also to conferences, academies, and distinctions attributed to Portuguese scientists.[2] The journal also contained a section dedicated to household matters, entitled "Useful Things," which welcomed articles with practical advice on a variety of household issues, including stain-cleaning, meat conservation, gold and bronze polishing, and soap manufacture. Sometimes, the Jesuits also gave medical advice on how to treat toothaches, coughs, or bronchitis. Readers seem to have particularly appreciated the culinary recipes published in this section.[3] In late 1917, for instance, an anonymous reader living in Porto wrote to the editor asking for the publication of more "tasteful and practical culinary recipes" as Christmas was approaching and *Vulgarização científica* had not published any recipes in the former three issues.[4] Because they wanted to attract young readers to their popular magazine, the Jesuits also published small inspirational novels in this series, including *Percy Wynn: Or Making a Boy of Him* (1893), one of the earlier works of American Jesuit Francis J. Finn (1859–1928).[5] By publishing technical articles dedicated to a learned reader alongside culinary recipes, novels, and household counsels, Silva Tavares and his companions hoped to broaden the appeal of their magazine and thus cover the expenses of the more specialized scientific series.

Because it was conceived as a popular magazine, *Vulgarização científica* covered a large variety of subjects, including history, sociology, ethnography, archeology, the history of science, and pedagogy.[6] However, the sections on agriculture, physics, geography, commerce and industry, and medicine and

---

2 Marta Mendonça, "*Brotéria* e a difusão da ciência em Portugal," in Rico and Franco, *Fé, ciência, cultura*, 255–76, here 267–69. The tradition of publishing book reviews can be traced back to the foundation of *Brotéria*. Between 1902 and 1906, Silva Tavares and his companions published 282 book reviews in the original series of *Brotéria*. After the restructuration, the three independent series continued to include book reviews in their issues. Overall, between 1907 and 1924, *Botânica*, *Zoologia*, and *Vulgarização científica* published 853 book reviews.

3 See especially Lena, "Arte culinária," *Brotéria: Vulgarização científica* 14 (1916): 108–10, 174–75, 227–28, 286–87, 334–35; Lena, "Arte culinária," *Brotéria: Vulgarização científica* 15 (1917): 31–32, 90–91, 171–72; Lena "Arte culinária," *Brotéria: Vulgarização científica* 16 (1918): 1–2.

4 The letter was included in a short reply by Silva Tavares, who signed under a pseudonym. See Dionel, "Ás leitoras da *Brotéria*," *Brotéria: Vulgarização científica* 16 (1918): 5.

5 The novel was published as a booklet between November 1915 and May 1916. Francis J. Finn: b. October 4, 1859, Saint Louis, Missouri (United States); e. March 24, 1879, Florissant, Missouri (United States); d. November 4, 1928, Cincinnati, Ohio (United States) (DHCJ, 2:1463–64).

6 See Table 8 in the appendix.

hygiene were clearly predominant, accounting for about sixty-nine percent of all popular articles. The vast majority of the lengthier articles were written by Silva Tavares and a few other Jesuits, namely Torrend, Azevedo Mendes, Oliveira Pinto, Raúl Sarreira (1889–1968), Manuel Rebimbas (1873–1944), Manuel Narciso Martins (1858–1940), and Artur Redondo (1873–1966).[7] Yet *Vulgarização científica* also welcomed articles by laymen, including the celebrated chemist Ferreira da Silva, and physicians José Pedro Dias Chorão (1853–1928) and Fernando de Almeida e Silva (1873–1942).

However, of all these authors, it was Silva Tavares who featured the most prominently. The founder and editor-in-chief of *Brotéria* was deeply invested in the popular series for many reasons, but the financial aspect seems to have been the most important. Ensuring *Vulgarização científica*'s commercial success was not a minor issue because it was a *sine qua non* condition for the publication of *Botânica* and *Zoologia*, which were already acknowledged as significant publications in their respective fields. This meant that, besides working systematically in the classification of plant galls around the globe, Silva Tavares dedicated a considerable part of his time to writing popular articles for the series. Between 1907 and 1924, for instance, he alone published 174 articles, which accounts for more than forty percent of all articles published in the popular series.[8]

In the opening pages, the first issue presented some details about its nineteen correspondents, including place of work, job, honorary affiliations, and main field of research (figure 18). Although the vast majority of these authors were Jesuit priests and scholastics, the first group also included five laymen: physicians Almeida e Silva and Dias Chorão, botanist José d'Ascenção Guimarães, Italian zoologist Mario Bezzi, and naturalist Thomaz Martins Mano (dates unknown).[9] In 1907, the journal's Jesuit correspondents from the Portuguese province were either teaching at the colleges of Campolide, São Fiel, and São Francisco, or pursuing their theological studies abroad in Ireland,

7 Raúl Sarreira: b. November 6, 1889, Ponte do Rol, Torres Vedras (Portugal); e. July 30, 1906, Barro (Portugal); d. April 11, 1968, Vila Coutinho (Mozambique) (*DHCJ*, 4:3506–7). Manuel Rebimbas Tavares: b. January 27, 1873, Pardelhas, Aveiro (Portugal); e. September 18, 1894, Barro (Portugal); d. December 26, 1944, Braga (Portugal) (*Mendizábal* 23,295). Manuel Narciso Martins: b. November 12, 1858, Labrunhal Cimeiro, Portalegre (Portugal); e. February 27, 1876, Barro (Portugal); d. July 23, 1940, Salvador da Bahia (Brazil) (*Mendizábal* 21,631). Artur Redondo: b. June 27, 1873, Vila Nova de Foscoa (Portugal); e. September 24, 1896, Barro (Portugal); d. March 17, 1966, Baturité, Ceará (Brazil) (*Mendizábal* 30,513).

8 Silva Tavares and Torrend, the two most prolific authors of *Vulgarização científica*, published a total of 213 articles, that is, about fifty percent of all articles. Silva Tavares sometimes published under the pseudonyms Dionel and Tessa.

9 "Redacção da *Brotéria*," *Brotéria: Vulgarização científica* 6 (1907): 10–11.

FIGURE 18 The network of *Brotéria*'s correspondents, 1907

the Netherlands, or Belgium. In addition, there were two other Jesuits, the Spaniard Longino Navás (1858–1938) and Austrian Johannes Rick, who were carrying out research in Spain and Brazil respectively.[10] Despite being listed in the first issue as correspondents, some authors did not write any articles for the popular series. It is possible that they accepted the invitation to contribute but that the collaboration eventually took another form, as was the case with Guimarães, Rick, Navás, and Bezzi. Although they did not publish any articles

10 Longino Navás: b. March 7, 1858, Cabacés, Tarragona (Spain); e. May 29, 1875; d. December 31, 1938, Gerona (Spain) (*Mendizábal*, 21,030).

for *Vulgarização científica*, they did publish in the scientific series.[11] During the following years, the contributors of the popular series increased to seventy-eight. However, the bulk of the magazine continued to depend on Silva Tavares and a few other authors.

Thus, based on the most representative sections of *Vulgarização científica*, and contrary to the aims of the magazine, the readership of the popular journal mostly consisted of learned men and women, regardless of their professional commitment to any particular branch of scientific inquiry. Sometimes, the articles in *Vulgarização científica* aimed to broadcast new discoveries or methods to peers, as in the particular case of medicine and hygiene. In such cases, the most relevant difference between an article in the popular magazine and an article in *Botânica* and *Zoologia* was simply a linguistic one. When considering both the intention behind its foundation and the actual contents of the magazine, it is clear that *Vulcarização científica* requires a more nuanced definition of the boundaries between a scientific journal, in the strict sense, and a popular science magazine.

1 Agriculture, Commerce, and Industry

The largest section in *Vulgarização científica* was on agriculture.[12] The first article, by Jesuit Manuel Martins, was published in 1908 and focused on a highly important topic, agricultural plagues. Given that corn was an essential cereal to feed the "dense population" of northern Portugal, Martins's intention was to provide the most effective means to combat the parasites that prevented corn from being harvested or conserved. For him, it was vital to eradicate such plagues because without corn "those good peasants would be forced to migrate" to "distant and insalubrious lands in search for their daily sustenance."[13] In subsequent years, *Vulgarização científica*'s correspondents also wrote

11 José d'Ascenção Guimarães, from the Lisbon Polytechnic School, published an important monograph on a group of parasitic plants. See José d'Ascenção Guimarães, "Monographia das Orobanchaceas," *Brotéria* 3 (1905): 5–208. Mario Bezzi, from Turin, published eight research articles on entomology (1909–22). Mycologist Johannes Rick and entomologist Longino Navás were two of the most prolific foreign authors of the scientific series of *Brotéria*. The former published thirty-two articles (1905–40) and the latter thirty-eight articles (1903–35).

12 Between 1908 and 1924, the popular series published seventy-two articles on agriculture (around 16.8 percent of all articles). See Table 8 in the appendix for more details.

13 Manuel Narciso Martins, "O milho grosso em Portugal e seus inimigos," *Brotéria: Vulgarização científica* 7 (1908): 68–83, here 73.

several articles on the production of wheat, soy, sugar, coffee, cocoa, wine, cotton, wool, and rubber.[14] The articles on wheat and wine focused on the Portuguese context. A few years later, the pursuit of food independence—in which wheat would play a key role—became one of the most pressing issues for the dictatorial Estado Novo regime.[15] In turn, the publication of articles on vegetal byproducts that were important for the Brazilian economy—particularly sugar, coffee, and cocoa—reflected the need to reach a wider readership during the Jesuits' exile, as while Silva Tavares and his companions were in Baía (1912–15), *Brotéria* depended almost exclusively on Brazilian subscribers.[16]

Like the agricultural section, the section dedicated to commerce and industry focused on the trade of Portuguese and Brazilian goods, namely fruit, rubber, cocoa, coffee, cereals, and dairy. Given its importance for the national economy, *Brotéria* also welcomed articles on the trade of cork and wine.[17] Among these articles, the most relevant was perhaps the piece by Ferreira da Silva, from the Polytechnic Academy of Porto.[18] From the 1890s, he was arguably the most prominent Portuguese chemist, partly due to his decisive role in the

---

14 See especially the following articles: Joseph Foulquier, "A borracha brazileira," *Brotéria: Vulgarização científica* 12 (1914): 5–31; Joaquim da Silva Tavares, "A producção do assucar brasileiro," *Brotéria: Vulgarização científica* 12 (1914): 282–95; Tavares, "Producção mundial de lã e particularmente nas colónicas britânicas," *Brotéria: Vulgarização científica* 12 (1914): 305–7; Florentino Barbosa, "O café na Parahyba," *Brotéria: Vulgarização científica* 12 (1914): 347–50; Camilo Torrend, "A cultura do algodão," *Brotéria: Vulgarização científica* 14 (1916): 5–12; 82–89; 137–45; J.M. da Cunha, "A produção de trigo em Portugal em 1923," *Brotéria: Vulgarização científica* 22 (1924): 135–6; J.M. da Cunha, "A colheita do vinho nos principais países produtores em 1923: O vinho português em 1923," *Brotéria: Vulgarização científica* 22 (1924): 137–39; António Monteiro Filipe, "Notas sobre a cultura do cacau em S. Tomé," *Brotéria: Vulgarização científica* 18 (1920): 174–80; Joaquim da Silva Tavares, "A cultura da soja," *Brotéria: Vulgarização científica* 22 (1924): 29–34; Camilo Torrend, "A cultura do cacau," *Brotéria: Vulgarização científica* 16 (1918): 12–19; 82–91; 128–35; 175–83; 274–80; Torrend, "A cultura do cacau," *Brotéria: Vulgarização científica* 17 (1919): 23–28, 41–46; 106–14; 128–38.

15 Tiago Saraiva, *Fascist Pigs: Technoscientific Organisms and the History of Fascism* (Cambridge, MA: MIT Press, 2016), 43–70.

16 Tavares, "A *Brotéria* nos primeiros 25 anos da sua existência," 35.

17 On the trade in wine, see especially "A producção vinicola em Portugal em 1913," *Brotéria: Vulgarização científica* 12 (1914): 313–15; "Exportação do vinho pela barra do Porto em 1913," *Brotéria: Vulgarização científica* 12 (1914): 315; Joaquim da Silva Tavares, "A producção vinicola em Portugal em 1915," *Brotéria: Vulgarização científica* 14 (1916): 188–89; Tavares, "O vinho português em 1916," *Brotéria: Vulgarização científica* 15 (1917): 91–95. On the production of and commerce in cork, see Joaquim da Silva Tavares, "A cortiça portuguesa," *Brotéria: Vulgarização científica* 15 (1917): 53–65; A. Vieira Novo, "A indústria corticeira," *Brotéria: Vulgarização científica* 21 (1923): 207–12.

18 António Ferreira da Silva, "Os vinhos do Porto," *Brotéria: Vulgarização científica* 13 (1915): 129–39; 181–95.

well-known controversies between the Brazilian and Portuguese governments over the adulteration of Port wine exported to Brazil in 1894 and again in 1900. The dispute was of paramount importance for both political and economic reasons. The Clinical Laboratory of Rio de Janeiro argued that the recently imported Port wine had been seriously tainted with salicylic acid, which sweetened the wine and artificially augmented its durability. This was indeed a very serious accusation, because it could hinder the trade of Port wine with Brazil and, most importantly, with England.[19] After analyzing the exported wine, Ferreira da Silva argued that the content of salicylates was residual and that it had not been caused by any kind of adulteration; instead, the change was a byproduct of the natural process of fermentation.[20] The crown acknowledged his services by making him peer of the realm and advisor to the king.

Ferreira da Silva briefly described the processes of Port wine manufacture in *Vulgarização científica* from harvest to aging and bottling, and presented and analyzed the most important figures in the trade of Port wine from 1678 onward. Interestingly, he did not address what had been the most pressing issue in the previous decades, the 1894 and 1900 controversies. Ferreira da Silva's article aimed to disseminate the most basic concepts on the manufacture and commerce of Port wine to a Portuguese readership. As a renowned expert in the field, his principal goal was to explain technical issues as simply as possible, without compromising the scientific objectivity of the concepts and processes at hand. By publishing an article in *Vulgarização científica* on this issue, Ferreira da Silva consequently lent his scientific credibility and expertise to the Jesuits' journal, with its publication testifying, once again, to the close friendship between the chemist and the Jesuit naturalists.

2 Geography

Between 1907 and 1917, the vast majority of the articles on geography aimed to disseminate historical and geographical facts about South American cities and

19 The trade in Port wine was of paramount importance for the Portuguese crown. In 1703, the Portuguese and English crowns had signed a commercial treaty, commonly referred to as Methuen Treaty, which favored both the export of Port wine and the import of British woolen cloth.

20 Ferreira da Silva published extensively on the controversy surrounding the adulteration of Port wine. Based on his personal notes, he published a very useful summary of the controversy. See António Ferreira da Silva, *A suposta salicilagem dos vinhos portugueses no Brasil (1900–1902): Memórias, notas e documentos* (Coimbra: Imprensa da Universidade, 1919). For more details, see especially Jorge Fernandes Alves and Rita C. Alves, *A.J. Ferreira da Silva: Nos caminhos da química* (Porto: Universidade do Porto, 2013).

provinces. The inclusion of a considerable number of photographs in these articles, as well as the simple prose, testified to the importance of geography in the broader program of popularization designed by the Jesuits.[21] As in the case of agriculture, Silva Tavares wrote several articles for this section based on his visits to urban and rural regions during his exile in Brazil, Argentina, and Spain, with his notes being published in the pages of *Vulgarização científica*.[22] By resourcefully combining his enthusiasm for photography with the situation in which he found himself, Silva Tavares offered his readers a copiously illustrated memoir of his exile. Along with essays on specific cities, the geography articles also addressed coetaneous economic issues, such as the construction of the Panama Canal and the financial implications of granting the monopoly on transatlantic communication to the United States.[23]

The successful relocation of *Brotéria* to Spain in 1914 led to some changes in the editorial policy of *Vulgarização científica*. Beginning in 1918, the geographical section shifted its focus from Brazil to Portugal, with articles on illiteracy rates, criminality, census data, and the cost of living. In other words, the essays began to address educational, economic, and demographic issues that were of particular interest for a learned Portuguese readership.[24] However, the topic that interested Silva Tavares the most in this period was fisheries, especially in the Minho River in northern Iberia.[25] While the first geography articles were

21 Between 1907 and 1924, *Vulgarização científica* published sixty-one articles on geography (around 14.2 percent of all articles). See Table 8 in the appendix for more details.

22 Joaquim da Silva Tavares, "A Argentina: Impressões de um excursionista," *Brotéria: Vulgarização científica* 11 (1913): 350–69; Tavares, "A cidade de Santos," *Brotéria: Vulgarização científica* 10 (1912): 101–38; Tavares, "O Corcovado," *Brotéria: Vulgarização científica* 10 (1912): 189–99; Tavares, "Uma excursão a Minas," *Brotéria: Vulgarização científica* 11 (1913): 126–62; Tavares, "Capital modelo: Bello Horizonte," *Brotéria: Vulgarização científica* 12 (1914): 65–107; Tavares, "Maceió, capital do Estado de Alagoas," *Brotéria: Vulgarização científica* 13 (1915): 324; Tavares, "Victória, capital do Estado do Espírito Santo," *Brotéria: Vulgarização científica* 14 (1916): 16; Tavares, "A cidade de Parahyba do Norte," *Brotéria: Vulgarização científica* 14 (1916): 81; Tavares, "Florianópolis, capital do Estado de Santa Catharina," *Brotéria: Vulgarização científica* 14 (1916): 145–46; Tavares, "Uma cidade encantada no Estado do Paraná," *Brotéria: Vulgarização científica* 14 (1916): 201–9; Tavares, "O salto de Iguaçu (Estado do Paraná)," *Brotéria: Vulgarização científica* 14 (1916): 258–71.

23 J. Alves Matheus, "O Canal de Panamá e as outras vias de comunicação entre os dois oceanos," *Brotéria: Vulgarização científica* 11 (1913): 178–80.

24 Joaquim da Silva Tavares, "O analphabetismo em Portugal," *Brotéria: Vulgarização científica* 16 (1918): 111–20; 151–64; 229–36; Tavares, "A criminalidade em Portugal," *Brotéria: Vulgarização científica* 17 (1919): 114–17; Tavares, "A carestia da vida em Portugal," *Brotéria: Vulgarização científica* 20 (1922): 142–49; Tavares, "O censo da população portuguesa feito em Dezembro de 1920," *Brotéria: Vulgarização científica* 22 (1924): 207–16.

25 Joaquim da Silva Tavares, "A pesca no Rio Minho," *Brotéria: Vulgarização científica* 18 (1920): 262–76; Tavares, "A pesca no Rio Minho," *Brotéria: Vulgarização científica* 19 (1921): 21–40; Tavares, "A pesca do atum em Portugal," *Brotéria: Vulgarização científica* 20 (1922):

mostly descriptive, the articles on fisheries included some suggestions on how to address specific problems. Rather than just disseminating knowledge in a simpler, abridged version to a less familiar audience, Silva Tavares engaged with the reader and presented his own conclusions.[26] After displaying and analyzing the official Portuguese and Spanish data on fisheries between 1920 and 1923, the editor-in-chief of *Brotéria* concluded that temperature and water purity were the two most decisive factors in determining the abundance or shortage of fish for a given year. In response to the shortage of salmon and allis shad in the Minho River, he urged for the creation of an Iberian laboratory of ichthyology—established either in Spain or Portugal—in order to reduce the costs involved in rearing these expensive species, promote their consumption by poorer rural populations, and improve the local economy.[27] Similar to the articles on agriculture, commerce, and industry, the later articles on Portuguese and Spanish geography insisted on the importance of food independence and economic sustainability. Considered together, these topics constituted a *basso continuo* in the Jesuit program of popularization.

3 Chemistry, Physics, and Seismology

Whereas the articles on geography, agriculture, and commerce were written in a fairly simple style, the earlier articles on chemistry, physics, and seismology addressed more technical issues.[28] As such, it is likely that the readership of those articles was more circumscribed, because the articles presumed at least a certain familiarity with scientific instruments, theories, and practices. Given

15–17; Tavares, "A pesca no Rio Minho em 1921," *Brotéria: Vulgarização científica* 20 (1922): 19–23; Tavares, "A pesca no Rio Minho em 1922," *Brotéria: Vulgarização científica* 21 (1923): 57–60; Tavares, "As pescarias no Rio Minho em 1923," *Brotéria: Vulgarização científica* 22 (1924): 89–95.

26 On the dominant view of popularization and the need to erase the distinction between the production and the transmission of knowledge for historical work, see especially Stephen Hilgartner, "The Dominant View of Popularization: Conceptual Problems, Political Issues," *Social Studies of Science* 20 (1990): 519–39; Roger Cooter and Stephen Pumfrey, "Separate Spheres and Public Spaces: Reflection on the History of Science Popularization and Science in Popular Culture," *History of Science* 32 (1994): 269–315; James Secord, "Knowledge in Transit," *Isis* 95 (2004): 654–72; and Bernadette Bensaude-Vincent, "A Genealogy of the Increasing Gap between Science and the Public," *Public Understanding of Science* 10 (2011): 99–113.

27 Tavares, "As pescarias no Rio Minho em 1923," 94–95.

28 Between 1907 and 1924, *Vulgarização científica* published fifty-nine articles on physics, twenty-five on chemistry, and five on seismology (around 20.7 percent of all articles). See Table 8 in the appendix for more details.

its increasing popularity and the recent ventures of Oliveira Pinto, it comes as no surprise that radioactivity emerged as one of the most discussed topics in the earlier chemistry articles (1907–17).[29] From 1918 onward, applied chemistry gradually replaced theoretical chemistry, with *Brotéria*'s correspondents, most notably Silva Tavares, paying more attention to matters of greater economic significance, namely the use of natural and artificial fuel in industry and in automobiles.[30]

The earlier physics articles addressed topics of particular interest for teachers of the natural sciences, namely X-rays and wireless telegraphy.[31] By privileging topics that had already proven to be of great interest to audiences at the colleges of Campolide and São Fiel, the Portuguese Jesuits aimed to bridge the gap between the practice and teaching of science.[32] The tight connection between the colleges and the journal seemed to have endured even after the 1910 expulsion. Presumably, the articles on X-rays and wireless telegraphy, by Raúl Sarreira and Oliveira Pinto respectively, replicated part of the physics curriculum for the scientific academies of Campolide and São Fiel. Given that Oliveira Pinto had performed some of the first experiments with wireless telegraphy in Portugal at Campolide in 1902, his lengthy article (eighty-nine pages) was probably well received by the Portuguese scientific community eight years later. After describing the historical contributions of Heinrich Hertz (1857–94), Michael Faraday (1791–1867), James Maxwell (1831–79), Édouard Branly (1844–1940), and Guglielmo Marconi (1874–1937), Oliveira Pinto delved into the explanation of technical concepts such as frequency, wavelength, and period. Most important, perhaps, was the thorough description of all equipment

29 António de Oliveira Pinto, "O estudo da radioactividade da materia: Laboratorio de Giff," *Brotéria: Vulgarização científica* 11 (1913): 51–68; Pinto, "A radioactividade das aguas medicinaes de fraca mineralisação," *Brotéria: Vulgarização científica* 13 (1915): 244–45; Manuel Rebimbas, "Radioactividade e gazes raros das fontes thermaes," *Brotéria: Vulgarização científica* 7 (1908): 181–84.

30 Joaquim da Silva Tavares, "A substituição da gasolina nos automóveis," *Brotéria: Vulgarização científica* 16 (1918): 73–77; Tavares, "A substituição da gasolina nos automóveis pelo álcool," *Brotéria: Vulgarização científica* 16 (1918): 239–40; Tavares, "O carvão líquido e a catálise: Vantagens industriais," *Brotéria: Vulgarização científica* 20 (1922): 66–70; Tavares, "Origem da energia industrial: Gás natural, petróleo e carvão," *Brotéria: Vulgarização científica* 21 (1923): 44–46.

31 Raúl Dias Sarreira, "Os raios X," *Brotéria: Vulgarização científica* 12 (1914): 155–67; Sarreira, "Raios e Pára-raios," *Brotéria: Vulgarização científica* 14 (1916): 215–21; Sarreira, "Os raios X," *Brotéria: Vulgarização científica* 15 (1917): 79–90; 118–24; 230–36; António de Oliveira Pinto, "Telegraphia sem fio," *Brotéria: Vulgarização científica* 9 (1910): 181–203; Pinto, "Telegraphia sem fio," *Brotéria: Vulgarização científica* 10 (1912): 14–33; 200–45.

32 On the solemn sessions of physics held at Campolide and São Fiel until 1910, see Chapter 4.

necessary for the transmission and reception of sound waves, because it provided the basic means to start using wireless telegraphy.

As well as publishing articles directly related to the teaching of experimental physics, the Portuguese Jesuits were also concerned with contemporary debates on modern physics. Shortly after the famous expedition of Arthur Stanley Eddington (1882–1944) and Andrew Crommelin (1865–1939) to Principe and Brazil in 1919, for instance, Torrend wrote a short article for the Jesuits' magazine expounding the intention of the excursion (i.e., the possible confirmation of Albert Einstein's [1879–1955] theory of general relativity).[33] In contrast to most previous expeditions, the focus of the observation was not the solar corona but a group of bright stars in the vicinity of the sun. The main objective of the British committee was to study the deflection of light emanating from those stars. By comparing two sets of photographs, before and after the eclipse of May 29, 1919, it would be possible to corroborate Einstein's theory.[34] As *Brotéria*'s article was written before the publication of Eddington's book on general relativity, Torrend presented and discussed the three possible outcomes. If there were no differences between the two sets of photographs, it would follow that light is not deflected by gravity. If, on the other hand, there was a deflection of seven-eighths of a second at the edge of the sun—0.87 according to Eddington—then light was deflected according to Newtonian principles. But, if the deflection of light at the edge of the sun was twice that figure—1.74 in Eddington's account—then Einstein's theory would be proven correct. As Torrend anticipated, the "conclusion of this study will constitute an important development in the physical knowledge of the globe."[35]

Beginning in 1920, the vast majority of articles published in the physics section concerned aviation.[36] This specific interest in aeronautics coincided

33 Camilo Torrend, "O eclipse total de 29 de Maio de 1919, no Brazil," *Brotéria: Vulgarização científica* 18 (1920): 40–41. Oddly miscategorized in the 1959 *Index of Brotéria* as a geography article, this article was in fact a popular article on physics: *Índices gerais da* Brotéria *científica (1902–2002)* (Lisbon: Brotéria, 2002), 144.

34 Arthur Stanley Eddington, *Space, Time and Gravitation: An Outline of the General Relativity Theory* (Cambridge: Cambridge University Press, 1920), 110–22.

35 Torrend, "O eclipse total de 29 de Maio de 1919, no Brazil," 41.

36 Dionel, "Os progressos da aviação," *Brotéria: Vulgarização científica* 18 (1920): 180–86; Dionel, "Aviação comercial," *Brotéria: Vulgarização científica* 22 (1922): 29–39; A. Lamego, "A aviação ao serviço do comércio," *Brotéria: Vulgarização científica* 20 (1922): 276–80; Lamego, "O voo à vela ou sem motor," *Brotéria: Vulgarização científica* 20 (1922): 283–89; Lamego, "Heroísmo e aventuras da aviação em 1922," *Brotéria: Vulgarização científica* 21 (1923): 131–34; Lamego, "Os progressos e serviços da aviação nos diferentes países," *Brotéria: Vulgarização científica* 21 (1923): 163–74; Lamego, "Os aviões comerciais," *Brotéria: Vulgarização científica* 21 (1923): 186–91; Lamego, "O voo à vela e suas causas," *Brotéria:*

with what historians have called the Golden Age of Aviation (1918–39), a period characterized by numerous technical developments in the manufacture of aircrafts and the increasing popularity of air racing and exploration.[37] In the Portuguese context, the highpoint of the golden era was undoubtedly the venture of Carlos Gago Coutinho (1869–1959) and Artur de Sacadura Cabral (1881–1924), who crossed the South Atlantic Ocean for the first time in 1922.[38] As was the case with the chemistry articles, the shift from theoretical physics to applied issues—developments in aviation, airplane commerce, and the first aerial circumnavigation—reflected the Jesuits' concern with reaching as wide a readership as possible.

Like the sections on physics and chemistry, the smaller section on seismology also sought to convey the latest discoveries in the field. With articles by Manuel Sánchez-Navarro Neumann, S.J. (1867–1941), director of the Astronomical, Geodynamical, and Meteorological Observatory of Cartuja (Granada, Spain), this section often presumed a learned reader.[39] In his first article for *Vulgarização científica*, the Spanish Jesuit delved into the study of earthquakes without scientific instruments.[40] With seventeen equations and mathematical deductions that allowed for the calculation of the depth of the hypocenter, the velocity of propagation of seismic waves, and the initial instant of the earthquake, this article was intended as an introduction to seismology for a reader at least familiar with elementary calculus and trigonometry. Unlike his description of the devastating earthquake in Messina, Italy (December 28, 1908), Sánchez-Navarro's article on the "latest discoveries in seismology" explained the

Vulgarização científica 21 (1923): 267–70; Lamego, "Os pára-quedas," *Brotéria: Vulgarização científica* 22 (1924): 10–15; Lamego, "A volta ao mundo em avião," *Brotéria: Vulgarização científica* 22 (1924): 74–77; Lamego, "Aeroplanos e suas aplicações," *Brotéria: Vulgarização científica* 22 (1924): 111–16; Lamego, "O esforço aeronáutico alemão," *Brotéria: Vulgarização científica* 22 (1924): 216–23; Lamego, "As grandes viagens aéreas," *Brotéria: Vulgarização científica* 22 (1924): 276–79.

37 For the American context of the golden age of aviation, see especially Joseph J. Corn, *The Winged Gospel: America's Romance with Aviation, 1900–1950* (New York: Oxford University Press, 1983), and Thomas Kessner, *The Flight of the Century: Charles Lindbergh and the Rise of American Aviation* (New York: Oxford University Press, 2010).

38 A. Vieira Novo, "A caminho do Brazil," *Brotéria: Vulgarização científica* 20 (1922): 188–97; A. Lamego, "Heroísmo e aventuras da aviação em 1922," *Brotéria: Vulgarização científica* 21 (1923): 131–34.

39 Manuel Sánchez-Navarro Neumann: b. January 23, 1867, Málaga (Spain); e. February 8, 1900, Granada (Spain); d. January 30, 1941, Puerto de Santa María, Cádiz (Spain) (*DHCJ*, 3:3493).

40 Manuel Sánchez-Navarro Neumann, "Os tremores de terra estudados sem instrumentos," *Brotéria: Vulgarização científica* 6 (1907): 217–31; 262–80.

functioning of a seismograph through the lenses of pendulum motion and introduced new trigonometric functions to calculate an earthquake's epicenter.[41]

The same section also contained two essays by a student at the Jesuit college of Campolide.[42] The first of these reproduced an article presented at a solemn session of the scientific academy of Campolide on March 19, 1909,[43] which was dedicated to seismology, partly in response to the recent catastrophe of Messina that had caused the death of tens of thousands. In front of an audience comprising the apostolic nuncio archbishop Giulio Tonti (1844–1918), as well as all the students and their families, Francisco Pereira Cabral (1893–?) expounded the three most accepted theories on the origin of earthquakes, introduced basic concepts of seismology, and delved into the functioning of a seismograph and the respective interpretation of a seismogram. At the end of his presentation, the audience viewed screen projections depicting "the destruction of Messina and Reggio."[44]

Shortly after the solemn session, on April 23, 1909, there was another severe earthquake in Benavente, Santarém. As he was unable to "personally check the damages" or "collect the necessary information *in loco*," Cabral wrote "only a few notes on the recent earthquake" for the Jesuits' popular magazine.[45] Unlike the first article, which was, above all, an overview of the most significant concepts, hypotheses, and instruments in seismology, this account instead focused on an analysis of a very recent case. Together with photographs of the disaster taken by Oliveira Pinto, the article also reproduced a copy of a seismogram, registered at the Spanish Jesuit Observatory of Ebro.[46] With the information

41 Manuel Sánchez-Navarro Neumann, "O recente terremoto de Messina," *Brotéria: Vulgarização científica* 8 (1909): 100–9; Neumann, "Os ultimos descobrimentos em sismologia," *Brotéria: Vulgarização científica* 12 (1914): 36–45; 108–14.

42 The articles were signed by F.P. Cabral, "academic and sixth grade student at the Colégio de Campolide." Hypothetically, the author could have been Francisco Pereira Cabral (1893–?). Son to Afonso do Vale Coelho Pereira Cabral (1857–1946) and Inês Van Zeller Guedes de Carvalho (1864–1950), he probably studied at Campolide, as his elder brothers Gonçalo (1883–1945) and Luís Gonzaga (1886–1964) did. If he studied at Campolide, then he was fifteen years old when the solemn session of the scientific academy dedicated to seismology took place (March 19, 1909). Because the articles state that the author was a member of the scientific academy and a sixth-grade student at Campolide (between fifteen and sixteen years old), it is possible (although not certain) that F.P. Cabral was short for Francisco Pereira Cabral.

43 F.P. Cabral, "Os terremotos: A proposito da catastrophe de Messina," *Brotéria: Vulgarização científica* 8 (1909): 110–28.

44 Cabral, "Os terremotos," 128.

45 F.P. Cabral, "Algumas notas sobre o tremor de terra de 23 de abril," *Brotéria: Vulgarização científica* 8 (1909): 182–90, here 182.

46 Cabral, "Algumas notas sobre o tremor de terra," 186.

he gathered, Cabral drew an isoseismal map and concluded that the epicenter had been in Samora Correia, a small city very close to Benavente. Both his map and the location of the epicenter seem to have been in conformity with the opinion of three renowned geologists, Alfredo Bensaúde (1856–1941), Salvador Calderón y Arana (1851–1911), and Paul (Paulo) Choffat (1849–1919), as Cabral himself noted.[47]

Although it turned out to be a relatively modest section in terms of the number of articles, the articles on seismology played a particularly important role in the history of *Vulgarização científica*. Like the earlier articles on chemistry and physics, the papers in this section introduced fundamental instruments, theories, and practices to a learned audience. But they also exemplified the successful intertwining between the Jesuit colleges of Campolide and São Fiel and the publication of *Brotéria*. Because it published highly mathematical and complex articles by a renowned expert alongside simpler texts by a fifteen-year-old student, this section defied the traditional, early twentieth-century understanding of the boundaries between a scientific journal and a popular magazine.

4 Medicine and Hygiene

The articles on medicine and hygiene in *Brotéria* were almost exclusively written by physicians. Whereas the articles on hygiene addressed public health issues—such as eating practices, nicotine dependence, and alcoholism—the medical articles focused on the characterization and treatment of infectious diseases, including flu, meningitis, pneumonia, typhoid fever, leprosy, and syphilis.[48] The sections on medicine and hygiene apparently followed

47 Cabral, "Algumas notas sobre o tremor de terra," 183.

48 Between 1907 and 1924, *Vulgarização científica* published forty-seven articles on medicine and hygiene (around eleven percent of all articles). See Table 8 in the appendix for more details. For the hygiene section, see especially Manuel Rebimbas, "Noções de hygiene alimentar," *Brotéria: Vulgarização científica* 6 (1907): 155–69; Paulino Vieilledent, "Perigos do tabaco," *Brotéria: Vulgarização científica* 7 (1908): 151–54; José Pedro Dias Chorão, "O tabaquismo e o alcoolismo," *Brotéria: Vulgarização científica* 8 (1909): 5–21; 41–57. For the medicine section, see José Pedro Dias Chorão, "Meningite cerebro-espinal epidémica," *Brotéria: Vulgarização científica* 6 (1907): 50—63; José Pedro Dias Chorão, "Gripe," *Brotéria: Vulgarização científica* 6 (1907): 239–61; Chorão, "Febre typhoide: Prophylaxia e therapeutica," *Brotéria: Vulgarização científica* 12 (1914): 115–22; 235–43; Chorão, "Um flagello social: A syphilis," *Brotéria: Vulgarização científica* 13 (1915): 65–74; Alejo Diz, "O tifo exantemático," *Brotéria: Vulgarização científica* 16 (1918): 140–44; José Pedro Dias Chorão, "A epidemia gripal de 1918 no concelho do Fundão," *Brotéria: Vulgarização científica* 17

a traditional model of popularization, because the articles were written by experts who sought to explain complex issues in the simplest prose. However, a closer inspection reveals that, once again, the boundaries between a popular magazine—that is, a journal directed to a broad, yet learned, audience—and a science journal—that is, a publication directed at peers—were rather fluid.

One of the most interesting examples of *Vulgarização científica*'s fluidity was perhaps the work of physician José Pedro Dias Chorão,[49] who published an important article in 1909 that reported the treatment of pneumonia in a student of São Fiel by means of an innovative procedure.[50] The fourteen-year-old student had manifested the first symptoms on January 18, 1909. For Dias Chorão, it appeared to be a minor flu. On January 21, however, the physician confirmed that it was pneumonia, and with the new diagnosis, he proceeded to administer "hypodermic injections of dissociated [colloidal] silver" every six hours.[51] The boy's general condition began to improve on January 22, and five days later, he was out of risk.[52] As this method had proven successful two years earlier, when he treated a sixty-one-year-old patient, Dias Chorão was mostly interested in communicating to his peers how to use colloidal silver in the treatment of pneumonia. His article went on to describe the procedure and instruments that had been used to obtain the valuable solution.[53] Later that year, Dias Chorão reported he had encountered three new cases of pneumonia and one case of an infection of the uterus, which had also been cured by means of colloidal silver.[54] He concluded that the "dosage of solute" could play a vital role in the treatment of pneumonia, as one infection only yielded after increasing the concentration of colloidal silver three times.[55] He was also able to identify at least two genera of bacteria that were responding to

---

(1919): 66–75; Júlio Marinho, "Tratamento da lepra," *Brotéria: Vulgarização científica* 22 (1924): 34–36.

49 There is not much written about this physician. Born in Capinha, a small village in the interior of Portugal, Dias Chorão practiced medicine in Fundão. As a local physician, he also served the Colégio de São Fiel. For a short biography, see Joaquim Candeias da Silva, "Evocação/Memória de alguns médicos notáveis da Beira Interior: Concelho do Fundão (v)," *Medicina na Beira Interior da Pré-História ao século XX: Cadernos de cultura* 20 (2006): 70–76.

50 José Pedro Dias Chorão, "A prata dissociada pela corrente electrica, no tratamento da pneumonia," *Brotéria: Vulgarização científica* 8 (1909): 97–99; 251–56.

51 Chorão, "A prata dissociada pela corrente electrica," 97.

52 Chorão, "A prata dissociada pela corrente electrica," 98.

53 Chorão, "A prata dissociada pela corrente electrica," 99.

54 Chorão, "A prata dissociada pela corrente electrica," 251–56.

55 Chorão, "A prata dissociada pela corrente electrica," 252.

this treatment, *Streptococcus* and *Staphylococcus*.[56] By applying the same procedure to the treatment of a severe infection of the uterus, which was seriously endangering a poor woman's life, Dias Chorão concluded that colloidal silver offered a promising method to fight infections, particularly "all infections of *Streptococcus* and *Staphylococcus*."[57] For its usefulness, this article could be seen as a popular article, in a traditional sense. However, Dias Chorão was trying to communicate with his peers, and, for that reason, he provided all data he collected, described in detail the experimental procedure to obtain colloidal silver, and exhorted the medical community to use this method in the treatment of specific infections.

Four years later, in 1913, Dias Chorão explained how he had arrived at the idea of using colloidal silver to treat infections[58] after reading a "very brief note" in the pages of *Revue de thérapique* (Review of therapeutics). However, the note was "so brief" that it did not give any details on how to obtain colloidal metals. Because he visited São Fiel frequently as a physician, he was able to ask professor of physics Valério Cordeiro about the procedure that had to be used to obtain the coveted solution.[59] After several failed attempts, Cordeiro decided to use an Oudin coil, an instrument frequently used in the solemn sessions of the scientific academy of São Fiel. Before using it for the first time, Dias Chorão had tried to scrutinize the medical benefits of colloidal silver, but it was only in 1912 that someone "courteously loaned" him a book on this particular subject.[60] In his 1913 article, Dias Chorão summarized the findings of two physicians, which confirmed the efficacy of using colloidal metals in the treatment of myriad diseases, including rheumatism, flu, smallpox, syphilis, meningitis, and diphtheria.[61] In addition to illustrating the tight connection between the teaching and practice of science at São Fiel, this example also reveals how particular medical practices could emerge in rural twentieth-century Portugal. Practical knowledge and experimentation superseded authoritative, though modern, theoretical knowledge.

56 Chorão, "A prata dissociada pela corrente electrica," 255.

57 Chorão, "A prata dissociada pela corrente electrica," 256.

58 José Pedro Dias Chorão, "Os metaes coloidaes electricos na therapeutica," *Brotéria: Vulgarização científica* 11 (1913): 40–50.

59 Chorão, "Os metaes coloidaes," 40. For biographical details on Valério Cordeiro, see Chapter 4.

60 Albert Robin, *Les ferments métalliques et leur emploi en thérapeutique* (Paris: J. Rueff Éditeur, 1907).

61 Louis Bousquet and H. Roger, *Les métaux colloïdaux en thérapeutique* (Paris: Masson, 1910).

Following in the footsteps of Dias Chorão, a few other physicians collaborated in the Jesuits' project of popularization. Possibly by suggestion of his close friend, former colleague at the University of Coimbra, and Jesuit alumnus Egas Moniz, Portuguese physician Fernando de Almeida e Silva applied for a position in Fundão in October 1899. In April 1900, after being officially appointed as a local physician, he moved to the village and established a clinic.[62] In 1908, he published an article on the process of baking bread and on the medical benefits of dark rye flour.[63] Although he was a member of the editorial team of *Vulgarização científica* and physician at São Fiel, this turned out to be his only contribution to the Jesuits' popular magazine. As a friend of the Jesuits, he was probably expected to have published a few more articles in *Brotéria*, but the expulsion of the Jesuits from Portugal in 1910 will almost certainly have hindered this venture.

During his first years in Brazil, Silva Tavares sought to find new authors for both the scientific and the popular series. Brazilian physicians Cândido de Mello Leitão (1866–1948) and Egas Moniz Barreto de Aragão (1870–1924) and Spanish physician Alejo Diz (1885–1936) willingly accepted the invitation to write a few articles for the medicine and hygiene sections. As with the earlier articles, the main aim of these papers was to provide practical solutions to everyday problems. When he wrote about baby-feeding, Mello Leitão included a possible schedule for breastfeeding as well as tables indicating the appropriate amount of milk during the baby's first fourteen months.[64] Similarly, when Jesuit scholastic Júlio Marinho (1891–1976) wrote about education, his main concern was to give useful advice to elementary and secondary teachers on topics such as the prevention of nearsightedness and the manufacture of student desks.[65]

---

62 Joaquim Candeias da Silva, "Evocação/Memória de alguns médicos notáveis da Beira Interior: Concelho do Fundão (IV)," *Medicina na Beira Interior da Pré-História ao século XX: Cadernos de cultura* 19 (2005): 83–88, here 85. Often referred to as Fernando d'Almeida (or de Almeida), Fernando de Almeida e Silva was son to papal count António de Almeida e Silva (1822–1900) and Maria Raquel Carvalho (1832–1911).

63 Fernando de Almeida, "O pão sob o ponto de vista hygienico," *Brotéria: Vulgarização científica* 7 (1908): 9–39.

64 Cândido de Mello Leitão, "Alimentação da Criança nos primeiros tempos da vida," *Brotéria: Vulgarização científica* 10 (1912): 173–88.

65 Júlio Marinho: b. November 16, 1891, Sidães, Fafe (Portugal); e. May 22, 1906, Barro (Portugal); d. January 4, 1976 Lisbon (Portugal) (*Verástegui*, 63). Júlio Marinho, "A miopia escolar," *Brotéria: Vulgarização científica* 15 (1917): 217–23; Marinho, "Higiene escolar," *Brotéria: Vulgarização científica* 13 (1915): 206–9; 247–52; Marinho, "Higiene escolar," *Brotéria: Vulgarização científica* 14 (1916): 47–49; Marinho, "Mobiliário escolar," *Brotéria: Vulgarização científica* 19 (1921): 211–16.

5 Science and Religion

In 1909, Silva Tavares created a section entirely dedicated to the natural sciences and religion.[66] This section aimed to scrutinize the conflict thesis between science and religion by addressing, among other things, the scientific plausibility of the theory of evolution and its epistemological value.[67] He wanted to provide a resolute answer to a fundamental question: "Can a Catholic admit evolution without compromising his faith?"[68] Later that year, German Jesuit Ambros Schupp (1840–1914) made a first attempt to answer the question.[69] He began by explaining the difference between Darwinism and evolution, because the two terms were often used interchangeably, and this "originated pitiful errors" for Christian "religion and morals."[70] After delving into the crucial concepts of species, constancy, evolution, and the survival of the fittest, Schupp asserted that Charles Darwin (1809–82) had written a "most remarkable work."[71] Although he was definitely not an enthusiastic supporter of a purely mechanistic explanation, Schupp believed that Darwinism had been "a powerful incentive for the progress of science."[72]

For Schupp, there were three arguments that supported the evolution of species: the evolution of the Solar System; the evolution of the Earth's crust; and the existence of fossils. Specifically favorable to Darwin's theory were the adaptation of species to climate changes and the artificial selection of species.[73] His main critique was not on the evolution of species but rather on

66 Joaquim da Silva Tavares, "As sciencias naturais e a religião," *Brotéria: Vulgarização científica* 8 (1909): 33–35.

67 The literature on the history of the relationship between science and religion is particularly extensive. See especially the classic study by John Hedley Brooke, *Science and Religion: Some Historical Perspectives* (Cambridge: Cambridge University Press, 1991), as well as the more recent Peter Harrison, *The Territories of Science and Religion* (Chicago: Chicago University Press, 2015). Because they offer a broad range of scholarly contributions on the relationship between science and religion, the following edited volumes can also be extremely helpful: Ronald L. Numbers, ed., *Galileo Goes to Jail and Other Myths about Science and Religion* (Cambridge, MA: Harvard University Press, 2010); Harrison, *Cambridge Companion to Science and Religion*; and Gary B. Ferngren, ed., *Science and Religion: A Historical Introduction* (Baltimore: John Hopkins University Press, 2017).

68 Tavares, "As sciencias naturais e a religião," 35.

69 Ambros Schupp: b. May 26, 1840, Montabaur, Hesse-Nassau (Germany); e. September 30, 1869 (Germany); d. November 13, 1914, Porto Alegre (Brazil) (*Mendizábal*, 13,414). Ambrósio Schupp, "A evolução e o homem," *Brotéria: Vulgarização científica* 8 (1909): 81–94; 147–63.

70 Schupp, "A evolução e o homem," 81.

71 Schupp, "A evolução e o homem," 84.

72 Schupp, "A evolução e o homem," 84.

73 Schupp, "A evolução e o homem," 87–88.

the origin of species. He claimed that there was no scientific basis to state that all species had originated from a few lifeforms, let alone from a single organism, as Ernst Haeckel (1834–1919) contended, because doing so contradicted the universally accepted axiom *omne vivum ex vivo* (all life is from life).[74] Because Schupp believed that the theory of natural selection did not explain animal symmetry, the development of sensorial organs, or the existence of two genders, among other crucial issues, he advocated for a theistic evolution, as proposed by many nineteenth-century Christian thinkers.[75] In Schupp's words, the belief in a "most intelligent being that governed the world wisely according to certain plans" easily conceded the "evolution of all organisms" in perfect "harmony with the doctrines of [Christian] faith."[76] While reminding his readers that evolution was not contrary to Christian faith, and that Darwinism had not been condemned by the Catholic Church, the German Jesuit insisted that man was not only "the end result of a phylogenetic or ontogenetic evolution of species" but the final cause of the evolution of nature.[77]

Although it initially seemed promising, this section of the journal fell short of its ambitious program. After Schupp's article, *Vulgarização científica* only published four other essays promoting a compatibilist vision of science and religion.[78]

---

74 Schupp, "A evolução e o homem," 89–90.

75 On the reception of Darwinism in the late nineteenth and twentieth century, see especially the classic works by Peter Bowler, James R. Moore, and Ronald L. Numbers: Peter J. Bowler, *The Eclipse of Darwinism: Anti-Darwinian Evolution Theories in the Decades around 1900* (Baltimore: Johns Hopkins University Press, 1983); James R. Moore, *The Post-Darwinian Controversies: A Study of the Protestant Struggle to Come to Terms with Darwin in Great Britain and America, 1870–1900* (New York: Cambridge University Press, 1979); Ronald L. Numbers, *The Creationists: From Scientific Creationism to Intelligent Design* (Cambridge, MA: Harvard University Press, 2006). For an introduction to the topic, see Peter J. Bowler and John Henry, "Evolution," in Ferngren, *Science and Religion*, 204–19; Jonathan Hodge, "Evolution," in Bowler and Pickstone, *Cambridge History of Science*, 6:242–64; James R. Moore, "Religion and Science," in Bowler and Pickstone, *Cambridge History of Science*, 6:541–62.

76 Schupp, "A evolução e o homem," 148.

77 Schupp, "A evolução e o homem," 153. On the reception of Darwinism by the Roman Catholic Church in the first decades after the publication of *On the Origin of Species* (1859), see especially Mariano Artigas, Thomas F. Glick, and Rafael A. Martínez, *Negotiating Darwin: The Vatican Confronts Evolution, 1877–1902* (Baltimore: Johns Hopkins University Press, 2006).

78 F. de Mello, "A propósito de aviação," *Brotéria: Vulgarização científica* 9 (1910): 36–41; Camilo Torrend, "Origem da cellula: Geração espontanea" *Brotéria: Vulgarização científica* 15 (1917): 247–53; Torrend, "Origem da cellula: Geração espontanea," *Brotéria: Vulgarização científica* 16 (1918): 103–10; Raúl Dias Sarreira, "A propósito de um livro recente: A. Emyieu,

6 The Foundation of a Cultural Journal

Because they were especially concerned with funding *Botânica* and *Zoologia*, the Portuguese Jesuits published a popular scientific journal for eighteen consecutive years, and the success of *Vulgarização científica* may have inspired other provinces to follow their example. In 1913, for example, the Spanish Jesuits from the Observatorio del Ebro emulated their Portuguese companions by publishing *Ibérica* (1913–36 and 1945–2004), a weekly magazine devoted to "general scientific culture."[79] The main objective of the Spanish magazine was to make "science accessible to everyone," as it subtitle clearly suggested: "El progreso de las ciencias y de sus aplicaciones" (The progress of science and its applications). *Ibérica* went on to become one of the most important popular scientific journals in Spain with articles on physics and the natural sciences, astronomy, meteorology, seismology, and geophysics.[80]

In 1925, Silva Tavares broadened the aims of *Vulgarização científica*, transforming it into a cultural, philosophical, and humanistic periodical. Like *Vulgarização científica*, the so-called cultural *Brotéria* was published alongside the scientific series (figure 12).[81] However, the main objective behind its creation was not the popularization of science, nor the funding of *Botânica* and *Zoologia*, but rather to communicate the Christian faith and Christian values to a broad, yet learned, audience. The Jesuits argued that the restructuration of *Vulgarização científica* and the creation of a monthly review was intended to accommodate their readers' pleas to address a larger number of topics, which would indeed contribute to a "greater social good."[82] Thus, in the pages of cultural

La part des croyants dans les progress de la science au XIXe siècle," Brotéria: Vulgarização científica 19 (1921): 184–87.

79 Ricard Cicera, "La ciencia al alcance de todos," *Ibérica* Spécimen A (1913): 1–5, here 1.

80 Although it shared certain similarities with *Vulgarização científica*, the journal was primarily aimed at a non-specialist readership, and hence differed from *Brotéria*'s research series *Botânica* and *Zoologia*. On the history of the Observatorio del Ebro and the journal *Ibérica*, see Josep Batlló and David Altadill, "The Ebre Observatory: Its Path to Ionospheric Research," *Advances in Space Research* 39 (2007): 941–46; Nestor Herran, "'Science to the Glory of God': The Popular Science Magazine *Ibérica* and Its Coverage of Radioactivity," *Science and Education* 21 (2012): 335–53; Pascual Bolufer, "Science and Technology in the 20th Century as Seen through the Journal *Ibérica* (1914–2003)," *Contributions to Science* 7 (2011): 185–90; Maria Genescà-Sitjes, "*Ibérica* Magazine (1913–2004) and the Ebro Observatory," *Contributions to Science* 9 (2013): 159–68.

81 Although it is simply referred to as cultural *Brotéria*, this journal was published under a variety of subtitles: *Fé: Ciências; Letras* (1925–31); *Revista contemporânea de cultura* (1932–64); *Revista de cultura* (1965–69); *Cultura e informação* (1970–99); *Revista de cultura* (1999–2000); *Cristianismo e cultura* (2001–). See figure 12 for more details.

82 "A servir de prólogo nesta segunda série," *Brotéria: Fé; Sciências; Letras* 1 (1925): 5–6, here 5.

Brotéria, the Jesuits sought to publish articles on literature, sociology, philosophy, theology, and history. Given the diversity of the subjects it covered, and its apologetic program, the cultural *Brotéria* shared many common features with the other learned journals published by the Society of Jesus all over the world. Whether Silva Tavares's decision represented the failure of his strategy to secure funding for the scientific series, or whether it was instead an effort to publish a magazine similar to other Jesuit periodicals, remains unclear.

Cultural *Brotéria* published works by Jesuits and lay authors, but the vast majority of essays were by Portuguese Jesuits such as Manuel Antunes (1918–85), António Leite (1911–2004), Domingos Maurício (1896–1978), and João Pereira Gomes (1917–2002), to name only a few.[83] Although cultural *Brotéria* had an exceptional longevity in the Portuguese context, only a few historians have delved into its history. However, the significance of the cultural series can easily be grasped by its popularity among both liberal and conservative thinkers. In the centennial celebration of *Brotéria*, thirty-eight personalities from the political, clerical, and academic spheres—including two former presidents, two bishops, and the patriarch of Lisbon—paid tribute to the Jesuit journal.[84] Regardless of creed or political stance, they celebrated the Jesuits' work in creating an exceptional cultural journal.

One year after the foundation of the cultural *Brotéria*, the political situation in Portugal changed once again when a military coup on May 28, 1926 overthrew the republican government, banned all political parties, and laid the groundwork for the foundation of the forty-year dictatorial regime of the Estado Novo. Because it published some speeches by Prime Minister António de Oliveira Salazar (1889–1970, in office 1932–68), the cultural *Brotéria* seemed to support the dictatorship, especially until the 1960s.[85] Unlike the First Republic, Estado Novo was particularly supportive of the Jesuits and their educational work. As well as reversing the Jesuits' exile, the new regime worked extensively on reestablishing diplomatic relations with the Holy See, the high point of which was the 1940 concordat that restored the Catholic Church some—but

83 For critical editions of the work of Manuel Antunes and João Pereira Gomes, see José Eduardo Franco, ed., *Obra completa do Padre Manuel Antunes*, 14 vols. (Lisbon: Fundação Calouste Gulbenkian, 2007–12); and Henrique Leitão and José Eduardo Franco, eds., *Jesuítas, ciência e cultura em Portugal: Obra selecta do Pe. João Pereira Gomes S.J.* (Lisbon: Esfera do Caos, 2012).

84 On the appraisal of the cultural *Brotéria*, see Rico and Franco, *Fé, ciência, cultura*, 19–86.

85 José Eduardo Franco, "A evolução editorial da *Brotéria*," in Rico and Franco, *Fé, ciência, cultura*, 103–26, here 116–17.

not all—of its former privileges.[86] In harmony with the propaganda of the Estado Novo, *Brotéria* also participated in the political celebrations of both the foundation of Portugal (1140) and the country's independence from Spanish rule (1640).[87] Between 1939 and 1940, the editorial team of *Brotéria* celebrated not only the national jubilees but also its own four-hundredth anniversary.[88]

In the following decades of the regime, the cultural *Brotéria* underwent a few additional changes. Beginning in 1945, the Jesuits created a section dedicated to philosophy that was later expanded into an independent magazine in 1947, the *Revista portuguesa de filosofia* (Portuguese journal of philosophy).[89] In the period following Vatican II (1962–65), the cultural *Brotéria* significantly broadened its scope and started publishing subtle critiques of the political, economic, social, and educational policies of the Estado Novo.[90] Twenty years later, Superior General Peter-Hans Kolvenbach (1928–2016, in office 1983–2008) sought to standardize the cultural journals published by the Society all over the world. In 1984, he addressed the editorial team of *Brotéria* and reminded them that the main objective of the journal should be "to deepen and announce the whole truth" to a learned Christian audience,[91] and because it played a crucial role in the "permanent formation of intellectual elites," *Brotéria* should place Christian faith and values at its core.[92]

---

86 On the history of the diplomatic relations between the Estado Novo and the Catholic Church, see especially Manuel Braga da Cruz, *O Estado Novo e a Igreja Católica* (Lisbon: Bizâncio, 1998).

87 The peak of the political celebrations was, undoubtedly, the design of a monumental exhibition in Lisbon in 1940. On the role of the Portuguese World Exhibition in commemorating the new regime, see especially Margarida Acciaiuoli, *Exposições do Estado Novo, 1934–1940* (Lisbon: Livros Horizonte, 1998); David Corkill and José Carlos Pina Almeida, "Commemoration and Propaganda in Salazar's Portugal: The 'Mundo Português' Exposition of 1940," *Journal of Contemporary History* 44 (2009): 381–99, and Ellen W. Sapega, *Consensus and Debate in Salazar's Portugal: Visual and Literary Negotiations of the National Text, 1933–1948* (University Park, PA: Pennsylvania State University Press, 2008).

88 For the celebration of the foundation of Portugal and independence from Spanish rule, see the issue of January 1939. For the celebration of the anniversary of the foundation of the Society of Jesus, see the issue of November 1940.

89 Lúcio Craveiro da Silva, "A Faculdade de Filosofia de Braga e a sua *Revista portuguesa de filosofia*: Mensagem e evolução," *Revista portuguesa de filosofia* 51 (1995): 281–94.

90 Franco, "A evolução editorial da *Brotéria*," 119.

91 Peter Hans Kolvenbach, "Função de uma revista de cultura," *Brotéria* 118 (1984): 483–84, here 483.

92 Kolvenbach, "Função de uma revista de cultura," 483–84.

CHAPTER 8

Taxonomy, Cytogenetics, and Plant Breeding in the Early Years of Estado Novo

Following Silva Tavares's death, Afonso Luisier was appointed *Brotéria*'s editor-in-chief (figure 19). Under his direction (1932–57), the editorship of the cultural and the scientific series was divided, and the scientific series *Zoologia* and *Botânica* were merged into a unique journal entitled *Brotéria: Ciências naturais* (Natural sciences) (figure 12). In this period, the Jesuit journal witnessed the consolidation of plant and animal taxonomy and the emergence of new research areas, especially cytogenetics and plant breeding. The collaboration between Luisier and some of the most distinguished Portuguese scientists served to reinforce the Jesuits' scientific prestige and political stability in the Estado Novo dictatorial regime.

Born in Fregnoley, Switzerland, on February 6, 1872, Luisier entered the Society of Jesus at the age of nineteen. Because he wanted to become a missionary to Portuguese Africa, he entered the novitiate in Portugal (1891–93). In contrast to the usual practice, Luisier completed his regency before studying philosophy. Between 1894 and 1897, he taught humanities at a Jesuit college

FIGURE 19
Afonso Luisier, S.J. (1872–1957)
AB

 | DOI:10.1163/9789004382367_009

in Lons-le-Saunier, France, and at the apostolic school in Guimarães. After his regency, he studied philosophy in Lyon, France (1898–99), Barro (1899–1900), Setúbal (1900–1), and Guimarães (1901–2). Between 1902 and 1906, he studied theology in Innsbruck, Austria, before finally being ordained on July 26, 1906. When he returned to Portugal, he was appointed professor of chemistry, physics, and natural history at Campolide, where he lived until 1910.[1]

Like his Jesuit companions, Luisier was not a broad-spectrum naturalist. He specialized in the study of bryophytes, especially in Portugal, Spain, Madeira, Azores, and Brazil, identifying and describing eighteen new species and thirteen new varieties of mosses over the course of his career. Although he published most of his research in *Brotéria*, he also wrote a few articles for other scientific periodicals, such as *Boletim da sociedade portuguesa de ciência naturais*, *Anais científicos da academia politécnica do Porto* (Scientific annals of the polytechnic academy of Porto), *Boletim da sociedade Broteriana* (Bulletin of the Sociedade Broteriana), and *Revista agronómica* (Agronomical journal).

While still a student of theology, Luisier started to correspond with some of the most respected Portuguese botanists, including Gonçalo Sampaio, Júlio Henriques, and Joaquim de Mariz (1847–1916). In 1902, his collaboration with Sampaio, Henriques, and Mariz allowed him to complete an important flora of the region of Setúbal and Serra da Arrábida, which comprised about one thousand different species.[2] As in the cases of Silva Tavares, Azevedo Mendes, Torrend, and Zimmermann, Luisier built his own network of correspondence, which in turn allowed for the circulation of countless specimens and, consequently, for the identification of new species and genera.[3] One of the examples that illustrates the importance of these networks of correspondence in Luisier's career is the description of the thistle *Centaurea luisier*. In 1915, while in Salamanca, Luisier collected what seemed to be an unknown species of thistle, and since he was not an expert on thistles, he sought outside help to identify this apparently new species. He sent the specimen to his friend Gonçalo

1 José Carvalhaes, "Padre Alphonse Luisier," *Brotéria: Ciências naturais* 54 (1958): 3–16; Carvalhaes, "Reverendo Padre Alphonse Luisier, S.J.: Homenagem ao cientista e ao mestre," *Boletim cultural de Santo Tirso* 5 (1957): 223–49. For more succinct biographies, see especially *DHCJ*, 3:2440–41 and *Grande enciclopédia portuguesa e brasileira*, 15:601; Franco, "Os directores da *Brotéria*," in Rico and Franco, *Fé, ciência, cultura*, 155–57.

2 Afonso Luisier, "Apontamentos sobre a flora da região de Setubal," *Boletim da sociedade broteriana* 19 (1902): 172–74; Luisier, "Catalogo das plantas vasculares de Setubal e da serra d'Arrabida," *Boletim da sociedade Broteriana* 19 (1902): 175–272; Luisier, "Appendice: Listas das plantas colhidas por Tournefort em Setubal e na serra d'Arrabida," *Boletim da sociedade Broteriana* 19 (1902): 272–77.

3 On the importance of these tailored networks for the circulation of botanical and zoological knowledge in twentieth-century Portugal, see Chapter 6.

Sampaio, a professor of botany at the University of Porto, but Sampaio was not an expert on thistles either, and despite being one of the most prestigious botanists of his generation, he was unable to provide a definite answer. The solution to what was a frequent issue in plant and animal taxonomy—the identification of a new species with the highest degree of certainty—consequently required further help. Sampaio then turned to Henriques, the dean of botany at the University of Coimbra, who finally confirmed that the thistle Luisier had collected near Salamanca was indeed a new species. Upon receiving this confirmation, Sampaio decided to describe the new plant in *Brotéria*. As had become a fairly common practice, Sampaio paid homage to its collector by naming it *Centaurea luisieri*.[4]

Like his Jesuit companions, Luisier was an enthusiastic collector, and before 1910, he had been able to gather a noteworthy collection of mosses. However, as we have seen, following the republican expulsion, his collection was confiscated by the new government. As was the case with Zimmermann and Torrend, he resorted to his scientific colleagues and made use of his nationality to retrieve it.[5] In 1913 (i.e., three years after being deprived of his mosses), the Swiss consul was able to negotiate the return of his compatriot's collection.[6] After the expulsion, Luisier traveled to the Netherlands, Belgium, and Spain. Between 1910 and 1912, he lived in Gemert, the Netherlands, and Alsemberg, Belgium. He returned to the Iberian Peninsula in 1912 and was appointed superior of the apostolic school in Salamanca, Spain (1912–15). From there, he went to Pontevedra (1916–18) and La Guardia (1918–32). During the exile, he resumed his work on the classification of mosses and published important works focused on the Iberian flora. Between 1917 and 1922, he wrote a series of articles on the mosses of Madeira.[7] And two years later, in 1924, he published a prize-winning monograph on the bryological flora of Salamanca.[8] Shortly after, the

4 Gonçalo Sampaio, "*Centaurea luisieri* (sp. n.)," *Brotéria: Botânica* 14 (1916): 104–5.

5 On the republican exile and the confiscation of the Jesuits' scientific collections, see Chapter 5.

6 Now at the Instituto Nun'Alvres, Santo Tirso, Luisier's collection is divided into three geographic sections: *Bryotheca Europaea*, *Bryotheca Atlantica*, and *Bryotheca exotica*.

7 Afonso Luisier, "Les mousses de Madère," *Brotéria: Botânica* 15 (1917): 81–98; *Brotéria: Botânica* 16 (1918): 29–71; *Brotéria: Botânica* 17 (1919): 17–142; *Brotéria: Botânica* 18 (1920): 5–22; 78–120; *Brotéria: Botânica* 20 (1922): 76–106.

8 Afonso Luisier, "Musci Salmanticenses: Descriptio et distributio specierum hactenus in provincia geographica Salmanticensi cognitarum; Brevi addito conspectu Muscorum totius Peninsulae Ibericae," *Memorias de la Real academia de ciencias exactas físicas y naturales de Madrid* 3 (1924): 1–280.

Jesuit botanist continued his lifelong goal, namely to describe and classify the mosses of the Portuguese archipelago of Madeira.[9]

When Azaña decreed the Jesuits' expulsion from Spain in 1932, Luisier was living in La Guardia. Together with most of his Portuguese companions, he traveled from La Guardia to Santo Tirso, in northern Portugal, and assisted in the relocation of the proclaimed heir of the colleges of Campolide and São Fiel, the wandering Instituto Nun'Alvres. Between 1932 and 1957, he taught natural sciences at the institute, where he also established a meteorological station. As in the case of São Fiel, the station was integrated into the national network of metrological observatories. During this period, Luisier continued to work on the systematic classification of mosses in Portugal, Azores, Madeira, and Brazil. He also wrote an extended biography of Félix de Avelar Brotero, which comprised a list of his most important works, including six original articles.[10]

After assuming the editorship of *Brotéria*, Luisier undertook the humanistic task of publishing a digest of a very important work on the entomological fauna of Madeira.[11] Between July and August 1935, Swedish naturalist Carl Olof Lundblad (1890–1970) led an expedition to Madeira and collected myriad insects, spiders, mites, and other arthropods. From 1938 to 1958, Lundblav and his collaborators published a series of articles describing the entomological fauna of Madeira in the Swedish journal *Arkiv för Zoologi* (Archive for zoology). Because he wanted to promote the study of the flora and fauna of the Atlantic islands, and the Swedish bulletin was not "easily found among us,"

9 Between 1927 and 1945, Luisier published a second series of articles on the mosses of Madeira: Afonso Luisier, "Les mousses de l'Archipel de Madère et en général des Îles Atlantiques," *Brotéria: Botânica* 23 (1927): 5–53; 129–145; *Brotéria: Botânica* 24 (1930): 18–47; 66–96; 119–40; *Brotéria: Botânica* 25 (1931): 5–20; 123–39; *Brotéria: Ciências naturais* 1 (1932): 164–82; *Brotéria: Ciências naturais* 7 (1938): 78–95; 110–31; *Brotéria: Ciências naturais* 11 (1942): 29–41; *Brotéria: Ciências naturais* 14 (1945): 78–94; 112–27; 156–76. Between 1936 and 1956, he also published a series of articles reporting recent discoveries of mosses in Madeira. See Afonso Luisier, "Recherches bryologiques récentes à Madère," *Brotéria: Ciências naturais* 5 (1936): 140–44; *Brotéria: Ciências naturais* 6 (1937): 88–95; *Brotéria: Ciências naturais* 8 (1939): 40–52; *Brotéria: Ciências naturais* 12 (1943): 135–44; *Brotéria: Ciências naturais* 16 (1947): 86–91; *Brotéria: Ciências naturais* 22 (1953): 178–91; *Brotéria: Ciências naturais* 25 (1956): 170–82.

10 Afonso Luisier, "Félix de Avelar Brotero," *Brotéria: Ciências naturais* 13 (1944): 145–58.

11 Afonso Luisier, "Artrópodes da Madeira, segundo as investigações do Dr. O. Lundblad," *Brotéria: Ciências naturais* 8 (1939): 18–39; 82–95; 101–12; *Brotéria: Ciências naturais* 9 (1940): 184–88; *Brotéria: Ciências naturais* 10 (1941): 61–69; 133–42; 179–84; *Brotéria: Ciências naturais* 11 (1942): 84–93; 137–44; 177–87; *Brotéria: Ciências naturais* 12 (1943): 29–36; 128–34.

Luisier decided to publish an annotated digest of Lundblav's work in the pages of *Brotéria*.[12]

The importance of Luisier's work on mosses led to him being recognized as one of the most distinguished botanists of his generation. He was a member of numerous societies, including the Sociedade Broteriana, the Portuguese Society of Natural Sciences, the Société valaisanne de sciences naturelles, and the Sullivant Moss Society. As Silva Tavares's successor as *Brotéria*'s editor-in-chief, Luisier was appointed corresponding fellow of the Lisbon Academy of Sciences in 1933, and almost a decade later, in 1942, the Faculty of Sciences, University of Porto, granted him an honorary doctorate. Despite having no formal scientific education, he thus became, by right, equal to university-trained botanists and zoologists.

Like Silva Tavares, who had stated his beliefs in the opening number of *Brotéria*, Luisier's calling to study the natural world was deeply embedded in his faith. On June 23, 1943, Luisier received his doctorate at a solemn session held at the Faculty of Sciences. There, he began by stating "loud and clear, that I have always loved the study of the natural sciences," and that "I have never found in my faith the slightest barrier to true progress in science," adding that his claim was far from original, because the Vatican Council had solemnly declared, several decades earlier, that there could never be a "true opposition between Faith and Reason."[13] The necessity of making such a statement in a university context reveals something important not only about Jesuit science and education but also about the history of science and religion in twentieth-century Portugal. When Luisier made his speech, three decades had elapsed since the Jesuits' expulsion, the dictatorial regime had celebrated its tenth anniversary, and diplomatic relations between the state and the Holy See were thriving.[14]

The decision to bestow such an honor was scientifically and politically motivated. As Américo Pires de Lima (1886–1966), director of the Faculty of Sciences, asserted, the honorary doctorate was, first and foremost, a public recognition of Luisier's "most relevant services to Portuguese culture." But he also hoped that this "just consecration" could "definitely neutralize any resentment" caused by the 1910 expulsion. By distancing himself from the "ingratitude" that had characterized the politicians of the First Republic (1910–26), Pires de Lima wanted to show that the Jesuits would not behave in ways that were contrary

12 Luisier, "Artrópodes da Madeira," 18.

13 Carvalhaes, "Padre Alphonse Luisier," 12.

14 On the history of the Catholic Church during this period, see especially Cruz, *O Estado Novo e a Igreja Católica*.

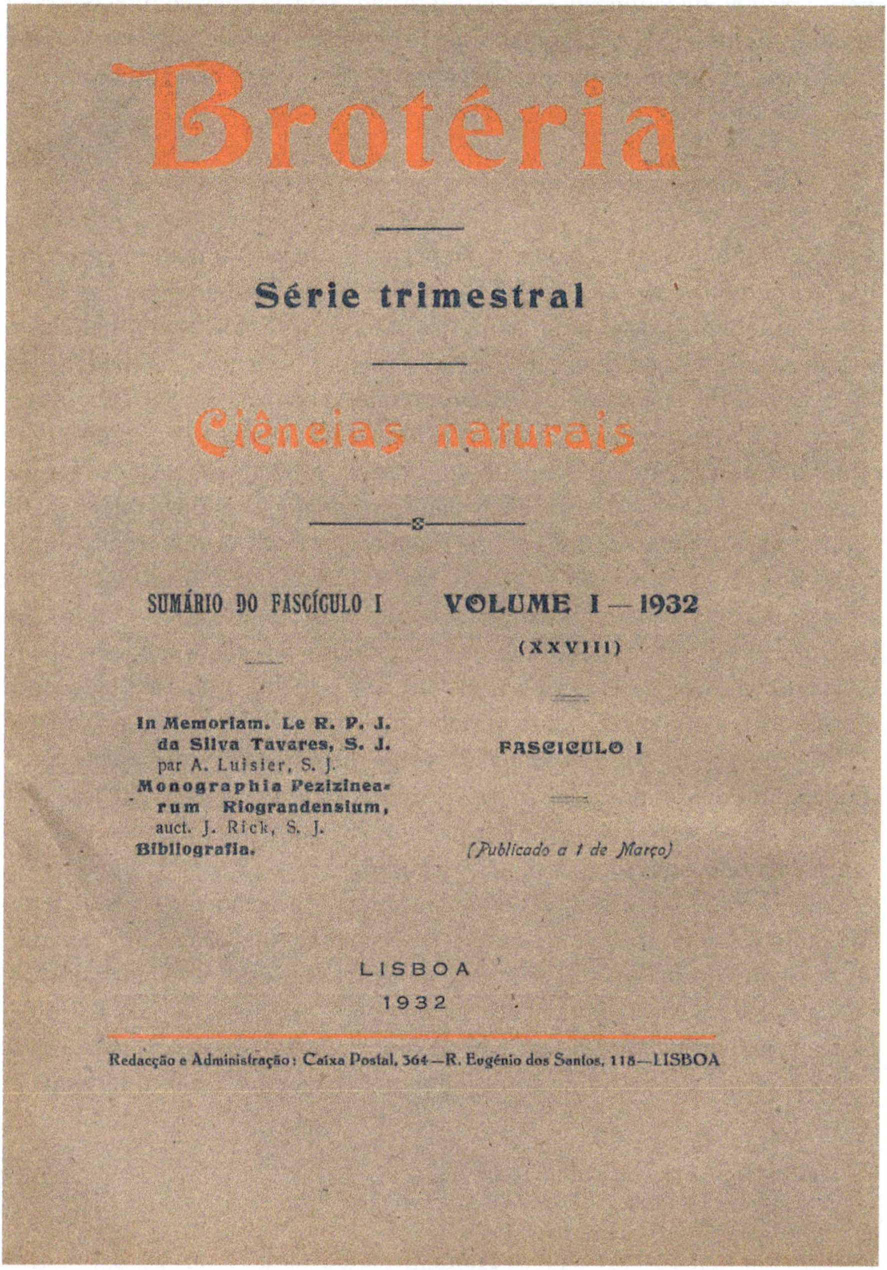

Brotéria

Série trimestral

Ciências naturais

SUMÁRIO DO FASCÍCULO I

In Memoriam. Le R. P. J. da Silva Tavares, S. J. par A. Luisier, S. J.
Monographia Pezizinearum Riograndensium, auct. J. Rick, S. J.
Bibliografia.

VOLUME I — 1932
(XXVIII)

FASCICULO I

(Publicado a 1 de Março)

LISBOA
1932

Redacção e Administração: Caixa Postal, 364—R. Eugénio dos Santos, 118—LISBOA

FIGURE 20 First issue of *Brotéria: Ciências naturais*, 1932
AB

to the new regime's interests.[15] While there had been an open conflict between the state and the Catholic Church under the First Republic, the Estado Novo was keen to ensure a harmonious relationship between both institutions.

1 Taxonomy and Evolution

After hearing about Silva Tavares's unexpected death on September 2, 1931, Provincial Cândido de Azevedo Mendes (in office 1919–24 and 1927–33) appointed Luisier as editor-in-chief of *Brotéria*. His decision was particularly significant because Azevedo Mendes embodied the original principles, methodologies, and scope of the Jesuit journal. As Zimmermann had left the Society of Jesus in 1921, Azevedo Mendes was by then the last representative of the initial group of naturalists behind the foundation of *Brotéria*. Serving as provincial for the second term, Azevedo Mendes knew that he could not steer the publication he had founded in 1902. Accordingly, he sought a Jesuit naturalist who had profound knowledge of *Brotéria*'s history and enjoyed an outstanding scientific reputation. Hence the nomination of Lusier came as no surprise, as he had been one of the most prolific authors to be published in the Jesuit journal and was particularly respected by his peers due to his contributions to the study of mosses. Aware of the difficulties of publishing a "journal entirely dedicated to scientific research," Luisier decided to merge the *Botânica* and *Zoologia* series into a unique magazine called *Brotéria: Ciências naturais* (1932–80) (figure 20).[16] Although he viewed this measure as temporary, in practice the unified journal continued to be published until 1980. (figure 12)

Under Luisier's direction, *Brotéria* continued to welcome articles on plant and animal taxonomy by former and new correspondents.[17] As in the earlier decades, the vast majority of articles were written by Jesuit naturalists, including Luisier, Rick, Navás, Jaime Pujiula (1869–1958), and Ignacio Sala de Castellarnau (1902–1973).[18] When Luisier took over the editorship of *Brotéria*, Navás was one of the most prolific correspondents of *Zoologia*; in fact, in terms of

15 Carvalhaes, "Padre Alphonse Luisier," 11.

16 Afonso Luisier, "Prólogo," *Brotéria: Ciências naturais* 1 (1932): 7–8.

17 See Table 9 in the appendix.

18 Longino Navás: b. March 7, 1858, Cabacés, Tarragona (Spain); e. May 29, 1875, Dussède, Haute-Garonne (France); d. December 31, 1938, Girona (Spain) (*DHCJ*, 3:2804). Jaime Pujiula: b. August 22, 1869, Besalú, Girona (Spain); e. August 13, 1887, Veruela, Zaragoza (Spain); d. December 15, 1958, Barcelona (Spain) (*DHCJ*, 4:3258–9). Ignacio Sala de Castellarnau: b. April 29, 1902, Esterri de Aneo, Lérida (Spain); e. May 29, 1917; d. February 23, 1973, Valencia (Spain) (*Verástegui* 133).

publications, he was second only to Azevedo Mendes. As a professor of natural history at the Jesuit college of Zaragoza, Spain, Navás worked on the description of neuropterans, bugs commonly known as net-winged insects. Early in his career, he published an important catalog of Iberian neuropterans in *Brotéria* in which he identified 408 different species.[19] Between 1903 and 1917, his research focused on the classification of net-winged insects collected in Africa, the Americas, Portugal, and Spain.[20] Beginning in the 1920s, he made a considerable effort to study the entomological collections of the natural history museums of Berlin and Paris,[21] identifying twenty-four species of insects in the former and describing 366 species of neuropterans, some of them firsthand, in the latter.

After the unification of *Zoologia* and *Botânica* in 1932, Navás continued to publish new descriptions of neuropterans in *Brotéria*.[22] For over three decades, he published forty-five articles on entomology and described more than four hundred new species of neuropterans. However, this gives only a glimpse of his taxonomical work, because throughout his career he identified 375 genera and more than two thousand species of net-winged insects. Like Luisier and other contemporary naturalists, Navás was a remarkable collector. Kept at the Jesuit college of Zaragoza and at the University of Barcelona, his collection of neuropterans comprised more than fifty thousand specimens, an impressive

---

19 Longino Navás, "Neurópteros de España y Portugal," *Brotéria* 5 (1906): 145–84; *Brotéria: Zoologia* 6 (1907): 43–100; *Brotéria: Zoologia* 7 (1908): 5–131.

20 See, for instance, Longino Navás, "Neurópteros de Egipto," *Brotéria: Zoologia* 8 (1909): 105; Navás, "Neurópteros de Zumbo (Africa Oriental, Mozambique) coleccionados por el P. Lopes," *Brotéria: Zoologia* 8 (1909): 106; Navás, "Neurópteros del Congo (Africa Occidental)," *Brotéria: Zoologia* 8 (1909): 106; Navás, "Neurópteros nuevos de América," *Brotéria: Zoologia* 10 (1912): 194–202; Navás, "Ascaláfidos (Ins. Neur.) sudamericanos," *Brotéria: Zoologia* 10 (1912): 203–33; Navás, "Neurópteros nuevos de América," *Brotéria: Zoologia* 11 (1913): 45–53; Navás, "Neurópteros sudamericanos," *Brotéria: Zoologia* 12 (1914): 45–56; 215–34; *Brotéria: Zoologia* 13 (1915): 5–13; *Brotéria: Zoologia* 14 (1916) 14–35; Navás, "Tricópteros nuevos de España," *Brotéria: Zoologia* 14 (1916): 5–11; 139–46; *Brotéria: Zoologia* 15 (1917): 5–17; 63–68.

21 Longino Navás, "Algunos insectos del Museo entomológico de Berlim 'Deutsches Entomologisches Museum,'" *Brotéria: Zoologia* 20 (1922): 87–92; Navás, "Algunos insectos del Museo de Paris," *Brotéria: Zoologia* 21 (1924): 99–114; *Brotéria: Zoologia* 22 (1925): 75–83; *Brotéria: Zoologia* 23 (1926): 85–115; *Brotéria: Zoologia* 24 (1927): 5–36; *Brotéria: Zoologia* 26 (1930): 5–24; 120–44; *Brotéria: Zoologia* 27 (1931): 101–36; *Brotéria: Ciências naturais* 3 (1934): 49–55.

22 Longino Navás, "Décadas de insectos nuevos," *Brotéria: Ciências naturais* 1 (1932): 62–85; 109–19; 145–55; *Brotéria: Ciências naturais* 2 (1933): 34–44; 101–10; *Brotéria: Ciências naturais* 3 (1934): 15–24; 133–44; *Brotéria: Ciências naturais* 4 (1935): 97–107; *Brotéria: Ciências naturais* 5 (1936): 161–70.

figure by any standard. Alas, during the Spanish Civil War (1936–39), tens of thousands of bugs were irretrievably lost after a severe bombardment in Barcelona.[23] Navás was arguably one of the most productive entomologists of his generation. By working closely with the Portuguese Jesuits, he played a key role in the consolidation of *Brotéria* as an international journal of entomology.

Whereas Navás and Azevedo Mendes worked mostly on insect taxonomy, Rick's research focused on the classification of fungi. Rick was born in Hatlerdorf, Austria, on January 19, 1869 and entered the Jesuit novitiate at the age of eighteen. He was ordained in 1901, in Valkenburg, the Netherlands, and in 1902 he completed his tertianship in the novitiate of Barro. In the following year, Rick moved to Rio Grande do Sul, southern Brazil, where he lived for the remainder of his life. Besides teaching natural history and mathematics at the Jesuit college of São Leopoldo, he was also deeply involved in social and spiritual apostolates—he founded two seminaries, established a leprosy asylum, and frequently went on popular missions.[24] Together with Torrend, Rick was arguably one of the most respected mycologists in twentieth-century Brazil. For over four decades, he dedicated his time to describing and classifying fungi, mostly collected in the federal state of Rio Grande do Sul. By the time of his death, he had amassed a particularly impressive collection—with more than eleven thousand specimens, Rick's collection is the largest of its kind in Brazil and the second largest in Latin America.

Rick started to publish his research in *Brotéria* shortly after his stay in Portugal as a tertian with an article in 1904 summarizing the fungi of Rio Grande do Sul,[25] which was followed in 1905 by a brief summary of the fungi he had collected in the region of Torres Vedras, where he had lived for ten months.[26] While at first he did not seek to describe new species or genera, but rather to summarize previous knowledge and correct the classification of some selected species, he dedicated the following three decades to studying the fungi of Rio Grande do Sul.[27] When Luisier took over the editorship of *Brotéria*, Rick was

23 Victor Monserrat, "Longino Navás, His Neuropterological Work and Collection," in *Recent Research in Neuropterology: Proceedings of the 2nd International Symposium on Neuropterology*, ed. Johann Gepp, Host Aspöck, and Herbert Hözel (Graz: Gedruckt von Druckhaus Thalerhof, 1986), 173–76.

24 For a biography, see Arthur Blásio Rambo and Arthur Rabuske, *Pe. João Evangelista Rick, S.J.: Cientista, colonizador, apóstolo social, professor* (São Leopoldo: Unisinos, 2004).

25 Johannes Rick, "Fungos do Rio Grande do Sul," *Brotéria* 3 (1904): 276–93.

26 Johannes Rick, "Fungos dos arredores de Torres Vedras," *Brotéria* 4 (1905): 159–63.

27 Between 1904 and 1931, Rick published thirteen articles on mycology in *Brotéria*. Most of these articles concerned fungi collected in the region of Rio Grande do Sul. See especially Rick, "Fungos do Rio Grande do Sul," 276–93; Rick, "Pilze aus Rio Grande do Sul (Brasilien)," *Brotéria* 5 (1906): 5–53; Rick, "Fungi gelatinosi et cartilaginosi rio-grandenses," *Brotéria:*

still collecting and cataloging Brazilian fungi, and the collaboration with the Portuguese journal continued for another eight years.[28]

Early in his career, Rick introduced a Jesuit scholastic to mycological studies. Ferdinand Theissen (1877–1919) was born in Krefeld, Germany, on July 27, 1877 and entered the Jesuit novitiate at the age of nineteen.[29] During his regency (1902–8), he served as teacher in the Jesuit college of São Leopoldo, Brazil, where he became Rick's disciple.[30] Following in his master's footsteps, Theissen studied the fungi of Rio Grande do Sul and published some of his works in *Brotéria*.[31] Before his premature death in 1919, he described four genera, thirty-eight species, and eleven new varieties of fungi in the Portuguese journal.[32] By publishing his own research in *Brotéria* and by promoting the journal among his naturalist companions, Rick played a crucial role in the consolidation of *Brotéria* as an international journal of plant taxonomy.

In the same way that Rick introduced Theissen to mycology, Navás introduced Sala de Castellarnau to the field of entomology. Born in Cataluña, Spain, on April 29, 1902, Sala de Castellarnau entered the Jesuit novitiate at the age of fourteen and went on to teach at the Jesuit college of Saint Joseph, Valencia,

Botânica 21 (1924): 131–41; Rick, "Monographia das Hypoxyleas riograndenses," *Brotéria: Botânica* 25 (1931): 21–50; Rick, "Monographia Pezizinearum Riograndensium," *Brotéria: Botânica* 25 (1931): 77–122.

28 Between 1932 and 1940, Rick published seventeen articles in *Brotéria: Ciências naturais*. As before, most of those articles focused on the description of fungi collected in Rio Grande do Sul. See especially Johannes Rick, "Monografia das *Rosellinias* riograndenses," *Brotéria: Ciências naturais* 1 (1932): 183–92; Rick, "Monografia das *Valsíneas* do Rio Grande do Sul," *Brotéria: Ciências naturais* 2 (1933): 83–99; Rick, "Polypori riograndenses," *Brotéria: Ciências naturais* 3 (1934): 180–89; *Brotéria: Ciências naturais* 4 (1935): 17–27; 84–94; Rick, "Polysticti riograndenses," *Brotéria: Ciências naturais* 4 (1935): 121–38; *Brotéria: Ciências naturais* 5 (1936): 171–78; *Brotéria: Ciências naturais* 6 (1937): 81–87; Rick, "Poriae riograndenses," *Brotéria: Ciências naturais* 6 (1937): 128–52; Rick, "Polyporeaceae tiograndenses," *Brotéria: Ciências naturais* 6 (1937): 153–68; Rick, "Genus *Stereum* riograndense," *Brotéria: Ciências naturais* 9 (1940): 42–47; 75–90; 139–48.

29 Ferdinand Theissen: b. July 27, 1877, Krefeld (Germany); e. April 28, 1897, Blikenbeek, Limburg (Netherlands); d. September 2, 1919, Gallenkirch, Voralberg (Austria) (*DHCJ*, 4:3788).

30 Afonso Luisier, "O P. Fernando Theissen," *Brotéria: Botânica* 18 (1920): 73–77, here 74.

31 Fernando Theissen, "*Marasmii* austro-brasilienses," *Brotéria: Botânica* 8 (1909): 53–65; Theissen, "*Perisporiales* riograndenses," *Brotéria: Botânica* 9 (1910): 5–44; Theissen, "*Hipocreaceae* riograndenses," *Brotéria: Botânica* 9 (1910): 121–49; Theissen, "*Hymenomycetes* riograndenses," *Brotéria: Botânica* 10 (1912): 5–28; Theissen, "Le genre *Asterinella*," *Brotéria: Botânica* 10 (1912): 101–24; Theissen, "Annotações á mycoflora brazileira I: O género *Vizella*," *Brotéria: Botânica* 12 (1914): 13–21; Theissen, "Annotações á mycoflora brazileira II: Especies novas," *Brotéria: Botânica* 12 (1914): 22–31; Theissen, "De hemisphaerialibus notae supplendae," *Brotéria: Botânica* 12 (1914): 73–96.

32 Luisier, "O P. Fernando Theissen," 74.

before heading the local natural history museum; he was also a member of several learned societies, including the Real Sociedad Española de Historia Natural and the Natural Geographic Society of Washington, DC.[33] In 1926, he went to Bombay, India, and, together with Navás, collected and classified hundreds of Indian bugs. As a result of this three-year stay in India, he published a catalog of net-winged bugs listing over 160 different species.[34] Unlike other Jesuit naturalists, Sala de Castellarnau worked on a variety of taxa, including not only neuropterans and other insects but also fishes, birds, mollusks, and arachnids.[35] He was also especially interested in the history of botany and zoology, and between 1934 and 1959 he wrote biographies of a few Jesuit naturalists, including Pujiula, and his "master of entomology" Longino Navás.[36] By welcoming these writings in the pages of *Brotéria: Ciências naturais*, Luisier was playing his part in a centennial Jesuit tradition, namely the publication of inspirational biographies. Yet while, in the old Society, hagiographies were meant to inspire Jesuit novices, scholastics, and priests to a life of heroism and sanctity, in the restored Society these biographies were intended to inspire not only Jesuits but also laymen to a life of scientific inquiry.

Although most Jesuit scientists dedicated themselves to plant and animal taxonomy, there were some notable exceptions. The Spanish Jesuit Jaime Pujiula, for instance, was not as interested in taxonomy as his companions.[37]

33 Juliá Masriera, "In memoriam: R.P. Ignacio Sala de Castellarnau S.J. (1903–1973)," *Miscel·lània zoológica* 3–4 (1974): 177–78.

34 Ignacio Sala de Castellarnau, "Catálogo de neurópteros de la Índia," *Brotéria: Ciências naturais* 15 (1946): 114–28.

35 Ignacio Sala de Castellarnau, "La invernación en los peces," *Brotéria: Ciências naturais* 12 (1943): 21–28; Castellarnau, "La estivación en los peces," *Brotéria: Ciências naturais* 12 (1943): 85–93; Castellarnau, "La invernación en los moluscos y arácnidos," *Brotéria: Ciências naturais* 14 (1945): 72–77; Castellarnau, "Catálogo de aves anilladas," *Brotéria: Ciências naturais* 18 (1949): 123–40; Castellarnau, "Binomia de los dípteros y ectoparásitos de quirópteros," *Brotéria: Ciências naturais* 25 (1956): 170–75.

36 Castellarnau, "Catálogo de neurópteros de la Índia," 114. See, especially, the following biographies: Ignacio Sala de Castellarnau, "El R.P. Henrique Dettmer (S.J.), 1873–1933," *Brotéria: Ciências naturais* 3 (1934): 66–77; Castellarnau, "Un ilustre botánico: El R.P. Ethelbert Blatter (S.J.)," *Brotéria: Ciências naturais* 4 (1935): 28–34; Castellarnau, "In Memoriam: R.P. Longino Navás," *Brotéria: Ciências naturais* 9 (1940): 131–38; Castellarnau, "El Embriólogo R.P. Jaime Pujiula," *Brotéria: Ciências naturais* 28 (1959): 95–104; Castellarnau, "El Biólogo Rev. P José Assmuth (S.J.)," *Brotéria: Ciências naturais* 27 (1958): 132–37; Castellarnau, "Rev. P. Herman Schmitz, S.J. (1878–1960)," *Brotéria: Ciências naturais* 30 (1961): 63–66.

37 On his life and work, see Francisco Teixidó Gómez, "El jesuita Jaime Pujiula Dilmé, científico clave de la investigación biológica barcelonesa," *Llull: Revista de la Sociedad española de historia de las ciencias y de las técnicas* 34, no. 73 (2011): 203–28.

Instead, as the director of the Institució Catalana d'Història Natural (1925–28), the oldest natural history society in Cataluña, Spain, Pujiula dedicated most of his time to studying plant physiology, cytology, and embryology.[38]

In the 1950s, after a lifetime spent studying the microstructure of pine leaves, common ivy, potatoes, sarsaparilla, fava beans, and blue periwinkles, Pujiula began to publish works outlining his concerns with the misuse of evolutionary theory. In 1955, for example, he criticized a purely "evolutionary and monistic" interpretation of some famous experiments conducted by Austrian botanist Gottlieb Haberlandt (1854–1945).[39] He was particularly concerned by the use psychobiologists were making of the discovery of sensorial organs in plants as proof of psychism (i.e., the belief that all living creatures have a soul). As a Jesuit naturalist, his concern was both scientific and theological. Rationally, Haberlandt's experiments did not endorse psychism. Besides, the belief that plants have a soul was completely absurd for a Catholic priest and theologian. He consequently reminded *Brotéria*'s readers that Haberlandt himself had disapproved of this interpretation and proceeded with a simple, physiological, explanation of photosensitivity and geotropism in plants. The "true scientific explanation" was simple, and perhaps most importantly, it was not contingent on any kind of eccentricity. There was no need to conceive a soul to achieve a perfectly sound scientific explanation. He thus concluded that "non sunt multiplicanda entia sine necessitate" (i.e., entities must not be multiplied without necessity).[40]

Two years later, in 1957, Pujiula further criticized the misuse of evolution by taxonomists, claiming that they had contributed to an indefinite "multiplication of species."[41] Determining what was indeed a new species, as opposed to a variety of a known species, was Pujiula's main concern. For him, the concept of species should be broadened so that a species could include many varieties

38 Jaime Pujiula, "Contribución anatómico-fisiológica al estudio de la caída de las hojas," *Brotéria: Botânica* 16 (1918): 71–83; Pujiula, "Algunas observaciones citológicas sobre todo en *Hedera helix* y *Solanum tuberosum*, y la primera variante del método tano-argéntico introducida por el. Sr. del Rio Hortega," *Brotéria: Botânica* 17 (1919): 67–96; Pujiula, "Contribución al conocimiento anatómico-fisiológico de los zarcillos de la zarzaparilla," *Brotéria: Botânica* 19 (1921): 66–72; Pujiula, "Dados citológicos sobre el meristemo radical de *Vicia faba* L.," *Brotéria: Botânica* 21 (1924): 90–96; Pujiula, "Aplicación de algunos métodos de tinción histológica animal al reino vegetal," *Brotéria: Botânica* 25 (1931): 51–64; Pujiula, "El hadroma de las hojas del pino," *Brotéria: Ciências naturais* 16 (1947): 173–91; Pujiula, "Observaciones en *Vinca maior*, v. variegata," *Brotéria: Ciências naturais* 22 (1953): 60–65.

39 Jaime Pujiula, "La biología filosófica," *Brotéria: Ciências naturais* 24 (1955): 191–94, here 192.

40 Pujiula, "La biología filosófica," 194.

41 Jaime Pujiula, "La autonomia de la vida," *Brotéria: Ciências naturais* 26 (1957): 194–96.

(i.e., the existence of "minor modifications" did not justify the multiplication of species). The distinction between "essential" and "non-essential modifications" was thus of paramount importance. Pujiula deemed it implausible that "God had created as many different species as taxonomists."[42] In "accordance with God's plan," life was essentially immutable, and the capacity of an organism to adapt did not support the endless proliferation of species. For Pujiula, plants and animals could adapt to different situations "without losing the essential."[43]

When the Spanish Jesuit criticized the misuse of evolution, and censured monism and psychism, *On the Origin of Species* (1859) was about to reach its first centenary.[44] Although the Catholic Church had initially been cautious, and vague, when it came to the question of evolution, in 1950 Pope Pius XII (1876–1958, r.1939–58) had formally allowed Catholics learned in "human sciences and sacred theology" to discuss "the doctrine of evolution, in as far as it inquires into the origin of the human body as coming from pre-existent and living matter."[45] The origin of the human soul was naturally off-limits. Learned in science and theology, as Pius XII desired, Pujiula was more than qualified to engage in such discussions. By publishing his thoughts on the misuse of evolution, Luisier echoed Silva Tavares's and Schupp's early concerns on the plausibility of evolutionary theory and its philosophical and theological implications.[46] Although these kinds of discussions were not a frequent occurrence in *Brotéria*, they reveal that, for Jesuit naturalists—as for Catholic scientists in general—a deeper understanding of the divine mysteries could only be achieved through synchronized knowledge of the Book of Nature and the Book of Revelation.

2 From Taxonomy to Cytogenetics and Plant Breeding

When Luisier took over the editorship of *Brotéria*, the Jesuit journal had established itself as a leading scientific publication in Portugal. Owing to its scientific reputation, wide scope, and broad dissemination, some of the most prominent Portuguese botanists and zoologists sought to publish their works in the journal's pages. Like their Jesuit correspondents, these scientists were

42 Pujiula, "La autonomia de la vida," 195.

43 Pujiula, "La autonomia de la vida," 196.

44 Charles Darwin, *On the Origin of Species by Means of Natural Selection* (London: John Murray, 1859).

45 Pius XII, *Humani generis*, no. 36.

46 Tavares, "As sciencias naturais e a religião," 33–35; Schupp, "A evolução e o homem," 81–94; 147–63.

mostly interested in animal and plant taxonomy. Occasionally, their research coincided with the Jesuits' scientific program, as in the case of Teodoro Monteiro (1919–95), who worked on the classification of Portuguese lepidopterans. While serving as abbot of the Benedictine monastery of Singeverga, Santo Tirso (in office 1969–77), he established an impressive collection of butterflies and moths comprising thirteen thousand specimens. A member of several learned societies, including the Société Entomologique de France, the Benedictine described part of his collection to *Brotéria*'s readers.[47] By welcoming his articles on the identification of Portuguese lepidopterans from 1956 onward, Luisier was able to fill the void left by the death of Azevedo Mendes thirteen years earlier.

Unlike Monteiro, the vast majority of Portuguese authors were not clerics. Instead, they were university-trained scientists employed at public laboratories and institutes, including the National Agronomic Station (Lisbon), the Institute of Botany (Lisbon), and the Institute of Botany Dr. Gonçalo Sampaio (Porto). Beginning in the 1940s, botanists working at the Institute of Botany in Lisbon started to publish their research in *Brotéria*.[48] In February 1942, while still a graduate student and teaching assistant at the Faculty of Sciences, Carlos das Neves Tavares (1914–72) published a short article describing fourteen species of little-known Portuguese lichens.[49] This was not his first incursion into these kinds of studies, as between 1939 and 1941 he had already published two brief notes in the *Bulletin de la Société portugaise de sciences naturelles* (Bulletin of the Portuguese Society of Natural Sciences) dedicated to the same subject.[50] Hence, in 1942, when he wrote his article for *Brotéria*, he considered it to be the third part of a larger work on the description of "novel or interesting" species of Portuguese lichens. Neves Tavares went on to become the head of

47 Teodoro Monteiro, "Contribuição para o conhecimento dos lepidópteros de Portugal," *Brotéria: Ciências naturais* 25 (1956): 49–70; *Brotéria: Ciências naturais* 26 (1957): 175–87; *Brotéria: Ciências naturais* 31 (1962): 119–30; Monteiro, "Pyralideos novos para Portugal," *Brotéria: Ciências naturais* 31 (1962): 181–99. For a short biography of Teodoro Monteiro, see Jorge Ferreira, "Teodoro Monteiro," in *Dicionário dos mais ilustres transmontanos e alto durienses*, ed. João Barroso da Fonte (Guimarães: Editora Cidade Berço, 2011), 2:232–33.

48 On the history of the Faculty of Sciences, see especially Ana Simões et al., *Uma história da Faculdade de Ciências da Universidade de Lisboa (1911–1974)* (Lisbon: Faculdade de Ciências da Universidade de Lisboa, 2013).

49 Carlos das Neves Tavares, "Notas liquenológicas," *Brotéria: Ciências naturais* 11 (1942): 42–48.

50 Carlos das Neves Tavares, "Notes lichenologiques," *Bulletin de la Societé portugaise de sciences naturelles* 13, no. 11 (1939): 53–60; *Bulletin de la Societé portugaise de sciences naturelles* 13, no. 28 (1941): 157–67.

the Institute of Botany (1967–72) and continued to publish some of his works in the Jesuit journal in the following decades.[51]

Ruy Telles Palhinha (1871–1957) also taught botany at the Faculty of Sciences, University of Lisbon. Born in Angra do Heroísmo, Azores, in 1871, Telles Palhinha was one of the most prestigious professors of botany of his generation.[52] He studied natural philosophy at the University of Coimbra and taught natural sciences at public high schools in Santarém and Lisbon before being appointed professor at the Faculty of Sciences. In 1921, he succeeded António Xavier Pereira Coutinho (1851–1939) as the director of the botanical garden and the Institute of Botany, heading both institutions until 1941. As he was also a fellow of numerous learned societies, including the Academy of Sciences, the Geographic Society, the Society of Natural Sciences, and the Société Botanique de France, Telles Palhinha was a very busy man. According to his contemporaries, he dedicated most of his time to teaching and directing the botanical facilities at the Faculty of Sciences; he also owned a pharmacy, where he "worked with enthusiasm."[53]

Although he did not have much time to carry out research, Telles Palhinha nevertheless made significant contributions to Portuguese botany in the 1930s. After three expeditions to the archipelago of Azores in 1934, 1937, and 1938, for instance, he produced the first systematic geobotanic survey of the Azorean flora.[54] He also wrote several biographies of Portuguese botanists, including Félix Avelar Brotero, Vandelli, Henriques, and Pereira Coutinho.[55] Around 1945, he was given a collection of letters addressed to the founder of the botanical garden, Francisco de Melo Breyner, count of Ficalho (1837–1903).[56] In the following years, Telles Palhinha studied Ficalho's correspondence and published

---

51 Carlos das Neves Tavares, "Líquenes da Serra da Estrela (Contribuição para o seu estudo)," *Brotéria: Ciências naturais* 14 (1945): 14–25, 49–61; Tavares, "Notas Lichénologiques," *Brotéria: Ciências naturais* 16 (1947): 145–57; Tavares, "Nótulas de Fanerogamia portuguesa," *Brotéria: Ciências naturais* 28 (1959): 69–75. For a short biography, see Fernando Catarino, "Carlos das Neves Tavares (1914–1972)," in *Memórias de professores cientistas*, ed. Ana Simões (Lisbon: Faculdade de Ciências da Universidade de Lisboa, 2001), 102–11.

52 Adriano Gonçalves da Cunha, "Prof. Doutor Ruy Telles Palhinha," *Boletim da sociedade Broteriana* 32 (1958): vii–xx; Aurélio Quintanilha, "In Memoriam: Ruy Telles Palhinha," *Boletim da sociedade portuguesa de ciências naturais* 14 (1972): 7–12.

53 Cunha, "Prof. Doutor Ruy Telles Palhinha," x.

54 Pedro Gouveia da Fonseca, "Darwin em Portugal (1910–1974): O Darwinismo e a evolução na produção científica de botânicos portugueses" (PhD diss., University of Coimbra, 2015), 222–42, here 233.

55 Cunha, "Prof. Doutor Ruy Telles Palhinha," x.

56 Francisco de Melo da Costa (1890–1945), marquis of Ficalho and Breyner's maternal grandson, gave him the letters shortly before passing away.

a selection of letters with the Portuguese Jesuits.[57] For the first time in its history, *Brotéria* welcomed the publication of primary sources on the history of botany.

While Silva Tavares and Telles Palhinha worked in Lisbon, Manuel Cabral Resende Pinto (1911–90), Arnaldo Rozeira (1912–84), and Joaquim Sampaio (1899–*c.*1965) worked at the Institute of Botany Dr. Gonçalo Sampaio, Porto. Like the French Jesuit Torrend, Resende Pinto was a prominent mycologist. Between 1942 and 1945, he published one of his most significant taxonomic works in *Brotéria*, a detailed catalog of *Hymenomycetes*.[58] Dedicated to Brotero, Henriques, Pereira Coutinho, and Torrend—his forerunners in the field of mycology—this catalog of bracket fungi and toadstools compiled information dispersed in more than fifty texts, most of them poorly known or difficult to find. In his introduction, Resende Pinto acknowledged Torrend, above all others, because he had identified "hundreds of species to our flora," adding that the study of Portuguese fungi could have been "completed, or almost completed" by Torrend himself long ago if the republican expulsion of 1910 had not "deprived us of his collaboration forever."[59] As had been the case with Pires de Lima's speech in 1942, Resende Pinto wanted to distance himself from the political, scientific, and anticlerical stance of the First Republic.

Rozeira, who served as head of the botanical garden and the Institute of Botany at the Faculty of Sciences, Porto, began publishing his works with the Portuguese Jesuits in the 1940s. In 1943, he described 151 new or little-known plants of the flora of northern Portugal and identified a new species of rockfoil, which he named *Saxifraga sampaioi* in homage to his "master of systematics," and patron of the Institute, Gonçalo Sampaio.[60] Three years later, in 1946, he again paid tribute to his professor of botany in a short article defending the morphological criteria employed by Sampaio in his works, which he claimed

57 Ruy Telles Palhinha, "Cartas de Alphonse de Candolle ao conde de Ficalho," *Brotéria: Ciências naturais* 17 (1948): 172–81; Palhinha, "Quatro cartas inéditas de Isaac Newton ao conde de Ficalho," *Brotéria: Ciências naturais* 17 (1948): 69–72; Palhinha, "Cartas de G. Schweinfurth para o conde de Ficalho," *Brotéria: Ciências naturais* 18 (1949): 167–72; Palhinha, "As Estufas do Jardim Botânico de Lisboa: Cartas do Dr. Goeze ao conde de Ficalho," *Brotéria: Ciências naturais* 24 (1955): 5–26.

58 Manuel Resende Pinto, "Hymeniales de Portugal," *Brotéria: Ciências naturais* 11 (1942): 5–20; 62–80; 106–36; 153–76; *Brotéria: Ciências naturais* 12 (1943): 5–20; 58–75; 116–27; 180–90; Pinto, "Hymeniales de Portugal: Hydnaceae," *Brotéria: Ciências naturais* 13 (1944): 25–36; Pinto, "Hymeniales de Portugal: Polyporaceae," *Brotéria: Ciências naturais* 13 (1944): 30–36; 131–40; Pinto, "Hymeniales de Portugal," *Brotéria: Ciências naturais* 14 (1945): 26–42.

59 Pinto, "Hymeniales de Portugal," 5–6.

60 Arnaldo Rozeira, "Plantas novas ou pouco citadas para Trás-os-Montes e Alto-Douro," *Brotéria: Ciências naturais* 12 (1943): 145–71, here 146.

was usually sufficient to determine the boundaries between different species.[61] In 1949, Rozeira published the dissertation he had presented in the tenure for professor of botany in the Faculty of Sciences (Porto).[62] The publication of this large monograph, unusual for a scientific periodical, testified to the close cooperation between *Brotéria* and Porto's academic community in this period.

Joaquim Sampaio also worked at the Institute of Botany, Porto. The son of Gonçalo Sampaio, patron of the institute, he wrote several articles identifying and classifying blue and green algae, having described several new species of the Portuguese flora.[63] Besides being an active taxonomist, Sampaio was also interested in the history of botany. Between 1946 and 1949, he analyzed the contributions of Romualdo Fragoso (1862–1928), Isaac Newton (1840–1906), and Gonçalo Sampaio to the history of Portuguese botany.[64] By publishing scientific biographies, as well as primary sources for the history of botany, *Brotéria* was playing a part in the consolidation of the discipline in twentieth-century Portugal.

One of the most interesting collaborations between *Brotéria* and public institutes was the journal's collaboration with the National Agronomic Station. In 1939, António Sousa da Câmara (1901–71), head of the Agronomic Station, published his first paper in *Brotéria* on the fragmentation of chromosomes caused by X-rays.[65] In the following years, Sousa da Câmara and his associates at the Department of Cytology and Cytogenetics continued to publish some of their works with the Portuguese Jesuits.[66] Although cytogenetics and plant breeding

61 Arnaldo Rozeira, "Gonçalo Sampaio como sistemata," *Brotéria: Ciências naturais* 15 (1946): 49–55.

62 Arnaldo Rozeira, "A secção *Stoechas Gingins*, do género *Lavandula Linn*," *Brotéria: Ciências naturais* 18 (1949): 5–84.

63 Joaquim Sampaio, "Sinopse das desmídias conhecidas na flora portuguesa," *Brotéria: Ciências naturais* 12 (1943): 97–115; *Brotéria: Ciências naturais* 13 (1944): 6–24; 88–130; *Brotéria: Ciências naturais* 14 (1945): 128–48; Sampaio, "Cianófitas da Serra da Estrela," *Brotéria: Ciências naturais* 16 (1947): 105–13; Sampaio, "As algas marinhas portuguesas e a extracção do iodo," *Brotéria: Ciências naturais* 16 (1947): 165–72; Sampaio, "Uma localidade nova para o *Ascophyllum nodosum* Le Jol., descoberta pelo Prof. G. Sampaio," *Brotéria: Ciências Naturais* 17 (1948): 56–58.

64 Joaquim Sampaio, "Subsídios para a história da botânica em Portugal I: O colector Isaac Newton e o estudo das criptogâmicas celulares portuguesas," *Brotéria: Ciências naturais* 15 (1946): 20–52; Sampaio, "Subsídios para a história da botânica em Portugal II: O Dr. Romualdo Fragoso, o Dr. Gonçalo Sampaio e a micologia portuguesa," *Brotéria: Ciências naturais* 17 (1948): 115–32; *Brotéria: Ciências naturais* 18 (1949): 86–92; 97–108.

65 António Sousa da Câmara, "Fragmentos cromosómicos produzidos pelos raios X," *Brotéria: Ciências naturais* 8 (1939): 166–79.

66 See especially António Sousa da Câmara, "A expulsão dos cromossomas pela centrifugação," *Brotéria: Ciências naturais* 11 (1942): 21–28; Miguel Noronha Wagner and Tristão

were the main focus of Sousa da Câmara's research group, it also worked on other subfields of biology, such as population genetics.[67] The publication of articles by researchers of the Agronomic Station, which was founded by Sousa da Câmara in 1936, was particularly timely. The station was modeled on the German Institute of Plant Breeding, and like its German counterpart, the Portuguese station promoted research on cytogenetics and plant breeding.[68]

In this period, research on the genetic improvement of wheat and rice was of paramount importance, because it supported one of the most pressing issues for Salazar's regime: the pursuit of food independence. The connection between the Portuguese Jesuits, the German Institute of Plant Breeding, and the scientific policies of the Estado Novo dated back to 1929 (i.e., seven years before the foundation of the Agronomic Station). With the objective of fulfilling national consumption needs, the government launched its first wheat campaign on August 21, 1929.[69] Shortly after, on November 29, 1929, Erwin Baur (1875–1933), head of German Institute (1928–33), went to Lisbon to give a lecture on the application of genetics to plant breeding[70] and later published his notes in the pages of *Brotéria: Botânica*.[71] By welcoming the works of Baur, one of the most prominent geneticists of his generation, and Sousa da Câmara

---

Mello Sampaio, "Nota acerca da movimentação dos univalentes em híbridos pentaplóides de trigo," *Brotéria: Ciências naturais* 19 (1950): 180–7; Augusto Simplício Duarte, "Notas sobre algumas problemas relacionados com o melhoramento do arroz," *Brotéria: Ciências naturais* 26 (1957): 28–33.

67 António Sousa da Câmara and Sarah Vasconcellos, "Contribuição para o estudo das populações de Drosophila melanogaster em Portugal," *Brotéria: Ciências naturais* 12 (1943): 49–57.

68 On the political, military, and economic reasons behind the foundation of state laboratories under the Nazi and fascist regimes in Germany, Italy, and Portugal, see Saraiva, *Fascist Pigs*. For the particular case of Portugal, see also Saraiva, "Laboratories and Landscapes: The Fascist New State and the Colonization of Portugal and Mozambique," *Journal for the History of Science and Technology* 3 (2009): 7–39; and Saraiva, "Fascist Labscapes: Genetics, Wheat and the Landscapes of Fascism in Italy and Portugal," *Historical Studies in the Natural Sciences* 40 (2010): 457–98.

69 *Diário do governo* 191 (August 21, 1929): 1875–77.

70 Before founding the German Institute of Plant Breeding, Baur wrote a compendium on human heredity with Fritz Lenz and Eugen Fischer. See Erwin Baur, Fritz Lenz, and Eugen Fischer, *Menschliche Erblichkeitslehre und Rassenhygiene* (München: J.F. Lehmann Verlag, 1921) (translated as *Human Heredity* [New York: Macmillan, 1931]). The book eventually won worldwide credit for the worst possible reasons, being Hitler's main source of inspiration for his racial and eugenic theories. See Hans-Walter Schmuhl, *The Kaiser Wilhelm Institute for Anthropology, Human Heredity and Eugenics, 1927–1935: Crossing Boundaries* (Dordrecht: Springer, 2008), 152.

71 Erwin Baur, "Aplicação da genética ao melhoramento das plantas cultivadas," *Brotéria: Botânica* 24 (1930): 49–57.

on cytogenetics and plant breeding, Silva Tavares and Luisier were displaying their public support for the official scientific policy pursued in the early years of the Estado Novo.

Most Portuguese scientists worked in one or two different institutions during their career. There were, however, some exceptions. The promising botanist and geneticist Flávio Resende (1907–67), for instance, worked at the National Agronomic Station and at the Institutes of Botany in Porto and Lisbon.[72] Resende obtained his doctorate from the University of Hamburg in February 1937. Although he had wanted to stay in Hamburg, there was a serious setback later that year, as in August 1938 the university dismissed his supervisor, Emil Heitz (1892–1965), for being one-quarter Jew,[73] and without Heitz—a pioneer in the emergent field of cytogenetics—Resende had to return to Portugal. Beginning in January 1938, he worked for seven months at the National Agronomic Station before traveling back to Germany to work on plant physiology and cytogenetics at the Kaiser Wilhelm Institut für Biologie. In 1941, after a three-year stay in Berlin, he returned to Portugal and began working at the Institute of Botany Dr. Gonçalo Sampaio. In 1942, he was appointed professor of botany at the University of Porto, and in 1943 professor of botany at the University of Lisbon. During his two-year stay in Porto, he published some of his research in *Brotéria*. Together with German botanist Karl von Poellnitz (1896–1945), he described in the Jesuit journal novel varieties of an African succulent plant.[74] After his appointment in Lisbon, he held several offices, including the direction of the botanical garden, the natural history museum, and the Institute of Botany (1944–67). He also founded a scientific journal in 1944 dedicated to biological sciences, the *Portugaliae acta biologica* (Portuguese biological acta), which remained his primary focus in the years that followed, as he did not publish again in *Brotéria*. However, his relationship with the Portuguese Jesuits did not end there. In the 1960s, he became a mentor to a young Jesuit biologist, the soon-to-be editor-in-chief of *Brotéria*.[75]

72 Carlos das Neves Tavares, "Vida e actividade científica do Prof. Dr. Flávio Ferreira Pinto Resende," *Portugaliae Acta Biologica: Serie B; Sistemática, ecologia, biogeografia e paleontologia* 9 (1967–1968): 1–63. See also his obituary in *Nature*: António Lima de Faria, "Professor Flávio Resende," *Nature* 214 (1967): 435. For a more recent account, see Fernando Catarino, "Flávio Ferreira Pinto Resende (1907–1967)," in *Memórias de professores cientistas*, ed. Ana Simões (Lisbon: Faculdade de Ciências da Universidade de Lisboa, 2001), 78–89.

73 Helmut Zacharias, "Emil Heitz (1892–1965): Chloroplasts, Heterochromatin, and Polytene Chromosomes," *Genetics* 141 (1995): 7–14, here 11.

74 Flávio Resende and Karl von Poellnitz, "Suculentas africanas: Variedades novas de *Haworthia tessellata* Haw," *Brotéria: Ciências naturais* 11 (1942): 49–55.

75 See Chapter 9.

After Silva Tavares's death, *Brotéria* extended its scientific program by publishing papers on subjects such as the history of botany, plant physiology, cytogenetics, and plant breeding, which was one of the major research priorities in the early years of the Estado Novo. The journal, which received financial support from the Portuguese government from 1930 onward, contributed to the diffusion of the work of some of the most prominent Portuguese botanists, namely Silva Tavares, Telles Palhinha, Resende Pinto, Rozeira, Sousa da Câmara, and Resende.

During this period, *Brotéria* renewed its importance as a scientific journal, and the Jesuits enjoyed a period of considerable political stability. Unlike the republican revolutionaries of 1910, the corporatist politicians of the Estado Novo glorified the Jesuits' contributions to science and education. As proclaimed heir of the Jesuit scientific tradition, Luisier was progressively honored by public institutions: the Academy of Sciences made him a fellow in 1933, and the University of Porto granted him an honorary doctorate in 1942. On February 6, 1957, on the official celebration of his eighty-fifth birthday, he was bestowed with the Military Order of Saint James of the Sword, the highest sate honor conferred on artists, scientists, and writers in Portugal. Six months after this celebration, on November 4, 1957, Luisier passed away in Caldas das Saúde, Santo Tirso. Once again, *Brotéria* needed a new editor.

CHAPTER 9

New Lenses to Read the Book of Nature: Biochemistry, Molecular Genetics, and Bioethics

In November 1957, *Brotéria* was again deprived of its editor-in-chief. Following Luisier's death, José Vilhena de Carvalhaes, S.J. (1912–2008) was entrusted with the publication of the Jesuit journal.[1] However, his term in office was rather brief. In 1961, Vilhena de Carvalhaes was appointed rector of the Instituto

FIGURE 21
Luís Archer, S.J. (1926–2011)
AB

1 José Carvalhaes: b. September 19, 1912, Lisbon (Portugal); e. September 14, 1932, Alpendurada (Portugal); d. January 17, 2008, Caldas da Saúde, Santo-Tirso (Portugal) (José Eduardo Franco, "Os directores da Brotéria: José Carvalhaes," in Rico and Franco, *Fé, ciência, cultura*, 168–70).

 DOI:10.1163/9789004382367_010

Nun'Alvres, the Jesuits' boarding school in northern Portugal, and passed on the direction of the journal.[2] It thus fell to his successor, Luís Archer (1926–2011), to steer the publication of *Brotéria* for forty consecutive years (figure 21).[3] During his editorship, he brought about the development of new research areas, especially biochemistry, molecular genetics, human genetics, and bioethics. One century after its foundation in a rural Jesuit college, *Brotéria* featured prominently on the stage of science publications in twentieth-century Portugal. Under Archer, the Jesuits' centennial investment in the teaching and practice of science in the colleges and the international scope of *Brotéria* were widely acknowledged by both the scientific community and the state.

Luís Jorge Peixoto Archer was born in Porto on May 5, 1926 into a distinguished Catholic family of Irish descent. Serene and introspective, Archer was a remarkable student. During his childhood, he became an enthusiast for classical music. He studied piano for nine years (1936–45), and his musical talent was widely acknowledged by his closest friends, colleagues, and companions in the Society of Jesus.[4] He attended the Liceu Rodrigues de Freitas, the town's first public high school, and then went on to study biology at the Faculty of Sciences, University of Porto. He graduated with honors and won the Rotary Club medal for best student of the Faculty of Sciences in 1947. After graduating, he was hired as a research assistant to Pires de Lima, former director of the Faculty of Sciences (in office 1929–32 and 1935–45) and one of the most respected botanists of his generation. A few months later, on December 7, 1947, he entered the Jesuit novitiate in Guimarães, causing general surprise and concern among his former professors at the University of Porto:[5] contrary to their expectations, it appeared that Archer was not to pursue a scientific career in biology.

Following the novitiate, Archer studied philosophy in Braga (1951–54) and theology in Frankfurt (1956–60). On July 31, 1959 (i.e., the feast day of Saint Ignatius), he was ordained in Frankfurt. When he returned to Portugal, Archer

2 Luís Archer, "R.P. José Carvalhaes S.J.," *Brotéria: Ciências naturais* 31 (1962): 54.

3 Luís Archer: b. May 5, 1926, Porto (Portugal); e. December 7, 1947, Guimarães (Portugal); d. August 8, 2011, Lisbon (Portugal) (Francisco Malta Romeiras, "The Emergence of Molecular Genetics in Portugal: The Enterprise of Luís Archer S.J.," *Archivum historicum Societatis Iesus* 164 [2013]: 501–12).

4 Francisco Malta Romeiras, "A última conversa com o P. Luís Archer: Um concílio entre Fé e Ciência?," *Brotéria* 173 (2011): 239–45.

5 Luís Archer, "Vinte anos de genética molecular em Portugal," *Brotéria: Genética* 10 (1989): 19–48, here 21. For a more recent edition of this autobiographical article, see Francisco Malta Romeiras and Henrique Leitão, eds., *Obra selecta do Padre Luís Archer, S.J.* (Lisbon: Fundação Calouste Gulbenkian, 2015), 1:319–50. See also Américo Pires de Lima, "Educação profissional," *O médico* 1 (1953): 85–91.

sought to deepen his studies of philosophy and psychology in the expectation that doing so would provide valuable tools for his apostolic ministries. However, his superiors insisted that he resumed his biological studies, because his talent for science would be of immense value for the Portuguese province. As Provincial Lúcio Craveiro da Silva, S.J. (1914–2007, in office 1960–66) put it shortly after his arrival, Archer was to be entrusted with a centennial, and most valuable, tradition in the Society of Jesus:

> I would not like to lose anything at all from the tradition followed by Fr. Luisier which has given our province such a good name. In our colleges, there has always been a tradition of having a group of men of science—which the father general says is the first and most important ministry of the Society—and it has been my desire that this tradition continues. Besides, given the constraints of the Portuguese milieu (and probably of all milieus), the prestige of science would confer a greater authority on all our apostolic activity. I have had direct experience of this in the social apostolate.[6]

Because science was "the first and most important ministry of the Society," and scientific prestige conferred "a greater authority on all apostolic activities," Archer started to look for a laboratory to pursue his doctoral studies in molecular genetics. He first turned to Flávio Resende for advice.[7] With his help, Archer was appointed as a research assistant in the laboratory of Francis J. Ryan (1916–63) at Columbia University.[8] However, in July 1963, shortly before embarking on his journey to New York, Archer received unsettling news: his future advisor had passed away unexpectedly.[9] Because he had already obtained permission to go to Columbia, he explored all options to purse his doctoral studies at the prestigious American university. Thus, after Ryan's passing, Archer contacted John A. Moore (1915–2002), chair of the Department of Zoology. In the beginning, the most pressing issue for Moore was to know if the Instituto de Alta Cultura (the national science foundation) would continue to fund the research conducted at Ryan's laboratory.[10] Two months later, in October 1963,

6 APSI, Lúcio Craveiro da Silva to Luís Archer (April 22, 1961). I would like to thank Claude Pavur for his precious help on the translation.

7 APSI, Luís Archer to Flávio Resende (March 25, 1963).

8 APSI, Luís Archer to Francis J. Ryan (April 26, 1963); APSI, Luís Archer to Francis J. Ryan (May 22, 1963); APSI, Luís Archer to Francis J. Ryan (May 24, 1963); APSI, Luís Archer to Flávio Resende (May 30, 1963); APSI, Luís Archer to Francis J. Ryan (June 12, 1963).

9 APSI, Viola Sugahara to Luís Archer (July 17, 1963).

10 APSI, John A. Moore to Luís Archer (July 26, 1963).

matters became "somewhat more clarified," and Moore informed Archer that the Instituto de Alta Cultura had agreed to fund Ryan's laboratory for another academic year. Although there was "no financial reason why you could not come," Moore felt that "it would be a mistake for you to join us this year," because Ryan's laboratory was being dismantled and there "were very few individuals working in the field of microbial genetics" at Columbia.[11] In November 1963, as Archer insisted on going to Columbia to study genetics under James Herbert Taylor (1916–98), Moore replied that his colleague was leaving Columbia in January 1964 for a position at Florida State University, and, for that reason, it would be impossible for Archer to pursue his doctoral studies at Columbia. Moore ended the epistolary correspondence with the Jesuit priest by expressing his regret that "Professor Ryan's death has so complicated your plans."[12]

When he heard about Archer's difficulties in going to the United States to pursue his doctorate, Portuguese biochemist Francisco Carvalho Guerra (1932–) wrote to Spanish scientist and Nobel laureate Severo Ochoa (1905–93) to put in a good word for his friend.[13] Ochoa, however, believed that his laboratory at the New York University School of Medicine was "extremely crowded." Besides, Ochoa said that his laboratory was not "an appropriate place for Dr. Archer" because he wanted to receive general training in genetics, and his work was "related to genetics in a very peripheral fashion." To work with the Nobel laureate, Archer would require "a firm foundation in chemistry, physical chemistry, and biochemistry." Ochoa reassured Guerra that Ryan's laboratory had indeed been "a good choice" because Ryan's research was "much more biologically oriented."[14]

Disappointed at not having yet found a laboratory where he would be able to work on genetics, Archer turned once again to Flávio Resende. The dean of botany at the Faculty of Sciences, Lisbon, suggested that he contact Ernst Freese (1925–90) at the National Institutes of Health (henceforth NIH), Maryland.[15] The Jesuit arrived in Washington, DC, on April 22, 1964, but, due to a misunderstanding, he only worked at Freese's laboratory for twelve days. Because

11 APSI, John A. Moore to Luís Archer (October 4, 1963).

12 APSI, John A. Moore to Luís Archer (November 22, 1963).

13 APSI, Francisco Carvalho Guerra to Severo Ochoa (August 28, 1963).

14 APSI, Severo Ochoa to Francisco Carvalho Guerra (October 28, 1963).

15 APSI, Luís Archer to Ernst Freese (February 8, 1964); APSI, Luís Archer to Ernst Freese (February 28, 1964); APSI, Luís Archer to Ernst Freese (March 6, 1964); APSI, Luís Archer to Ernst Freese (March 17, 1964); APSI, Luís Archer to Ernst Freese (April 12, 1964); APSI, Luís Archer to Ernst Freese (April 14, 1964); APSI, Luís Archer to Ernst Freese (April 16, 1964); APSI, Ernest Freese to Luís Archer (March 12, 1964); APSI, Ernest Freese to Luís Archer (April 2, 1964).

the letters of reference sent from Portugal always mentioned the Jesuit priest as "Dr. Luís Archer," Freese thought that Archer had already obtained his doctorate. Unfortunately, Freese was not interested in hosting Archer as a doctoral candidate and told him to return after finishing his studies.[16] He was formally dismissed on May 4, 1964. On the eve of his thirty-eighth birthday, Archer was facing a critical crossroad. He had forfeited a scientific career to become a Jesuit, and seventeen years later he was trying to resume his biological studies in obedience to his superiors, but a series of contingencies kept hindering his plans. Without a laboratory in which to work, Archer could not comply with the national funding agency (the Instituto de Alta Cultura) that had awarded him a fellowship to complete his doctorate. He needed to find another institution and start all over again.

In the fall of 1964, Archer was finally admitted as a free student at Georgetown University. As well as studying biochemistry and molecular biology, he also started to work on a research project on the effects of anthrax toxin on oxidative metabolism of rat lung. On May 4, 1966, exactly two years after being dismissed from the NIH, the Jesuit delivered a paper at a scientific meeting in Los Angeles.[17] And in the following months he published the results of his research in the *American Journal of Physiology*, the *Journal of Bacteriology*, and *Brotéria*.[18] His early achievements as a biochemist did not go unnoticed in Lisbon and Washington, and in 1967 the Instituto de Alta Cultura awarded Archer a prize for his article in the *Journal of Bacteriology*. At Georgetown, and after earlier turning him down, professor of microbial genetics Otto Landman (1925–2003) approached the Jesuit and offered to become his advisor. As Archer later quipped in an autobiographical essay, the anthrax toxin had "brought him back to life."[19]

In 1966, Archer took his comprehensive exam and was formally admitted as a doctoral candidate at Georgetown University. Under Landman's guidance, he started to work on bacterial genetics and soon found that his results

16 Romeiras, "A última conversa com o P. Luís Archer," 240.

17 Irving Gray and Luís Archer, "Effect of Anthrax Toxin on Oxidative Metabolism of Rat Lung," *Bacteriological Proceedings* 67 (1966): 52.

18 Irving Gray and Luís Archer, "The Effect of Anthrax Toxin on Oxidative Metabolism of Lung Tissue," *American Journal of Physiology* 210 (1966): 1313–15; Archer and Gray, "Alterations in the Activity of Some Oxidative Enzymes by the Action of Anthrax Toxin," *Brotéria: Ciências naturais* 35 (1966): 169–77; Gray and Archer, "Metabolic Changes in Nicotinamide Adenine Dinucleotide in Response to Anthrax Toxin," *Journal of Bacteriology* 93 (1967): 36–39.

19 Luís Archer, "Vinte e cinco anos de genética molecular," *Memórias da Academia das Ciências de Lisboa: Memórias da classe de ciências* 30 (1989): 129–76, here 136. For a more recent edition, see *Obra selecta do Padre Luís Archer, S.J.*, 1:369–410.

contradicted the work of German geneticist and Stanford fellow Walter Bodmer (1936–).[20] In 1967, he presented his research on DNA-mediated transformation in *Bacillus subtilis* in Washington, New York, and Denver, and finally received his coveted doctorate in molecular biology from Georgetown.[21] His research soon found its way to Stanford, and Bodmer himself ended up confirming Archer's results.[22]

In the 1960s, molecular genetics was *terra incognita* for Portuguese biologists, and so Archer's return raised some very high expectations. As Resende bluntly put it in 1965, the Jesuit's forthcoming arrival would provide a "great opportunity" for the emergence of the discipline in Portugal.[23] Resende was soon proven right. In 1967, Dutch microbiologist Nicolaas van Uden (1921–91) invited Archer to teach a summer course in molecular biology at the Gulbenkian Institute of Science, a new and vibrant research center in Oeiras, a small town in the outskirts of Lisbon.[24] The course, which took place between June 3 and July 16, 1968, was an instant success. While the first edition of the summer course was, above all, a theoretical introduction to the novel field of genetics, Archer was able to include experimental classes in 1969. His efforts to establish a hands-on workshop were highly praised by Portuguese biologists. At the age of seventy-seven, Aurélio Quintanilha (1892–1987), one of the most respected geneticists of his generation, came from Mozambique to participate in Archer's course and was astonished by the molecular turn in biology and by the Jesuit's pedagogical skills.[25]

20 Walter Bodmer, "Recombination and Integration in *Bacillus subtilis*: Involvement of DNA Synthesis," *Journal of Molecular Biology* 14 (1965): 534–57.

21 Luís Archer, "Transport of Transforming DNA into the Cell Interior in Thymine-Starved *Bacillus subtilis* with Chromosomes Arrested at the Terminus" (PhD diss., Georgetown University, 1967).

22 Charles D. Laird, Louise Wang, and Walter Bodmer, "Recombination and DNA Replication in *Bacillus subtilis* Transformation," *Mutation Research* 6 (1968): 205–9.

23 APSI, Flávio Resende to Luís Archer (May 16, 1965).

24 See especially APSI, Luís Archer to Nicolaas van Uden (July 6, 1967); APSI, Luís Archer to Nicolaas van Uden (February 9, 1968); APSI, Luís Archer to Nicolaas van Uden (March 23, 1968); APSI, Luís Archer to Nicolaas van Uden (March 26, 1968); APSI, Luís Archer to Nicolaas van Uden (April 2, 1968); APSI, Nicolaas van Uden to Luís Archer (June 23, 1967); APSI, Nicolaas van Uden to Luís Archer (August 7, 1967); APSI, Nicolaas van Uden to Luís Archer (February 13, 1968); APSI, Nicolaas van Uden to Luís Archer (March 25, 1968); APSI, Nicolaas van Uden to Luís Archer (March 27, 1968); APSI, Nicolaas van Uden to Luís Archer (April 23, 1968); APSI, Nicolaas van Uden to Luís Archer (May 13, 1968).

25 APSI, Aurélio Quintanilha to Luís Archer (January 20, 1970). See also Luís Archer, "Mestre Quintanilha faz-se aluno," *Brotéria: Genética* 14 (1993): 29–32.

Between 1968 and 1970, Archer worked on the translation of his dissertation and taught molecular biology at the Faculty of Sciences, University of Porto.[26] In 1971, he established a laboratory of molecular genetics at the Gulbenkian Institute of Science, which he directed for twenty consecutive years. As director of the newly founded laboratory, one of his greatest concerns was to train the next generation of molecular geneticists. Hence, beginning in the 1970s, he undertook a series of initiatives with the aim of promoting the science of molecular genetics in his homeland. In 1972, he organized the First International Meeting on Bacterial Transformation, thus launching a network of scientific correspondence that would benefit him and his students.[27] As more powerful techniques to analyze, cut, and engineer DNA began to emerge, the Jesuit felt the need to return to the United States. In 1973, he worked for eight months at the Massachusetts Institute of Technology (MIT) with geneticist Maurice Sanford Fox (1924–), and in 1976 he worked for five months at the University of Rochester under physician and microbiologist Frank Young (1931–), with whom he learned the pioneering techniques of genetic engineering.

After returning from New York, Archer organized a ground-breaking hands-on workshop on genetic manipulation in Oeiras. During the workshop, the participants performed, for the first time, innovative techniques for genetic engineering, including the isolation of a restriction enzyme, the construction of a plasmid map, and the realization of a southern blot.[28]

In the late 1960s and into the 1970s, Archer was particularly concerned with the consolidation of molecular genetics. Besides establishing the first university courses in Lisbon, Porto, Coimbra, and Braga, he also founded the Portuguese Society of Genetics in 1974, and wrote the first textbook of the discipline in Portuguese in 1976.[29] One of his most important contributions was perhaps the establishment of a research-oriented school of thought. Since the late 1970s, and especially in the 1980s, some of the most promising Portuguese molecular biologists worked under his supervision. When they completed their doctorates, they started their own research groups and supervised a few more doctoral students. As a result, by the end of his scientific career, Archer had directly or indirectly influenced four generations of graduate biologists (a total of ninety-one researchers in 2009).

---

26 Luís Archer, "Transformação genética em células sincronizadas de *Bacillus subtilis*" (PhD diss., Universidade do Porto, 1969).

27 The proceedings were published the following year. See Luís Archer, ed., *Bacterial Transformation* (London: Academic Press, 1973).

28 Archer, "Vinte anos de genética molecular em Portugal," 33.

29 Luís Archer, *Genética molecular* (Lisbon: Editorial Brotéria, 1976).

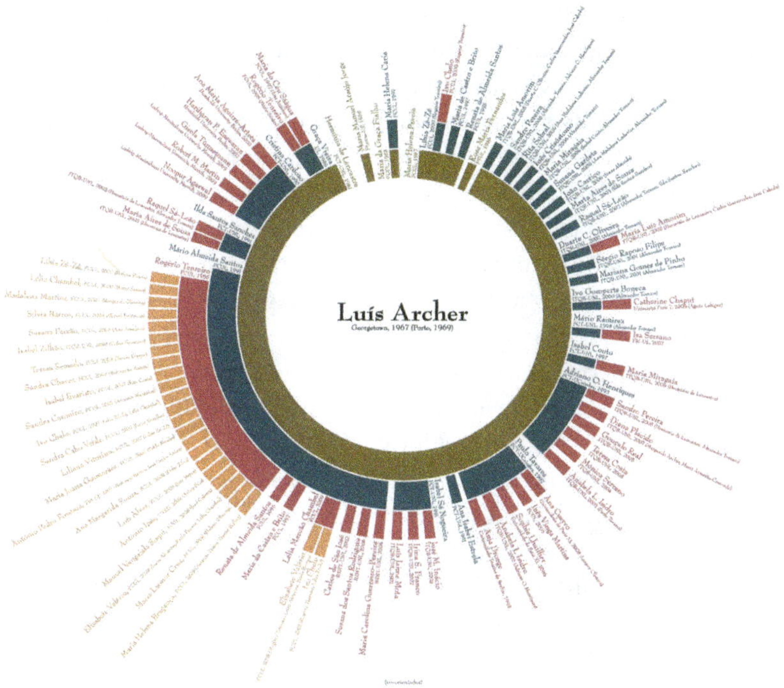

FIGURE 22 Luís Archer's school of research, 2009
CREDITS: JOSÉ ARTUR MARTINHO SIMÕES, ADRIANO O. HENRIQUES, ANA SANCHEZ, LUÍS MORGADO. © ITQB

1 The Molecular Turn

In addition to teaching and doing research in molecular genetics, Archer was particularly invested in the publication of *Brotéria*, which he directed for forty consecutive years (1962–2002).[30] When Provincial Craveiro da Silva appointed him to steer the publication of the journal, Archer was not yet a scientist but rather a teacher and *kapellmeister* at the Instituto Nun'Alvres, the Jesuit boarding school in Santo Tirso. The nomination as editor-in-chief of *Brotéria* in 1962 therefore represented his first mission as an apostle of science, "the first and most important ministry of the Society."[31] As editor-in-chief, Archer brought

30 He was also editor-in-chief of the cultural *Brotéria* between 1972 and 1975, and again between 1993 and 2000.

31 APSI, Lúcio Craveiro da Silva to Luís Archer (April 22, 1961).

about the development of new research topics in the Jesuit journal, namely biochemistry, molecular genetics, human genetics, and, finally, bioethics.

In the last issue of 1961, that is, the last issue before Archer's official appointment, *Brotéria* published its first article on biochemistry. Written by William D. Sullivan, S.J. (1918–96) and James V. Coyle from Boston College, the article focused on the effect of radiation on the activity of a metabolic enzyme at different states of division of the ciliate *Tetrahymena*.[32] In the following years, Sullivan continued to publish his works on the activity of metabolic enzymes with the Portuguese Jesuits.[33] Although twenty-five years had elapsed since Hans Krebs (1900–81) had discovered the citric acid cycle in 1937—the very same discovery that would earn him the Nobel Prize in Physiology or Medicine in 1953—the study of metabolism and metabolic enzymes still occupied most biochemists at the time. Most importantly, Sullivan and Coyle's article announced that significant changes lay ahead for the Jesuit journal. In the 1960s, *Brotéria* abandoned taxonomy and systematics, gradually replacing these subjects with biochemistry, cell biology, and molecular genetics.[34] As in previous decades, the journal published shorter research papers as well as lengthier essays. Between 1965 and 1969, Francisco Carvalho Guerra, Roberto Salema (1932–), and Luís Archer published their doctoral dissertations on mitochondrial tumefaction, biogenesis of starch, and DNA-mediated transformation in *Brotéria: Ciências naturais*.[35]

---

32 William D. Sullivan and James V. Boyle, "The Effect of X-radiation on the Succinic Dehydrogenase Activity on the Dividing Stages of *Tetrahymena pyriformis* GL.," *Brotéria: Ciências naturais* 30 (1961): 77–82. For Sullivan's obituary, see "Sonntag Institute Director William Sullivan, S.J., Dies," *Boston College Chronicle* 4, no. 14 (1996): 6.

33 William D. Sullivan and Adam J. von Knobelsdorff, "The *In vitro* and *In vivo* Effects of Fluoride on Succinic Dehydrogenase Activity," *Brotéria: Ciências naturais* 31 (1962): 3–13; William D. Sullivan, "The Spectrophotometric Determination of Malic Dehydrogenase and 'Malic' Enzyme in Normal Populations of *Tetrahymena pyriformis* GL.," *Brotéria: Ciências naturais* 33 (1964): 143–58.

34 Levi Guerra and F. Edmund Hunter, "Sucrose Inhibition of Gramicidin Induced Swelling of Isolated Rat Liver Mitochondria," *Brotéria: Ciências naturais* 34 (1965): 227–46; Elinor O'Brien and William D. Sullivan, "Electrophoretic Patterns of Proteins Constituents in Tissues of Tumor-bearing and Non-tumor-bearing Animals," *Brotéria: Ciências naturais* 43 (1974): 3–14; Carlos Azevedo, "Nucléolo: Estrutura, citoquímica e aspectos funcionais," *Brotéria: Ciências naturais* 48 (1978): 3–52.

35 Francisco Carvalho Guerra, "Tumefacção mitocondrial: Estudo comparativo das mitocôndrias do cérebro e do fígado," *Brotéria: Ciências naturais* 34 (1965): 3–226; Luís Archer, "DNA-Mediated Transformation," *Brotéria*: *Ciências naturais* 36 (1967): 107–76; Roberto Salema, "Amido: Estudo ultraestrutural da sua biogénese em plantas superiores," *Brotéria: Ciências naturais* 38 (1969): 1–127.

Following the publication of his doctoral dissertation in 1968, Archer started to look for other potential contributions in genetics, especially in the subfields of cytogenetics and molecular genetics. Without a proper laboratory to do research and teach the next generation of biologists, he initially turned to foreign biologists.[36] By publishing doctoral dissertations, as well as shorter articles, exploring the novel fields of biochemistry and molecular genetics, *Brotéria* thus became the leading journal of modern biology in Portugal.[37]

Archer was particularly keen on modernizing *Brotéria* and transforming it into one of the vectors for the promotion of genetics at an academic level. As with Silva Tavares and Luisier before him, he combined his own scientific interests with the editorial program of *Brotéria*, thus fulfilling the need for a specialized journal on genetics in Portugal. In 1980, after twenty years as editor-in-chief of *Ciências naturais*, Archer redefined the journal, creating *Brotéria: Genética* (figure 23).[38] He argued that while in the first decades of the twentieth century "the natural sciences were perfectly united around systematics and taxonomy," the variety and complexity of approaches to study the natural world in the preceding years made it "more difficult to support a unified vision of life phenomena."[39] For Archer, genetics provided a solution to reach a more comprehensive understanding of biology. Because the survival of all species required diversity and evolution, genetics "was not only the science of perpetuity, but also the science of innovation."[40] Since it was already producing changes in the world economy, industry, demography, society, and politics,

---

36 See, for instance, the following articles published between 1970 and 1974: R.B. Ghosh, "Karyomorphological Studies of Somatic Chromosomes in *Ailanthus excelsa* Roxb. An Ornamental and a Road-Side Plant," *Brotéria: Ciências naturais* 39 (1970): 3–8; Syed Shafi Ashgar, D.K. Khawaja, A.K. Jafri, "5'-Nucleotidase Activity in the Tissues of the Cat-fish, *Heteropnuestes fossilis* Bloch.," *Brotéria: Ciências naturais* 41 (1972): 3–7; S. Bagchi and R.M. Datta, "On the Nature of Chromosome in Meiotic Mechanism of a Natural Tetraploid *Sesbania benthamiana* Domin.," *Brotéria: Ciências naturais* 42 (1973): 31–37; Asit Kumar Banerjee and Archana Sharma, "Chromosome Studies on Some Indian Members of *Compositae*. I. *Tribe Inuloideae*," *Brotéria: Ciências naturais* 43 (1974): 15–32; Bhagwan Din Chaurasia and Vijay K. Sharma, "Karyological Studies in *Phaseolus mngo* Linn.," *Brotéria: Ciências naturais* 43 (1974): 33–34; Bhagwan Din Chaurasia and Vijay K. Sharma, "Karyological Studies in *Asphodelus tenuifolius* Cav.," *Brotéria: Ciências naturais* 43 (1974): 35–37.

37 On the history of biochemistry in Portugal, see especially Isabel Amaral, *A emergência da bioquímica em Portugal: As escolas de investigação de Marck Athias e de Kurt Jacobsohn* (Lisbon: Fundação Calouste Gulbenkian-Fundação para a Ciência e Tecnologia, 2006).

38 On the history of *Brotéria: Genética*, see especially Miguel Mota, "A contribuição da *Brotéria* para o desenvolvimento da genética," in Rico and Franco, *Fé, ciência, cultura*, 517–27.

39 Luís Archer, "*Brotéria*: Das ciências naturais à genética," *Brotéria: Genética* 1 (1980): 5–6, here 5.

40 Archer, "*Brotéria*: Das ciências naturais à genética," 5.

FIGURE 23 First issue of *Brotéria: Genética*, 1980
AB

he believed that, in the near future, genetics would become the foundation for "great changes in the living and social worlds."[41]

41 Archer, "*Brotéria*: Das ciências naturais à genética," 6.

Although there were a few Portuguese periodicals that published works of genetics, such as the *Boletim da sociedade Broteriana*, the *Revista agronómica*, the *Agronomia lusitana* (Portuguese agronomy), and the *Revista de biologia* (Journal of biology), *Brotéria: Genética* was the first national journal entirely dedicated to the new science. For twenty-two consecutive years, *Genética* was funded by some non-profit organizations, including the Portuguese Society of Genetics (1980–2002), the National Institute for Scientific Research and the Foundation for Science and Technology (1983–99), and the National Institute for Agricultural Research (1989–91). The association between the Society of Genetics and the Jesuit journal was particularly treasured, because *Brotéria: Genética* was, from the start, the official publication of the Society of Genetics. This was a rather peculiar situation, because the journal belonged to a religious order and was, at the same time, the official organ of a scientific organization with members of all creeds and political stances. Naturally, both the Society of Jesus and the Society of Genetics benefited from this close relationship. In the 1980s, *Brotéria*, and by extension the Society of Jesus, lent its scientific tradition and prestige to the newly created Society of Genetics (1974), and received, in return, sufficient funds for its publication. However, this inevitably raised a few constraints. Unlike other scientific publications, *Genética* only had one editor-in-chief during its existence, most likely because Archer was the only member of the Society of Genetics who was also a Jesuit.

Between 1980 and 2002, *Genética* published 330 research articles, review essays, and short notes on bacterial genetics, animal breeding, plant breeding, human genetics, and bioethics. The publication of articles on plant and animal breeding seemed to follow the same editorial program as *Ciências naturais*.[42] The publication of works of molecular and human genetics, on the other hand, was one of the key innovations of *Genética*. These articles focused on the pathology, incidence, and diagnosis of severe monogenic diseases and of chromosomal disorders, such as cystic fibrosis, sickle-cell anemia, and Down

42 The articles on animal breeding focused on the study and betterment of hamsters, pigs, goats, and cats. See Jorge Antunes-Correia and Tristão Mello Sampaio, "Somatic Association of Syrian Hamster Chromosomes," *Brotéria: Genética* 1 (1980): 43–51; Jorge Antunes-Correia, "Genética da sensibilidade ao halotano: Uma situação com grande interesse actual no melhoramento porcino," *Brotéria: Genética* 1 (1980): 131–33; Jorge Antunes-Correia and H. Krug, "Frequência genética das hemoglobinas em caprinos portugueses," *Brotéria: Genética* 1 (1980): 181–84; Jorge Antunes-Correia, "Frequências génicas da pelagem em gatos portugueses," *Brotéria: Genética* 4 (1983): 157–65. In turn, the articles on plant breeding focused mostly on wheat and triticale. See Tristão Mello Sampaio, "Bactérias fixadores de azoto atmosférico em trigo," *Brotéria: Genética* 1 (1980): 83–85; Maria do Céu Barradas, Tristão Mello Sampayo, and Zaida Cunha, "Gene Activation for High Molecular Mass Glutenins in Tetraploid Wheat: The Case of Resende," *Brotéria: Genética* 12 (1991): 69–71; Francisco Bagulho, "Avanço no tratamento de triticale," *Brotéria: Genética* 5 (1984): 34–47.

syndrome.[43] By publishing these essays, the Jesuit journal was taking part in the dissemination of clinical results to a wider readership. Most of the research articles on microbial genetics were written by members of Archer's laboratory, including Hermínia Lencastre (1947–), Helena Paveia (1945–2009), and Graça Vieira, who worked on the location and utilization of the genes for arabinose in *Bacillus subtilis*, and on the molecular characterization of bacteriophages.[44] Besides publishing the results of her research, Hermínia Lencastre—who was Archer's first graduate student (1974–81) at the Gulbenkian Institute of Science—also wrote an important review essay on genetic engineering in bacteria.[45]

To celebrate its tenth anniversary, the European Meeting on Bacterial Transformation and Transfection convened, once again, in Portugal.[46] The conference was held at the Gulbenkian Foundation, in Lisbon, between August 30 and September 2, 1982 and Luís Archer served as chairman.[47] The conference opened with a plenary lecture entitled "Searching a *Haemophilus influenzae* Library for Genes Induced during Competence Development," by American physician and biologist Hamilton O. Smith (1931–), who had recently been awarded the Nobel Prize in Physiology or Medicine for his pioneering

43 Jorge Sequeiros and Paula Coutinho, "Genetic Aspects of Machado-Joseph Disease," *Brotéria: Genética* 2 (1981): 137–47; Maria Guida Boavida, "Anomalias cromossómicas humanas," *Brotéria: Genética* 4 (1983): 79–80; Maria Guida Boavida et al., "Um estudo citogenético em recém-nascidos com síndroma malformativo," *Brotéria: Genética* 4 (1983): 127–45; Ângela Duarte et al., "Fibrose quística em Portugal: Patologia molecular e diagnóstico pré-natal," *Brotéria: Genética* 14 (1993): 87–93.

44 See especially Luís Archer and Helena Paveia, "Location of Genes for Arabinose Utilization in *Bacillus subtilis* Chromosome," *Brotéria: Genética* 1 (1980): 169–76; Archer and Paveia, "Genes for L-arabinose Utilization in *Bacillus subtilis*," *Brotéria: Genética* 13 (1992): 149–59; Luís Archer, Rosa M. Fernandes, and Hermínia Lencastre, "Two Newly Isolated Temperate Phages of *Bacillus subtilis*," *Brotéria: Genética* 4 (1983): 27–33; Luís Archer et al., "Location of Prophage IG3 in *Bacillus subtilis* Chromosome," *Brotéria: Genética* 6 (1985): 57–63; Luís Archer, Graça A. Vieira, and Hermínia Lencastre, "Relatedness of Bacteriophages PBS1, AR9, 3NT and I10 of *Bacillus subtilis*," *Brotéria: Genética* 6 (1985): 179–89.

45 Hermínia Lencastre, "Engenharia genética em *Bacillus subtilis*: I; Plasmídeos nativos," *Brotéria: Genética* 3 (1982): 15–38; Lencastre, "Engenharia genética em *Bacillus subtilis*: II; Plasmídeos compostos," *Brotéria: Genética* 4 (1983): 81–125; Lencastre, "Engenharia genética em *Bacillus subtilis*: III; Bacteriófagos," *Brotéria: Genética* 6 (1985): 25–55.

46 Luís Archer, "Regresso a Portugal do European Meeting on Bacterial Transformation," *Brotéria: Genética* 1 (1980): 191. The previous meetings convened in Oeiras, Portugal (1972), Kraków, Poland (1974), Granada, Spain (1976), York, England (1978), and Florence, Italy (1980).

47 Luís Archer, "Volta a Portugal um congresso europeu de origem portuguesa," *Brotéria: Genética* 3 (1982): 81.

discoveries in genetic engineering (1978).[48] With seventy participants from Spain, Italy, the United Kingdom, Belgium, Germany, the Netherlands, Poland, Czechoslovakia, Sweden, Norway, Russia, Japan, and the United States, the meeting was a tremendous success. Given their obvious interest for the Portuguese Society of Genetics, the proceedings were published in *Brotéria* in 1982.[49] As editor of *Genética*, head of the laboratory of molecular genetics at the Gulbenkian Science Institute, and convener of international congresses, Luís Archer promoted research in his field of expertise while at the same time providing an outlet for the publication of the most important works of his students and peers.

2 The History of Genetics

Along with review essays, research articles, and abstracts, *Brotéria: Genética* also published shorter biographies of Portuguese scientists. Between 1980 and 2002, the journal published thirty-seven biographies of contemporary geneticists, including António Sousa da Câmara, Abílio Fernandes (1906–94), Miguel Pereira Coutinho (1915–98), and Tristão de Melo Sampaio (1923–97).[50] It also paid tribute to Aurélio Quintanilha and José Antunes Serra (1914–90), leading figures in modern genetics during the Estado Novo. Professor of botany at the University of Coimbra, and an expert on fungal genetics, Quintanilha became known for his works on the sexuality and reproduction of mushrooms in the 1930s. In 1935, however, his career at the University of Coimbra came to an abrupt end. On May 13, 1935, Prime Minister António de Oliveira Salazar (1889–1970, in office 1932–68) discharged all public officers that had opposed—or showed signs of opposing—the "fundamental principles of the Constitution."[51] Although he did not engage in politics, he was forced to step down from the university and was forbidden to work at private and public institutions from then on.

48 David B. Danner and Hamilton O. Smith, "Searching a *Haemophilus influenzae* Library for Genes Induced during Competence Development," *Brotéria: Genética* 3 (1982): *19.

49 Luís Archer, ed., "6th European Meeting on Bacterial Transformation and Transfection: Scientific Program and Abstracts," *Brotéria: Genética* 3 (1982): *3–106.

50 Miguel Pereira Coutinho, "Homenagem à memória do Prof. António de Sousa da Câmara," *Brotéria: Genética* 2 (1981): 7–9; Luís Archer, "Prof. Doutor Abílio Fernandes," *Brotéria: Genética* 16 (1995): 5–6; Henrique Guedes-Pinto, "O Eng.° Tristão de Mello-Sampayo," *Brotéria: Genética* 18 (1997): 5–8; Maria Luísa Neves, "Recordando o Padre Luisier: Nos 40 anos do seu falecimento," *Brotéria: Genética* 18 (1997): 99–101; Henrique Guedes-Pinto, "O Sr. Professor Miguel Pereira Coutinho: Recordações dispersas," *Brotéria: Genética* 19 (1998): 5–11.

51 *Diário do governo* 108 (May 13, 1935): 649–50.

After two years living on a meager pension, Quintanilha went to Paris, where he worked at the Muséum National d'Histoire Naturelle (1937–39) and served as a volunteer in the French army during the Second World War (1939–40). Following the German occupation, he left Paris for Lisbon to work with Sousa da Câmara at the National Agronomic Station (1941–43). Because he was a retired public officer, he could not receive a salary. So, in the hope of being readmitted to public service, he ended up working for two years without receiving any pay. In 1943, Salazar intervened again and discharged him from his duties at the Agronomic Station: he was to be sent to Mozambique to serve as director of the Centro de Investigação Científica Algodoeira (1943–62), a center that promoted research on the betterment of cotton in the colonies.[52] In the 1970s, things eased somewhat for the Portuguese geneticist. He was bestowed the Military Order of Saint James of the Sword in 1972 and was reinstated as professor of botany at the University of Coimbra in 1974. While the former event happened in the final years of the dictatorship, the latter happened after the Carnation Revolution (i.e., the military uprising that put an end to the dictatorship on April 25, 1974).

For his pioneering works on genetics, for training a new generation of researchers—including Abílio Fernandes, Flávio Resende, and José Antunes Serra—and for being mistreated during the dictatorship, Quintanilha's return to Portugal was surrounded with joyfulness and hope. The Portuguese Society of Genetics honored him by making him its first honorary fellow in 1974. As a founding member of the Society of Genetics, Archer was particularly involved in the tribute paid to the Portuguese dean of genetics. In 1975, the Jesuit published in *Brotéria: Ciências naturais* the papers read by José Antunes Serra, Vitorino Nemésio (1901–78), and Quintanilha at the homage held at the Gulbenkian Foundation on November 11, 1974.[53] In the following years, Archer paid two other tributes to Quintanilha. In 1988, one year after his passing, *Brotéria: Genética* published a series of testimonies from his peers,[54] and

52 The Estado Novo was particularly interested in promoting the betterment of cotton in the colonies. For a historical context, see especially Saraiva, "Laboratories and Landscapes," 7–39, and Saraiva, "Fascist Labscapes," 457–98.

53 José Antunes Serra, "Aurélio Quintanilha: Algumas palavras acerca da sua obra como professor e investigador," *Brotéria: Ciências naturais* 44 (1975): 157–74; Vitorino Nemésio, "Perfil de Aurélio Quintanilha," *Brotéria: Ciências naturais* 44 (1975): 175–88; Aurélio Quintanilha, "História da genética em Portugal," *Brotéria: Ciências naturais* 44 (1975): 189–208.

54 Henrique Guedes-Pinto, "Aurélio Quintanilha: fragmentos para um esboço," *Brotéria: Genética* 9 (1988): 5–9; José Antunes Serra, "Professor Aurélio Quintanilha: Impressões e recordações pessoais de homenagem," *Brotéria: Genética* 9 (1988): 9–17; Miguel Pereira Coutinho, "Como conheci o Prof. Aurélio Quintanilha," *Brotéria: Genética* 9 (1988): 19–21;

five years later, the journal again showed its support by printing a few more eulogies on the occasion of Quintanilha's centenary.[55] This threefold homage to Quintanilha suggests that Archer wanted to emphasize the Jesuits' detachment from the controversial measures taken against the dean of genetics during the Estado Novo.

In the 1990s, *Brotéria: Genética* paid an important tribute to Antunes Serra, one of Quintanilha's most talented students in Coimbra.[56] Professor in Coimbra (1936–52) and Lisbon (1953–87), Antunes Serra's research covered a huge variety of biological topics. Although he was mostly admired for his contributions to ovine genetics, his publications covered an immense variety of topics, ranging from the constitution of melanins to cell division, and from the concept of genes to evolution. In 1944, while teaching in Coimbra, he founded the journal *Portugaliae acta biologica* with Flávio Resende, where he published some of his works. His research also started to feature in high-profile journals such as *Nature* and *Science* as early as 1945. This was a significant achievement for a scientist working in the European periphery under an authoritarian regime.[57] Although he was not dismissed from the University of Coimbra

---

Luís Archer, "Quintanilha: O velho mestre feito jovem aprendiz," *Brotéria: Genética* 9 (1988): 23; Abílio Fernandes, "Lembrando o Prof. Doutor Aurélio Quintanilha," *Brotéria: Genética* 9 (1988): 135–50.

55 Luís Archer, "Centenário de Aurélio Quintanilha (1892–1992)," *Brotéria: Genética* 14 (1993): 5–6; Archer, "Mestre Quintanilha faz-se aluno," *Brotéria: Genética* 14 (1993): 29–32; António Viveiros Bettencourt, "As minhas memórias do Professor A. Quintanilha," *Brotéria: Genética* 14 (1993): 43–45; Fernando Catarino, "No centenário de Aurélio Quintanilha," *Brotéria: Genética* 14 (1993): 19–22; Miguel Pereira Coutinho, "Como vejo a figura do Professor Quintanilha," *Brotéria: Genética* 14 (1993): 9–13; Vitorino Nemésio, "Perfil de Aurélio Quintanilha," *Brotéria: Genética* 14 (1993): 55–65.

56 Quintanilha, "História da genética em Portugal," 194–95. On Serra's early life and work, see especially Maria do Mar Gago, "The Emergence of Genetics in Portugal: J.A. Serra at the Crossroads of Politics and Biological Communities (1936–1952)," (MSc. diss. University of Lisbon, 2009); Maria do Mar Gago, "Things of Darkness: Genetics, Melanins and the Regime of Salazar (1936–1952)," *Centaurus* 57 (2015): 1–27.

57 José Antunes Serra and A. Queiroz Lopes, "Chemical Constitution of Nucleolar Inclusions in Growing Oocyte Cells," *Nature* 155 (1945): 792; José Antunes Serra, "Constitution of Hair Melanins," *Nature* 157 (1945): 771–2; Serra, "A Possible Explanation of the Relations between Hereditary Anemias and Depigmentation in the Mouse," *Nature* 159 (1947): 504–5; Serra and Arturo Pereira, "Quantitative Microdetermination of Amino Acids after Paper Chromatography," *Science* 113 (1951): 387–88; Serra, "Interpretation of Nucleolar Inclusions," *Nature* 181 (1958): 1544–45; Serra, "Cytochemical Demonstration of Masked Lipids," *Science* 128 (1958): 28–29; Serra, "A Position Effect Explanation of the Gene Conversion," *Nature* 183 (1959): 700; Serra and Paulo Picciochi, "Demonstration of Canonic Gonial Mitosis and Meiosis in *Parascaris equorum*," *Science* 132 (1960): 1400–1.

in the purges of 1935 and 1947, Antunes Serra was forbidden to travel abroad from 1947 onward. This prohibition had profound implications for his career, because he had to turn down many invitations either to attend international conferences or to work at prestigious laboratories. Antunes Serra was finally able to leave Portugal in 1965. He initially worked at the University of Alberta, Canada (1965–66) and then at the Pacific Northwest Research Foundation, Seattle (1968–70).

In the 1960s, Antunes Serra published his *magnum opus*: a three-volume handbook of modern genetics.[58] With more than 1,900 pages, his textbook was intended to be a guide for postgraduate studies of genetics. He returned to Lisbon in 1970 and went on to create a genetics laboratory at the Faculty of Sciences, Lisbon, which he directed until 1987.

Like Quintanilha, Antunes Serra was highly esteemed by his peers: in 1984, he was made honorary fellow of the Portuguese Society of Genetics, and in 1989 he was bestowed the Military Order of Saint James of the Sword. On March 15, 1991, nine months after his passing, his friends organized a final tribute at the Faculty of Sciences, Lisbon. As with Quintanilha, Archer seized the opportunity to honor Antunes Serra and published the speeches given at the ceremony in *Brotéria: Genética*.[59] By publishing short biographies and eulogies of geneticists, and, occasionally, lengthier essays on the history of the discipline, *Brotéria* thus came to play a significant role in the construction of a public image of genetics and the science's leading practitioners in twentieth-century Portugal. The cases of Quintanilha and Antunes Serra, in particular, seemed to serve another purpose, to testify to the Jesuits' close participation in the rehabilitation of scientists mistreated during the dictatorship.

58 José Antunes Serra, *Modern Genetics*, 3 vols. (London: Academic Press, 1965–68). This work expanded *Moderna genética: Geral e fisiológica*, published in Portuguese twenty years earlier (José Antunes Serra, *Moderna genética: Geral e fisiológica*, 2 vols. [Coimbra: Author edition, 1949]).

59 Luís Archer, "Homenagem ao Prof. Doutor José Antunes Serra," *Brotéria: Genética* 13 (1992): 5; Carlos Almaça, "As primeiras fases da obra científica do Prof. J.A. Serra," *Brotéria: Genética* 13 (1992): 33–40; Jorge Antunes-Correia, "A contribuição fundamental do Professor José Antunes Serra na genética do melhoramento dos ovinos," *Brotéria: Genética* 13 (1992): 15–18; João Côrte-Real, "Como conheci o Senhor Professor Serra: Uma experiência de treze anos," *Brotéria: Genética* 13 (1992): 27–31; Miguel Pereira Coutinho, "O Professor Antunes Serra, um verdadeiro mestre," *Brotéria: Genética* 13 (1992): 9–12; Maria de Lourdes Sampaio Silva, "Algumas palavras e recordações de homenagem ao Prof. José Antunes Serra," *Brotéria: Genética* 13 (1992): 19–21. Following Serra's passing in 1990, Rolanda Albuquerque de Matos published a comprehensive list of Serra's publications. See Rolanda Albuquerque de Matos, "Professor José Antunes Serra," *Brotéria: Genética* 12 (1991): 5–44.

3 The Emergence of Bioethics

Bioethics was one of the most pressing topics covered by *Brotéria: Genética*. The vast majority of the articles were written by Archer who was keen on promoting the debate on the ethical questions related to the life sciences. In 1980, when he created *Genética*, Archer was already an accomplished geneticist. In the first decade after his return to Portugal, he had created an international laboratory of molecular genetics (1971), participated in the foundation of a Society of Genetics (1974), introduced genetic engineering techniques to his peers (1976), and completely redefined the editorial program of *Brotéria* (1980) to better serve the community of Portuguese geneticists. Although he did not forsake his research and teaching duties, in the late 1970s he began to dedicate a considerable part of his time to bioethics. The new discipline combined his scientific interests with his early desire to study philosophy, and it soon became a second career for him.

Between 1980 and 2007, Archer published extensively on the bioethical implications of human artificial reproduction, prenatal screening, gene therapy, and cloning.[60] While the Catholic Church disapproved of artificial reproduction in every possible case, Archer openly defended the legitimacy of assisted reproductive technology for infertile heterosexual couples—which he considered not only a lesser evil but also an acceptable extension of the spouses' sterile conjugal act. In harmony with the church's teachings, he condemned surrogacy and the use of artificial reproduction techniques in fertile spouses, homosexual couples, and single parents. He was also particularly concerned with the ethical conundrum raised by the creation of spare embryos to guarantee a successful pregnancy.[61] Using his knowledge as a molecular geneticist

60 In this period, Luís Archer published five books and more than one hundred articles on bioethics. See Luís Archer, *Temas biológicos e problemas humanos* (Lisbon: Editorial Brotéria, 1981); Archer, *Desafios da nova genética* (Lisbon: Editorial Brotéria, 1992); Luís Archer, Jorge Biscaia, and Walter Osswald, eds., *Bioética* (Lisbon: Verbo, 1996); Luís Archer et al., *Novos desafios à bioética* (Porto: Porto Editora, 2001); Archer, *Da genética à bioética* (Coimbra: Gráfica de Coimbra 2, 2006). His papers on bioethics were recently published in a modern edition of his selected works. See Francisco Malta Romeiras and Henrique Leitão, eds., *Obra selecta do Padre Luís Archer, S.J.*, vol. 3, *Bioética (1980–1997)* (Lisbon: Fundação Calouste Gulbenkian, 2016), and Francisco Malta Romeiras and Henrique Leitão, eds., *Obra selecta do Padre Luís Archer, S.J.*, vol. 4, *Bioética (1998–2007)* (Lisbon: Fundação Calouste Gulbenkian, 2017).

61 For Archer's views on artificial reproduction, see especially Luís Archer, "Reprodução humana artificial," *Brotéria: Genética* 6 (1985): 93–96; Archer, "Reprodução humana artificial: Um grito de alarme," *Brotéria: Genética* 7 (1986): 109–14; Archer, "O bebé-proveta que me falou em sonhos," *Brotéria* 122 (1986): 569–72; Archer, "Alarme sobre reprodução

who cloned genes in bacteria on a daily basis, Archer worked hard to debunk popularly misconstrued notions of cloning and gene therapy.[62] As a bioethicist, he frequently stressed that, while somatic gene therapy posed no particular ethical problems, this was not the case for germ-line gene therapy and prenatal screening, because, at the core, they promoted eugenics.[63]

Because he was a forerunner in the emerging field of bioethics, Archer participated in some of the first national and European committees dedicated to the ethical implications of biotechnology. At first, however, the Holy See called into question his role as an advisor to the Council of Europe. In December 1984, one year after his nomination to the ad hoc Committee of Experts on Ethical and Legal Problems Relating to Human Genetics, the apostolic nuncio in Lisbon, Archbishop Salvatore Asta (1914–2004, in office 1984–89), reached out to the Jesuit provincial urging Archer's dismissal.[64] In his view, Archer could not serve as an advisor to the Council of Europe because canon 285§3 asserted that "clerics are forbidden to assume public offices which entail a participation in the exercise of civil power."[65] Although Superior General Peter-Hans Kolvenbach and Cardinal António Ribeiro (1928–98), patriarch of Lisbon, believed that his role did not entail "participation in the exercise of civil power," and argued that he was just a scientific advisor, Archer eventually stepped down, in obedience to the Holy See.[66] To avoid similar situations in the future,

artificial," *Brotéria* 124 (1987): 91–99; Archer, "Gerar ou produzir vida humana?," *Brotéria* 124 (1987): 519–32; Archer, "Para uma ética da reprodução humana," *Brotéria* 145 (1997): 218–25; Archer, "Treze anos de reflexão sobre procriação medicamente assistida," *Brotéria: Genética* 20 (1999): 125–55; Archer, "Procriação medicamente assistida: Evolução do pensamento ético de 1986 a 1999," *Brotéria* 150 (2000): 337–67. For the official position of the Catholic Church, see John Paul II, *Evangelium vitae* (Washington, DC: United States Catholic Conference, 1995), no. 14.

62 See especially Luís Archer, "Clonagem: Mito e ciência," *Brotéria* 149 (1999): 75–90; Archer, "Clonagem: Verdade científica e sonho mítico," *Brotéria: Genética* 20 (1999): 101–24; Archer, "Clonagem não reprodutiva: Usos terapêuticos e objecções éticas," *Cadernos de bioética* 22 (2000): 65–81.

63 Luís Archer, "Genetic Engineering and Human Freedom," *Brotéria: Genética* 10 (1989): 49–62; Archer, "Engenharia genética: Esperanças e medos," *Cadernos de bioética* 15 (1997): 49–54; Archer, "Ethical Aspects of Gene Therapy," *Brotéria: Genética* 19 (1998): 13–26; Archer, "Terapia génica: Dez anos depois," *Brotéria: Genética* 21 (2000): 53–66.

64 APSI, Salvatore Asta to Manuel Vaz Pato (December 14, 1984).

65 *Code of Canon Law* (London: Collins, 1983), no. 285§3.

66 APSI, Manuel Vaz Pato to Peter-Hans Kolvenbach (January 10, 1985); APSI, Peter-Hans Kolvenbach to Manuel Vaz Pato (January 30, 1985); APSI, Manuel Vaz Pato to Salvatore Asta (April 26, 1985); APSI, Manuel Vaz Pato to Peter-Hans Kolvenbach (June 10, 1985); APSI, Salvatore Asta do Manuel Vaz Pato (March 7, 1986); APSI, Manuel Vaz Pato to Salvatore Asta (April 4, 1986); APSI, Manuel Vaz Pato to Salvatore Asta (April 5, 1986); APSI, Manuel Vaz Pato to Peter-Hans Kolvenbach (April 6, 1986); APSI, Salvatore Asta to Manuel

Kolvenbach recommended that the provincial disclose, in advance, Archer's nominations to the patriarch of Lisbon.[67] Witnessing the leading expert on bioethics stepping down from an advisory position at the Council of Europe, in obedience to the Holy See, was puzzling for the Portuguese government. Yet, it did not prevent further nominations. In 1991, he was one of the founding members—and then chairman (1996–2001)—of the Conselho Nacional de Ética para as Ciências da Vida, a council created to reflect and give advice to the government on ethical issues related to the life sciences. And in 1993 he was appointed to the Group of Advisors to the European Commission on the Ethical Implications on Biotechnology (1993–97).

Luís Archer's contributions to molecular genetics and bioethics earned him a reputation as one of the leading geneticists both in Portugal and abroad. He was a senior fellow of numerous learned societies, including the Lisbon Academy of Sciences (1985), the European Academy of Sciences and Arts (1993), and the New York Academy of Sciences (1995). He was also an honorary fellow of the Portuguese Society of Genetics (1993), the Portuguese Society of Human Genetics (1998), and the Portuguese Society of Microbiology (2003), to name only a few. The Catholic University of Portugal awarded him an honorary doctorate in 2007, with the New University of Lisbon following suit in 2009. More importantly, in 1991 Luís Archer became the first molecular geneticist, the first bioethicist, and the first Jesuit, to receive the Grand Cross of the Military Order of Saint James of the Sword, the greatest state honor conferred on a writer, artist, or scientist in Portugal.[68]

Vaz Pato (April 11, 1986); APSI, Peter-Hans Kolvenbach to Manuel Vaz Pato (May 12, 1986); APSI, Manuel Vaz Pato to Peter-Hans Kolvenbach (June 4, 1986); APSI, Peter-Hans Kolvenbach to Manuel Vaz Pato (July 25, 1986).

67 APSI, Peter-Hans Kolvenbach to Manuel Vaz Pato (June 18, 1986); APSI, Manuel Vaz Pato to Cardinal António Ribeiro (June 30, 1986).

68 The Military Order of Saint James of the Sword comprises six different grades: Grand Collar, Grand Cross, Grand Officer, Commander, Officer, Knight or Dame. The Grand Collar is usually conferred upon foreign heads of state. The only exception to date is José Saramago, who received the Grand Collar after winning the Nobel Prize in Literature (1998). Being the second highest grade, the list of Grand Crosses is rather short. Among its recipients are some of the most distinguished twentieth-century writers (Sophia de Mello Breyner Andresen [1998], António Lobo Antunes [2004]), painters (Columbano Bordallo Pinheiro [1920], Carlos Reis [1940], Paulo Rego [2004]), musicians (Amália Rodrigues [1990], Maria João Pires [1998]), and scientists (Egas Moniz [1945], José Antunes Serra [1989]). In 2008, Walter Osswald, Daniel Serrão, and Jorge Biscaia also received the Grand Cross for their pioneering work in bioethics.

Conclusion

Science and education were among the most valued and distinctive ministries of the Society of Jesus in the early modern period. Between its foundation in 1540 and the order's universal suppression in 1773, the Society of Jesus created colleges and universities throughout the world and established long-distance networks of communication that proved fundamental for the circulation of knowledge between Europe, Asia, and the Americas. In this period, the constant observation of the heavens, the drawing of accurate maps, and the description of new species of animals and plants reflected the Jesuits' practical, Ignatian spirituality. In the Portuguese Empire, the Jesuits ran one university and thirty-seven colleges and were responsible for tens of thousands of students. The Jesuits' colleges made important contributions to the teaching and practice of mathematics, astronomy, cosmography, cartography, natural sciences, and even astrology and chiromancy. In the 1750s, the all-powerful marquis of Pombal deployed an array of political, diplomatic, and economic actions against the Society that ultimately led to the Jesuits' expulsion in 1759. After banning the Jesuits, Pombal presented them as the main obstacle to the country's progress and modernization. Although this accusation was unable to withstand even a cursory level of scrutiny, it nevertheless proved extremely resilient, deeply affecting the order's return in the nineteenth century.

After a first unsuccessful attempt during the Liberal Wars (1829–34), the Society of Jesus was effectively restored in Portugal by Carlos Rademaker. The restoration of 1858 was contingent on a series of events, including Rademaker's return to Lisbon, the wide support from liberal governments and conservative aristocrats, and the chaotic status of the Padroado. As education had been one of the most significant ministries of the "old" Society of Jesus, and as the accusations of obscurantism lingered, nineteenth-century Jesuits made significant efforts to promote an education grounded on an "alliance between religion and science."[1] In this period, the promotion of field trips to observe the heavens, the systematic collection of meteorological data, the curation of natural history collections, attendance at international conferences, and the publication of original research works all testify to the importance of the Jesuits' informal training as modern scientists and teachers.

Besides intertwining the teaching and practice of science in their colleges, the Jesuits were particularly invested in drawing their students to science.

1 Colégio de Campolide, *Prospecto* (Lisbon: Typographia Universal, 1886), 1–2, here 1.

 | DOI:10.1163/9789004382367_011

The students were invited to engage in activities ranging from participation in scientific academies to going on field trips to observe eclipses or to collect new specimens of animals and plants for the natural history museum. By resourcefully investing in a hands-on experimental teaching style in the natural sciences, the colleges of Campolide and São Fiel soon became the most noteworthy boarding schools in nineteenth-century Portugal. The Jesuits put the teaching and practice of science at the top of their educational priorities while simultaneously projecting a public image as modern scientists and pedagogues, thus regaining their credibility and prestige among the Portuguese royal family, the aristocracy, and the bourgeoisie. The rehabilitation of the Jesuits' public image was of paramount importance for the approval of new apostolic missions and the diffusion of Ignatian spirituality.

At the turn of the nineteenth century, the Jesuit colleges enjoyed a highly positive reputation, and even freemasons, republicans, and anticlerical politicians chose Campolide and São Fiel for the education of their children. However, when the usual accusations of obscurantism resurfaced, the Jesuits' situation became rather fragile, and they were once again expelled from Portugal on October 8, 1910. The republicans viewed the Jesuits' expulsion, the closure of their colleges, and the confiscation of their books, collections, and instruments as a necessary first step toward achieving social, cultural, and political stability in the new regime. Exiled to Spain, Belgium, and Brazil, the Jesuits established new missions and colleges and continued to work on the classification of new species of animals and plants and on the publication of *Brotéria*.

Founded in 1902 as the taxonomy journal of a seemingly minor boarding school in the countryside, *Brotéria* was probably the Jesuits' most important scientific endeavor in twentieth-century Portugal. Unlike the other journals published by the Society worldwide, *Brotéria* was conceived as a scientific journal and published extensively on a variety of subjects, ranging from taxonomy to biochemistry, molecular genetics, and bioethics. At first, the journal sought to contribute to the progress of botany and zoology through the identification of new species of animals and plants. As well as corresponding to the scientific interests of Joaquim da Silva Tavares, Cândido de Azevedo Mendes, and Carlos Zimmermann, the publication of a journal of taxonomy also followed a trend regarding the practices of the Society of Jesus in the eighteenth century, when the natural sciences became one of the most popular disciplines for Jesuit scholars worldwide. Under the direction of Silva Tavares (1902–31), the journal played a significant role in the consolidation of botany and zoology in Portugal. In this period, a relatively small group of Jesuit naturalists—including Silva Tavares, Azevedo Mendes, Torrend, Zimmermann, and Luisier—built tailored networks of scientific correspondence that allowed for the circulation, description, and classification of hundreds of gall wasps, moths, butterflies, mushrooms, algae, lichens, and mosses collected worldwide.

Brotéria was enthusiastically received by Portuguese botanists and zoologists. Yet because the number of paying subscribers was initially insufficient to cover its publication, Silva Tavares restructured the journal in 1907, dividing *Brotéria* into three separate magazines: *Vulgarização científica*, *Botânica*, and *Zoologia*. Published until 1924 and written in the vernacular, *Vulgarização* was designed to make a profit. By publishing articles on agriculture, geography, physics, commerce, and medicine alongside culinary recipes, novels, and household advice, Silva Tavares and his companions broadened the audience of their magazine and were able to cover the expenses of the more specialized scientific series *Botânica* and *Zoologia*. In 1925, the editorial program of the Jesuit journal was redefined once again, and *Vulgarização* gave way to the so-called cultural *Brotéria*. Published alongside the scientific series, the journal sought to propagate the Catholic faith and values to a broad readership.

Following the passing of its first editor, *Brotéria* witnessed the consolidation of taxonomy and the emergence of new research areas, namely plant physiology, cytogenetics, and plant breeding, one of the major research lines in the early years of the Estado Novo. Under the direction of Afonso Luisier (1931–57), *Brotéria* became a leading biology journal. It featured works by celebrated botanists and zoologists, and its publication was subsidized through national funds. After the nomination of Archer as editor-in-chief in 1962, *Brotéria* became a leading journal in biochemistry, molecular genetics, genetic engineering, human genetics, and bioethics. In this period, the Jesuits' efforts to promote the teaching and practice of science were widely recognized, and Luís Archer was bestowed the highest scientific and state honors.

During the Thirty-Second General Congregation (1974–75), Superior General Pedro Arrupe (1907–91, in office 1965–83) urged the Jesuits to aim their apostolates at the service of faith and the promotion of justice.[2] Portuguese Jesuits responded to Arrupe's calling to be at the service of faith by creating centers of Ignatian spirituality for university students in Coimbra (1976), Braga (1979), Lisbon (1984), and Porto (1989). They responded to Arrupe's calling to be promoters of justice by establishing noteworthy non-profit organizations that worked either in developing countries or with minorities in Portugal, such as the Development NGO Leigos para o Desenvolvimento, in 1986, and the Fundação Gonçalo da Silveira, in 2004. They also supported the creation of a national food bank in 1991.

2 Arrupe, "Our Mission Today: The Service of Faith and the Promotion of Justice," 298–316.

By the turn of the twentieth century, it had become clear that science was no longer "the first and most important ministry of the Society," as Provincial Lúcio Craveiro declared in 1961.[3] In 2002, seventy-six-year-old Luís Archer published the final issue of *Brotéria: Genética*—a comprehensive table of contents listing all research articles, essays, and reviews published in the series *Botânica*, *Zoologia*, *Vulgarização científica*, *Ciências naturais*, and *Genética*. One century after its modest creation as a taxonomy journal, *Brotéria* had served its founding purpose: to "disseminate the taste for the natural sciences in our country," while unveiling, at the same time, the "greatness of God" stamped in the "vast book" of nature.[4]

3 APSI, Lúcio Craveiro da Silva to Luís Archer (April 22, 1961).
4 "Duas palavras de introdução," v.

Appendix

TABLE 1 Jesuits in Portugal, 1860–1910[a]

	Priests	Scholastics	Coadjutors	Total
1859–60	4	4	2	10
1860–61	6	8	7	21
1861–1862	6	10	6	22
1862–63	5	14	8	27
1863–64	7	13	8	28
1864–65	10	16	12	38
1865–66	10	24	12	46
1866–67	10	28	14	52
1867–68	20	32	17	69
1868–69	10	35	21	66
1869–70	10	35	18	63
1870–71	14	38	17	69
1871–72	15	36	19	70
1872–73	16	38	20	74
1873–74	19	42	20	81
1874–75	18	46	22	86
1875–76	23	45	26	94
1876–77	26	46	26	98
1877–78	28	49	28	105
1878–79	32	49	29	110
1879–80	34	50	32	116
1880–81	36	58	34	128
1881–82	40	60	36	136
1882–83	40	68	39	147
1883–84	42	68	41	151
1884–85	44	69	42	155
1885–86	47	69	45	161
1886–87	52	69	48	169
1887–88	54	72	49	175
1888–89	60	67	52	179
1889–90	70	68	54	192
1890–91	77	72	56	205

 | DOI:10.1163/9789004382367_012

TABLE 1 Jesuits in Portugal, 1860–1910a (*cont.*)

	Priests	Scholastics	Coadjutors	Total
1891–92	81	74	58	213
1892–93	92	86	62	240
1893–94	95	83	63	241
1894–95	98	96	67	261
1895–96	99	105	70	274
1896–97	104	98	70	272
1897–98	104	99	74	277
1898–99	110	97	82	289
1899–1900	111	105	85	301
1900–1	114	108	94	316
1901–2	119	107	91	317
1902–3	123	107	94	324
1903–4	132	99	96	327
1904–5	135	97	100	332
1905–6	140	98	102	340
1906–7	138	100	106	344
1907–8	139	94	106	339
1908–9	150	93	113	356
1909–10	147	101	112	360

a Sources: *Annual Catalogues of the Portuguese Province* (*1858–1910*).

TABLE 2 Jesuits, servants, and students at the Colégio de Campolide, 1858–1910[a]

	Priests	Scholastics	Coadjutors	Servants	Students
1857–58	2	–	1	–	–
1858–59	2	2	1	–	–
1859–60	2	3	–	–	–
1860–61	5	4	–	–	–
1861–62	5	1	4	–	–
1862–63	6	3	4	–	–
1863–64	7	2	5	–	–
1864–65	8	3	5	–	–
1865–66	5	4	4	7	43
1866–67	7	5	7	6	48

	Priests	Scholastics	Coadjutors	Servants	Students
1867–68	6	6	8	6	58
1868–69	8	7	8	7	72
1869–70	9	6	6	9	62
1870–71	7	7	8	10	64
1871–72	7	7	8	12	80
1872–73	7	7	8	14	85
1873–74	8	8	10	14	85
1874–75	8	7	10	17	98
1875–76	9	7	10	18	115
1876–77	7	7	9	–	*c.*100
1877–78	7	8	10	–	–
1878–79	8	8	8	–	130
1879–80	6	10	8	–	160
1880–81	7	11	10	–	156
1881–82	7	12	9	–	153
1882–83	8	9	9	–	130
1883–84	8	9	9	–	130
1884–85	8	9	10	–	130
1885–86	8	10	9	–	140
1886–87	10	7	11	30	150
1887–88	11	6	12	22	137
1888–89	14	4	11	20	120
1889–90	10	6	11	–	–
1890–91	12	7	12	24	180
1891–92	13	6	12	–	220
1892–93	12	8	12	30	240
1893–94	12	9	13	45	256
1894–95	11	11	11	45	278
1895–96	11	11	13	45	300
1896–97	11	13	13	–	320
1897–98	11	13	13	–	333
1898–99	11	14	14		–
1899–1900	12	10	16	–	–
1900–1	15	12	16	–	–
1901–2	19	9	16	–	250
1902–3	21	7	16	–	–
1903–4	21	8	15	–	–

TABLE 2 Jesuits, servants, and students at the Colégio de Campolide, 1858–1910 (*cont.*)

	Priests	Scholastics	Coadjutors	Servants	Students
1904–5	16	11	15	–	–
1905–6	16	10	16	–	–
1906–7	14	11	16	–	290
1907–8	14	11	16	–	–
1908–9	17	10	19	–	250
1909–10	–	–	–	–	280

a Sources: *Annual Catalogues of the Portuguese Province (1858–1910)*. Manuel Borges Grainha, *História do colégio de Campolide* (Coimbra: Imprensa da Universidade, 1913).

TABLE 3 Jesuits, servants, and students at the Colégio de São Fiel, 1863–1910[a]

	Priests	Scholastics	Coadjutors	Servants	Students
1863–64	2	–	1	–	80
1864–65	2	–	1	–	–
1865–66	3	–	2	–	–
1866–67	3	2	2	–	–
1867–68	3	1	2	–	–
1868–69	3	2	2	–	–
1869–70	4	1	3	–	–
1870–71	5	1	3	–	–
1871–72	4	1	3	–	–
1872–73	4	2	4	–	–
1873–74	5	3	5	–	–
1874–75	6	4	6	–	–
1875–76	7	4	7	–	–
1876–77	10	9	8	–	–
1877–78	8	8	9	–	–
1878–79	7	9	9	–	–
1879–80	7	9	9	–	138
1880–81	8	9	10	–	–
1881–82	10	9	10	–	–
1882–83	9	10	11	–	–
1883–84	11	10	11	–	–
1884–85	13	7	13	–	–

	Priests	Scholastics	Coadjutors	Servants	Students
1885–86	12	9	14	–	–
1886–87	15	8	13	–	–
1887–88	14	9	14	–	–
1888–89	14	8	16	–	–
1889–90	12	10	16	–	–
1890–91	13	10	15	–	–
1891–92	11	8	16	–	–
1892–93	12	8	14	–	–
1893–94	20	29	12	–	–
1894–95	14	24	12	–	–
1895–96	16	20	12	–	–
1896–97	16	21	11	–	–
1897–98	19	25	12	–	–
1898–99	13	11	12	–	–
1899–1900	15	11	13	–	–
1900–1	15	12	12	–	–
1901–2	20	14	13	–	–
1902–3	21	12	14	–	–
1903–4	22	11	15	–	–
1904–5	21	11	14	–	–
1905–6	22	9	14	–	–
1906–7	23	8	15	–	–
1907–8	22	8	18	–	–
1908–9	22	8	19	–	–
1909–10	22	8	20	–	367

a Sources: *Annual Catalogues of the Portuguese Province (1858–1910)*; Joaquim Augusto de Sousa Refóios, *O Collegio de São Fiel no Louriçal do Campo e o de Nossa Senhora da Conceição na Covilhã: Apontamentos sobre o jesuitismo no districto de Castello-Branco* (Coimbra: Imprensa da Universidade, 1883); Manuel Borges Grainha, *História do colégio de Campolide* (Coimbra: Imprensa da Universidade, 1913); Francisco Rodrigues, *A formação intelectual do jesuíta: Leis e factos* (Porto: Livraria Magalhães e Moniz, 1917).

TABLE 4 Professors of mathematics and natural sciences at the Colégio de Campolide, 1865–1910[a]

	Mathematics	Physics, chemistry, natural history
1865–66	António Tomás	–
1866–67	Joannes B. Meli	–
1867–68	António Cordeiro	–
1868–69	António Cordeiro	–
1869–70	António Cordeiro, Francisco Borges	Francisco Borges
1870–71	António Cordeiro, Francisco Borges	Francisco Borges
1871–72	António Cordeiro, Francisco Borges	Francisco Borges
1872–73	João Arraiano, Francisco Borges	Francisco Borges
1873–74	João Arraiano, Francisco Borges	Francisco Borges
1874–75	João Arraiano, Francisco Borges	Francisco Borges
1875–76	Abel Carvalhão	João Arraiano
1876–77	João Arraiano	João Arraiano
1877–78	João Arraiano	João Arraiano
1878–79	José de Magalhães	Manuel de Moraes
1879–80	José de Magalhães	Manuel de Moraes
1880–81	Francisco Borges, Manuel de Moraes	José de Magalhães
1881–82	Manuel de Moraes, José de Magalhães	José de Magalhães (P., C.), Francisco Borges (NH)
1882–83	Francisco Borges, Manuel de Moraes, Francisco Xavier da Cunha	Francisco Borges
1883–84	Francisco Borges, Francisco Xavier da Cunha, João Justino	Francisco Borges
1884–85	Francisco Borges, Francisco Xavier da Cunha, João Justino	Francisco Borges
1885–86	Francisco Borges, Francisco Xavier da Cunha	Francisco Borges
1886–87	Francisco Borges, Luís Mendes, António Azevedo	Francisco Borges
1887–88	Francisco Borges, Luís Mendes. António Azevedo, Manuel Graça	Francisco Borges
1888–89	Manuel Graça	–
1889–90	Manuel Graça	Joaquim da Silva Tavares
1890–91	António Oliveira Pinto	António Vaz
1891–92	António Azevedo	António Vaz

	Mathematics	Physics, chemistry, natural history
1892–93	João Justino, João Baptista Gonçalves	António Vaz
1893–94	João Justino	António Vaz
1894–95	Luís Alves, Salústio dos Santos, Fernando de Macedo	Luís Lopes
1895–96	Salústio dos Santos, António Gonçalves, António Nunes, Mâncio Moraes	João Nazaré, Joaquim Cunha (botany and zoology)[b]
1896–97	Luís Alves, João Rodrigues, José Ribeiro, Salústio dos Santos, Serafim Nazaré	João Nazaré, Joaquim Cunha (botany and zoology)
1897–98	Joaquim Cunha, João Madureira, Salústio dos Santos, Francisco Zamith	João Nazaré, Joaquim Cunha (botany and zoology)
1898–99	Luís Maria de Almeida, Manuel Fernandes, António Gonçalves, Joaquim Cunha, João Frias, José Ribeiro Laia	João Nazaré, José Ribeiro Laia (botany)
1899–1900	Manuel Fernandes, António Gonçalves, Joaquim Cunha, José Beirão, Manuel Elvas	João Nazaré. José Beirão (botany and zoology), José Ribeiro Laia (botany and zoology)
1900–1	Luís Maria de Almeida, Luís Gonzaga de Azevedo, Manuel Fernandes, Manuel da Fonseca, Manuel Guimarães Pestana, José Beirão, José Ribeiro Laia	João Nazaré, Manuel Guimarães Pestana, José Aparício, Manuel da Fonseca, José Beirão, José Ribeiro Laia[c]
1901–2	Luís Alves, Luís Gonzaga de Azevedo, António Oliveira Pinto, Avelino de Miranda, Domingos da Silva, Manuel Pestana, Rudolfo Chorão	João Nazaré, António Oliveira Pinto, Manuel Pestana
1902–3	Luís Maria de Almeida, Luís Alves, Luís Baecher, Avelino Miranda, António Oliveira Pinto, Rudolfo Chorão	João Nazaré, António Oliveira Pinto, Rudolfo Chorão

TABLE 4 Professors of mathematics and natural sciences at the Colégio de Campolide, 1865–1910 (*cont.*)

	Mathematics	Physics, chemistry, natural history
1903–4	Luís Maria de Almeida, Luís Alves, Avelino Miranda, Salústio dos Santos, Rudolfo Chorão	António Oliveira Pinto
1904–5	Luís Maria de Almeida, Luís Alves, António Oliveira Pinto	António Oliveira Pinto, Manuel Elvas, Rudolfo Chorão
1905–6	Luís Maria de Almeida, Luís Alves, António Antunes, António Oliveira Pinto, Artur Redondo, António Monteiro	António Oliveira Pinto, João Nazaré, Manuel Elvas, Frederico de Menezes
1906–7	Luís Maria de Almeida, Luís Alves, António Oliveira Pinto, António Monteiro, Manuel Elvas	Afonso Luisier, António Oliveira Pinto, António da Cruz. Manuel Elvas, Frederico de Menezes
1907–8	Luís Alves, António Oliveira Pinto, Manuel Elvas, Manuel dos Santos, António Neto	António Oliveira Pinto, Manuel Elvas, Frederico de Menezes, Leo Meyer
1908–9	António Oliveira Pinto, António Monteiro, Frederico de Menezes	Afonso Luisier, António Oliveira Pinto, Frederico de Menezes, Alfredo Costa, José de Magalhães
1909–10	Salústio dos Santos, Manuel dos Santos, Francisco de Miranda	Afonso Luisier, António Oliveira Pinto, João Frias, João Nazaré, Alfredo Costa, José de Magalhães

a Sources: *Annual Catalogues of the Portuguese Province (1858–1910).*

b In this year, for the first time, the catalogs mention a professor of "elements of botany and zoology." This course was designed as a propedeutics to natural history, which was only taught during the high-school years.

c In this year, Manuel Guimarães Pestana taught elements of botany and zoology, natural history, physics, and chemistry; João Nazaré taught natural history, physics, and chemistry, while José Aparício, Manuel da Fonseca, José Beirão, and José Ribeiro Laia taught elements of botany and zoology.

TABLE 5 Professors of mathematics and natural sciences at the Colégio de São Fiel, 1872–1910[a]

	Mathematics	Physics, chemistry, natural history
1872–73	João dos Remédios	–
1873–74	João dos Remédios	–
1874–75	Jacob Pereira, Luís Campo Santo	–
1875–76	Francisco Antunes, Jacob Pereira, Luís Campo Santo	–
1876–77	Francisco Lucas, Jacob Pereira, José Maria da Silva	José Borges Grainha
1877–78	José Borges Grainha, José Maria da Silva	José Borges Grainha
1878–79	José Borges Grainha, José Maria da Silva	José Borges Grainha
1879–80	José Borges Grainha	José Borges Grainha
1880–81	José Borges Grainha, José Pereira Dias, José Maria da Silva	José Borges Grainha
1881–82	José Borges Grainha, José Pereira Dias, José Maria da Silva, César Augusto da Silva	José Borges Grainha
1882–83	João Arraiano, Alexandre Castelo	João Arraiano
1883–84	João Arraiano, Alexandre Castelo	João Arraiano
1884–85	João Arraiano, Alexandre Castelo	João Arraiano
1885–86	João Arraiano, Alexandre Castelo	João Arraiano
1886–87	João Arraiano, Alexandre Castelo, Alexandre Moreira	João Arraiano
1887–88	Alexandre Moreira, João Nazaré	João Nazaré
1888–89	Manuel Moraes	João Nazaré
1889–90	Manuel Fernandes	João Nazaré
1890–91	Manuel Fernandes	Joaquim da Silva Tavares
1891–92	Manuel Fernandes	Joaquim da Silva Tavares
1892–93	António de Matos	Joaquim da Silva Tavares

TABLE 5 Professors of mathematics and natural sciences at the Colégio de São Fiel, 1872–1910 (*cont.*)

	Mathematics	Physics, chemistry, natural history
1893–94	António de Matos	Joaquim da Silva Tavares, António Oliveira Pinto
1894–95	António de Matos, José Pereira Dias	António Oliveira Pinto
1895–96	António de Matos, José Pereira Dias	António de Matos
1896–97	Manuel de Almeida, José Pereira Dias	António de Matos, Cândido Azevedo Mendes
1897–98	Manuel Fernandes, José Pereira Dias, Filipe Pinheiro	António de Matos, Cândido Azevedo Mendes
1898–99	Luís Alves, João Baptista Barreto, Cândido Azevedo Mendes, Carlos Zimmermann, José Bernardo, José Pereira Dias	João Baptista Barreto, Cândido Azevedo Mendes, Carlos Zimmermann, Manuel Pacheco, José Bernardo, José Pereira Dias
1899–1900	Luís Alves, Manuel Martins, Cândido Azevedo Mendes, Carlos Zimmermann, Manuel Rebimbas, José Bernardo	Luís Lopes, Cândido Azevedo Mendes, Carlos Zimmermann
1900–1	Luís Lopes, Cândido Azevedo Mendes, Carlos Zimmermann, Manuel Pacheco, José Bernardo	Luís Lopes, Manuel Martins, Joaquim da Silva Tavares, Cândido Azevedo Mendes, Carlos Zimmermann, Manuel Pacheco, Miguel Barcelos
1901–2	Luís Lopes, Luís Maria de Almeida, Carlos Zimmermann, Jorge de Brito e Cunha, José Bernardo, Valério Cordeiro	Joaquim da Silva Tavares, Cândido Azevedo Mendes, Jorge de Brito e Cunha, Valério Cordeiro

	Mathematics	Physics, chemistry, natural history
1902–3	Manuel Martins, Manuel Rebimbas, Jorge de Brito e Cunha, José Bernardo, José Simas	Luís Lopes, Manuel Rebimbas, Joaquim da Silva Tavares
1903–4	Luís Lopes, Manuel Fernandes, Manuel Martins, Jorge de Brito e Cunha, José Bernardo, José Simas	Manuel Rebimbas, Joaquim da Silva Tavares
1904–5	Carlos Moreira, Manuel Martins, José Simas	Manuel Rebimbas, Joaquim da Silva Tavares, Valério Cordeiro
1905–6	Joaquim Lima, José Pereira Dias, Anacleto Pereira Dias, João Pacheco, José Simas	Manuel Martins, Joaquim da Silva Tavares, Anacleto Pereira Dias, João Pacheco, Valério Cordeiro
1906–7	Joaquim Lima, José Pereira Dias, José Simas	Joaquim da Cunha, Joaquim da Silva Tavares, João Pacheco, Valério Cordeiro
1907–8	António Gonçalves, Cândido Azevedo Mendes, Carlos Moreira, Joaquim Lima, José Pereira Dias, Jorge de Brito e Cunha, João Pacheco, José Simas, Albino Teixeira	Cândido Azevedo Mendes, João Nazaré, Jorge de Brito e Cunha, Alfredo Costa, João Azevedo
1908–9	António Gonçalves, Cosmas Rodrigues, Joaquim Lima, José Pereira Dias, Manuel Sampaio, Albino Teixeira, Jorge de Brito e Cunha	Manuel Rebimbas, João Nazaré, Augusto de Paiva, Manuel Sampaio
1909–10	Luís Alves, António Gonçalves, José Pereira Dias, Manuel Sampaio, Albino Teixeira	Carlos Zimmermann, Rudolfo Chorão, Augusto de Paiva, João Ferreira

a Sources: *Annual Catalogues of the Portuguese Province (1858–1910)*. José Beirão taught Latin, Portuguese, zoology, and mathematics; José Ribeiro Laia taught Latin, Portuguese, history, geography, botany, zoology, and mathematics; João Nazará taught physics, chemistry, and natural history.

TABLE 6 Professors of mathematics and natural sciences at the Colégio de São Francisco, 1877–1908[a]

	Mathematics	Physics, Chemistry, Natural History
1877–78	Benedito Rodrigues	–
1881–82	Avelino Miranda	–
1884–85	Luís Maria de Almeida	–
1886–87	–	João Nazaré
1887–88	José da Lapa	–
1888–89	José da Lapa	António Vaz
1889–90	José Lapa	–
1890–91	–	António Azevedo
1891–92	–	António Oliveira Pinto
1892–93	António Oliveira Pinto	–
1898–99	José Lapa	–
1899–1900	–	Joaquim da Silva Tavares[b]
1900–1	–	António Oliveira Pinto[c]
1901–2	Thomas Dignam	Thomas Dignam
1902–3	António Cordeiro, José Beirão	Anacleto Pereira Dias
1903–4	–	Anacleto Pereira Dias
1904–5	Anacleto Pereira Dias	–
1905–6	Manuel Pestana	Joaquim da Cunha
1906–7	–	Anacleto Pereira Dias
1907–8	Manuel Pestana	Manuel Pestana

a Sources: *Annual Catalogues of the Portuguese Province (1858–1910)*.
b Joaquim da Silva Tavares was professor of physics, chemistry, and mechanics.
c In this year, António Oliveira Pinto was also professor of physics, chemistry, and mechanics.

TABLE 7 Articles published in *Brotéria*, 1902–2002[a]

	Articles	%
Zoologia	189	10.8
Botânica	203	11.6
Vulgarização científica	429	24.6
Ciências naturais	595	34.1
Genética	330	18.9
Total	1746	

a Sources: *Índices gerais da* Brotéria científica (Lisbon: Brotéria, 2002).

TABLE 8 Articles published in *Brotéria: Vulgarizaçao científica*, 1907–24[a]

	1907–17	1918–24	Total	%
Agriculture	39	33	72	16.8
Physics	25	34	59	13.8
Geography	32	29	61	14.2
Commerce and Industry	25	27	52	12.1
Hygiene	17	8	25	5.8
Chemistry	16	7	23	5.4
Useful and Harmful Animals	23	–	23	5.4
Medicine	13	9	22	5.1
Arboriculture	14	6	20	4.7
Physiology	8	2	10	2.3
Culinary	8	–	8	1.9
Archeology	–	7	7	1.6
Bacteriology	7	–	7	1.6
History of Science	7	–	7	1.6
Natural Sciences and Religion	3	2	5	1.2
Plant Pathology	5	–	5	1.2
Seismology	5	–	5	1.2
Others	–	5	5	1.2
Sociology	–	4	4	0.9
History	–	4	4	0.9
Ethnography	2	–	2	0.5
Microscopy	2	–	2	0.5
Pedagogy		1	1	0.2
Total	251	178	429	

a Sources: *Índices gerais da* Brotéria científica (Lisbon: Brotéria, 2002).

TABLE 9 Articles published in *Brotéria: Ciências naturais*, 1932–59[a]

	Articles	%
Botany	195	53.4
Zoology	165	45.2
Geology	5	1.4
Total	365	

a Sources: *Índices gerais da* Brotéria científica (Lisbon: Brotéria, 2002).

TABLE 10 Articles published in *Brotéria: Ciências naturais*, 1960–79[a]

		Articles	%
Zoology		124	53.9
	Biographies	2	0.9
	Biochemistry	47	20.4
	Evolution	3	1.3
	Physiology	30	13.0
	Genetics	4	1.7
	Morphology, anatomy, histology	16	7.0
	Taxonomy	16	7.0
	Varia	6	2.6
Botany		92	40.0
	Biographies	5	2.2
	Physiology and biochemistry	17	7.4
	Genetics	19	8.3
	Morphology, cytology	21	9.1
	Pathology	5	2.2
	Taxonomy	22	9.6
	Varia	3	1.3
Bacteriology		9	3.9
Molecular biology		5	2.2
Total		230	

a Sources: *Índices gerais da* Brotéria científica (Lisbon: Brotéria, 2002).

TABLE 11 Articles published in *Brotéria–Genética*, 1980–2002[a]

		Articles	%
Research		105	31.8
	Bacterial genetics	10	3.0
	Animal breeding	19	5.8
	Plant breeding	52	15.8
	Human genetics	24	7.3
Notes and reviews		102	30.9
	Bacterial genetics	15	4.5
	Genetics	34	10.3
	Animal breeding	8	2.4
	Plant breeding	15	4.5
	Human genetics	30	9.1
Ethics		123	37.3
	Bioethics	27	8.2
	Biographies	37	11.2
	News and notes	58	17.6
	Total	330	

a Sources: *Índices gerais da* Brotéria científica (Lisbon: Brotéria, 2002).

Bibliography

"À los lectores." *Razón y fé* 1 (1901): 1–5.

Abreu, Luís Machado de. *Ensaios anticlericais*. Lisbon: Roma Editora, 2004.

Abreu, Luís Machado de, and José Eduardo Franco, eds. *Ordens e congregações religiosas no contexto da I República*. Lisbon: Gradiva, 2010.

Acciaiuoli, Margarida. *Exposições do Estado Novo, 1934–1940*. Lisbon: Livros Horizonte, 1998.

Agra, Leandro, Andrea Bezerra, and David Barbosa. "URM Mycological Herbarium: Revision of the Myxomycetes Collection." *Brazilian Journal of Botany* 37 (2014): 299–313.

Albuquerque, Luís de. "A 'Aula da Esfera' do Colégio de Santo Antão no século XVII." *Anais da Academia portuguesa de história* 21 (1972): 337–91.

Alden, Dauril. "The Gang of Four and the Campaign against the Jesuits in Eighteenth-Century Brazil." In *The Jesuits II: Cultures, Sciences and the Arts, 1540–1773*, edited by John W. O'Malley, Gauvin Alexander Bailey, Steven J. Harris, and T. Frank Kennedy, 707–24. Toronto: Toronto University Press, 2006.

Alden, Dauril. *The Making of an Enterprise: The Society of Jesus in Portugal, Its Empire and Beyond, 1540–1750*. Stanford: Stanford University Press, 1996.

Almaça, Carlos. "As primeiras fases da obra científica do Prof. J.A. Serra." *Brotéria: Genética* 13 (1992): 33–40.

Almeida, André Ferrand de "A formação do espaço brasileiro e o projecto do novo Atlas da América portuguesa (1713–1748)." Lisbon: Comissão Nacional para as Comemorações dos Descobrimentos Portugueses, 2001.

Almeida, Fernando de. "O pão sob o ponto de vista hygienico." *Brotéria: Vulgarização científica* 7 (1908): 9–39.

Almeida, Fortunato de. *História da igreja em Portugal*. 8 Vols. Coimbra: Imprensa Académica, 1910–22.

Almeida, José Veríssimo de. "Bibliografia: Primeira contribuição para o estudo dos fungos da região setubalense." *Revista agronómica* 1 (1903): 368–69.

Alorna, Marquês de. *As prisões da Junqueira durante o ministério do marquês de Pombal*. 2nd ed. Lisbon: Typographia Universal, 1882 [1857].

Alves, Jorge Fernandes, and Rita C. Alves. *A.J. Ferreira da Silva: Nos caminhos da química*. Porto: Universidade do Porto, 2013.

Amaral, Isabel. *A emergência da bioquímica em Portugal: As escolas de investigação de Marck Athias e de Kurt Jacobsohn*. Lisbon: Fundação Calouste Gulbenkian–Fundação para a Ciência e Tecnologia, 2006.

Andrade, António Alberto Banha de. *A reforma pombalina dos estudos secundários (1759–1771): Contribuição para a história da pedagogia em Portugal*. Coimbra: Universidade de Coimbra, 1981.

Antunes-Correia, Jorge. "A contribuição fundamental do Professor José Antunes Serra na genética do melhoramento dos ovinos." *Brotéria: Genética* 13 (1992): 15–18.

Antunes-Correia, Jorge. "Frequências génicas da pelagem em gatos portugueses." *Brotéria: Genética* 4 (1983): 157–65.

Antunes-Correia, Jorge. "Genética da sensibilidade ao halotano: Uma situação com grande interesse actual no melhoramento porcino." *Brotéria: Genética* 1 (1980): 131–33.

Antunes-Correia, Jorge, and H. Krug. "Frequência genética das hemoglobinas em caprinos portugueses." *Brotéria: Genética* 1 (1980): 181–84.

Antunes-Correia, Jorge, and Tristão Mello Sampaio. "Somatic Association of Syrian Hamster Chromosomes." *Brotéria: Genética* 1 (1980): 43–51.

Antunes, João Lobo. "Ciência e fé." *Brotéria* 152 (2001): 331–48.

Antunes, Manuel, ed. *Como interpretar Pombal? No bicentenário da sua morte*. Lisbon: Brotéria, 1982.

Antunes, Manuel. "O marquês de Pombal e os jesuítas." In *Como interpretar Pombal? No bicentenário da sua morte*, edited by Manuel Antunes, 25–142. Lisbon: Brotéria, 1982.

Araújo, Ana Cristina, ed. *O marquês de Pombal e a universidade*. Coimbra: Universidade de Coimbra, 2000.

Araújo, António. *Jesuítas e antijesuítas no Portugal republicano*. Lisbon: Roma Editora, 2004.

Archer, Luís, ed. "6th European Meeting on Bacterial Transformation and Transfection: Scientific Program and Abstracts." *Brotéria: Genética* 3 (1982): *3–106.

Archer, Luís. "Alarme sobre reprodução artificial." *Brotéria* 124 (1987): 91–99.

Archer, Luís, ed. *Bacterial Transformation*. London: Academic Press, 1973.

Archer, Luís. "O bebé-proveta que me falou em sonhos." *Brotéria* 122 (1986): 569–72.

Archer, Luís. "*Brotéria*: Das ciências naturais à genética." *Brotéria: Genética* 1 (1980): 5–6.

Archer, Luís. "Centenário de Aurélio Quintanilha (1892–1992)." *Brotéria: Genética* 14 (1993): 5–6.

Archer, Luís. "Clonagem não reprodutiva: Usos terapêuticos e objecções éticas." *Cadernos de bioética* 22 (2000): 65–81.

Archer, Luís. "Clonagem: Mito e ciência." *Brotéria* 149 (1999): 75–90.

Archer, Luís. "Clonagem: Verdade científica e sonho mítico." *Brotéria: Genética* 20 (1999): 101–24.

Archer, Luís. *Da genética à bioética*. Coimbra: Gráfica de Coimbra 2, 2006.

Archer, Luís. *Desafios da nova genética*. Lisbon: Editorial Brotéria, 1992.

Archer, Luís. "DNA-Mediated Transformation." *Brotéria: Ciências naturais* 36 (1967): 107–76.

Archer, Luís. "Engenharia genética: Esperanças e medos." *Cadernos de bioética* 15 (1997): 49–54.

Archer, Luís. "Ethical Aspects of Gene Therapy." *Brotéria: Genética* 19 (1998): 13–26.

Archer, Luís. "Genetic Engineering and Human Freedom." *Brotéria: Genética* 10 (1989): 49–62.

Archer, Luís. *Genética molecular*. Lisbon: Editorial Brotéria, 1976.

Archer, Luís. "Gerar ou produzir vida humana?" *Brotéria* 124 (1987): 519–32.

Archer, Luís. "Homenagem ao Prof. Doutor José Antunes Serra." *Brotéria: Genética* 13 (1992): 5.

Archer, Luís. "Mestre Quintanilha faz-se aluno." *Brotéria: Genética* 14 (1993): 29–32.

Archer, Luís. "Para uma ética da reprodução humana." *Brotéria* 145 (1997): 218–225.

Archer, Luís. "Procriação medicamente assistida: Evolução do pensamento ético de 1986 a 1999." *Brotéria* 150 (2000): 337–67.

Archer, Luís. "Prof. Doutor Abílio Fernandes." *Brotéria: Genética* 16 (1995): 5–6.

Archer, Luís. "Quintanilha: O velho mestre feito jovem aprendiz." *Brotéria: Genética* 9 (1988): 23.

Archer, Luís. "Regresso a Portugal do European Meeting on Bacterial Transformation." *Brotéria: Genética* 1 (1980): 191.

Archer, Luís. "Reprodução humana artificial." *Brotéria: Genética* 6 (1985): 93–96.

Archer, Luís. "Reprodução humana artificial: Um grito de alarme." *Brotéria: Genética* 7 (1986): 109–14.

Archer, Luís. "R.P. José Carvalhaes S.J." *Brotéria: Ciências naturais* 31 (1962): 54.

Archer, Luís. *Temas biológicos e problemas humanos*. Lisbon: Editorial Brotéria, 1981.

Archer, Luís. "Terapia génica: Dez anos depois." *Brotéria: Genética* 21 (2000): 53–66.

Archer, Luís. "Transformação genética em células sincronizadas de *Bacillus subtilis*." PhD diss., Universidade do Porto, 1969.

Archer, Luís. "Transport of Transforming DNA into the Cell Interior in Thymine-Starved *Bacillus subtilis* with Chromosomes Arrested at the Terminus." PhD diss., Georgetown University, 1967.

Archer, Luís. "Treze anos de reflexão sobre procriação medicamente assistida." *Brotéria: Genética* 20 (1999): 125–55.

Archer, Luís. "Vinte anos de genética molecular em Portugal." *Brotéria: Genética* 10 (1989): 19–48.

Archer, Luís. "Vinte e cinco anos de genética molecular." *Memórias da Academia das Ciências de Lisboa: Memórias da classe de ciências* 30 (1989): 129–76.

Archer, Luís. "Volta a Portugal um congresso europeu de origem portuguesa." *Brotéria: Genética* 3 (1982): 81.

Archer, Luís, and Irving Gray. "Alterations in the Activity of Some Oxidative Enzymes by the Action of Anthrax Toxin." *Brotéria: Ciências naturais* 35 (1966): 169–77.

Archer, Luís, Rosa M. Fernandes, and Hermínia Lencastre. "Two Newly Isolated Temperate Phages of *Bacillus subtilis*." *Brotéria: Genética* 4 (1983): 27–33.

Archer, Luís, Ilda M. Santos, Rosa M. Fernandes, and Hermínia Lencastre. "Location of Prophage IG3 in *Bacillus subtilis* Chromosome." *Brotéria: Genética* 6 (1985): 57–56.

Archer, Luís, Graça A. Vieira, and Hermínia Lencastre. "Relatedness of Bacteriophages PBS1, AR9, 3NT and I10 of *Bacillus subtilis*." *Brotéria: Genética* 6 (1985): 179–89.

Archer, Luís, Jorge Biscaia, and Walter Osswald, eds. *Bioética*. Lisbon: Verbo, 1996.

Archer, Luís, and Helena Paveia. "Genes for L-arabinose Utilization in *Bacillus subtilis*." *Brotéria: Genética* 13 (1992): 149–59.

Archer, Luís, and Helena Paveia. "Location of Genes for Arabinose Utilization in *Bacillus subtilis* Chromosome." *Brotéria: Genética* 1 (1980): 169–76.

Archer, Luís, Jorge Biscaia, Walter Osswald, and Michel Renaud, eds. *Novos desafios à bioética*. Porto: Porto Editora, 2001.

Arenas, Mar García. "La Compañía de Jesús en la *Deducción cronológica y analítica* pombalina," *Revista de historia moderna-anales de la Universidad de Alicante* 21 (2003): 315–48.

Arrupe, Pedro. "Our Mission Today: The Service of Faith and the Promotion of Justice." In *Jesuit Life and Mission Today: The Decrees and Accompanying Documents of the 31st–35th General Congregations of the Society of Jesus*, edited by John W. Padberg, 298–316. St. Louis, MO: Institute of Jesuit Sources, 2009.

Artigas, Mariano, Thomas F. Glick, and Rafael A. Martínez. *Negotiating Darwin: The Vatican Confronts Evolution, 1877–1902*. Baltimore: Johns Hopkins University Press, 2006.

Ashgar, Syed Shafi, D. K. Khawaja, A. K. Jafri. "5'-Nucleotidase Activity in the Tissues of the Cat-fish, *Heteropnuestes fossilis* Bloch." *Brotéria: Ciências naturais* 41 (1972): 3–7.

Assumpção, Tomás Lino de, ed. *História geral da Companhia de Jesus, desde a fundação até aos nossos dias*. Lisbon: Empreza da História de Portugal, 1901; reprinted as: *História geral dos jesuítas*. Rio de Janeiro: Moraes Editores, 1982.

Asúa, Miguel de. *Science in the Vanished Arcadia: Knowledge of Nature in the Jesuit Missions of Paraguay and Río de la Plata*. Leiden: Brill, 2014.

Azevedo, Carlos. "Nucléolo: Estrutura, citoquímica e aspectos funcionais." *Brotéria: Ciências naturais* 48 (1978): 3–52.

Azevedo, Luís Gonzaga de. *Proscritos: Notícias circunstanciadas do que passaram os religiosos da Companhia de Jesus na revolução de Portugal de 1910*. 2 Vols. Valladolid: Florencio de Lara, 1911.

Bagchi, S., and R. M. Datta. "On the Nature of Chromosome in Meiotic Mechanism of a Natural Tetraploid *Sesbania benthamiana* Domin." *Brotéria: Ciências naturais* 42 (1973): 31–37.

Bagulho, Francisco. "Avanço no tratamento de triticale." *Brotéria: Genética* 5 (1984): 34–47.

Baldini, Ugo. "The Academy of Mathematics of the Collegio Romano from 1553 to 1612." In *Jesuit Science and the Republic of Letters*, edited by Mordechai Feingold, 47–98. Cambridge, MA: MIT Press, 2003.

Baldini, Ugo. "As assistências ibéricas da Companhia de Jesus e a actividade científica nas missões asiáticas (1578–1640): Alguns aspectos culturais e institucionais." *Revista portuguesa de filosofia* 54 (1998): 195–245.

Baldini, Ugo. "Giovanni Paolo Lembo's Lessons in Lisbon: A Partial Content Analysis." In *Proceedings of the International Conference History of Astronomy in Portugal: Institutions, Theories, Practices*, edited by Luís Saraiva, 123–81. Lisbon: Sociedade Portuguesa de Astronomia, 2014.

Baldini, Ugo. "L'insegnamento della matematica nel Collegio di S. Antão a Lisbona, 1590–1640." In *A Companhia de Jesus e a missionação no Oriente*, edited by Nuno da Silva Gonçalves, 275–310. Lisbon: Fundação Oriente-Brotéria, 2000.

Baldini, Ugo. *Legem impone subactis: Studi su filosofia e scienza dei gesuiti in Italia, 1540–1632*. Rome: Bulzoni, 1992.

Baldini, Ugo. "The Portuguese Assistancy of the Society of Jesus and Scientific Activities in Its Asian Missions until 1640." In *História das ciências, Portugal e o Oriente*, edited by Luís Saraiva, 49–104. Lisbon: Fundação Oriente, 2000.

Baldini, Ugo. "The Teaching of Mathematics in the Jesuit Colleges of Portugal, from 1640 to Pombal." In *The Practice of Mathematics in Portugal*, edited by Luís Saraiva and Henrique Leitão, 293–465. Coimbra: Acta Universitatis Conimbrigensis, 2004.

Banerjee, Asit Kumar, and Archana Sharma. "Chromosome Studies on Some Indian Members of *Compositae*: I; *Tribe Inuloideae*." *Brotéria: Ciências naturais* 43 (1974): 15–32.

Bangert, William V. *A History of the Society of Jesus*. St. Louis, MO: Institute of Jesuit Sources, 1986.

Barbosa, Florentino. "O café na Parahyba." *Brotéria: Vulgarização científica* 12 (1914): 347–50.

Barradas, Maria do Céu, Tristão Mello Sampayo, and Zaida Cunha. "Gene Activation for High Molecular Mass Glutenins in Tetraploid Wheat: The Case of Resende." *Brotéria: Genética* 12 (1991): 69–71.

Battló, Josep, and David Altadill. "The Ebre Observatory: Its Path to Ionospheric Research." *Advances in Space Research* 39 (2007): 941–46.

Baur, Erwin. "Aplicação da genética ao melhoramento das plantas cultivadas." *Brotéria: Botânica* 24 (1930): 49–57.

Baur, Erwin, Fritz Lenz, and Eugen Fischer. *Menschliche Erblichkeitslehre und Rassenhygiene*. Munich: J. F. Lehmann Verlag, 1921; reprinted as: *Human Heredity*. New York: Macmillan, 1931.

Becker, V. "Un duel darwiniste (Haeckel et His)." *Études* 21 (1877): 672–88.

Beneke, Everett Smith. "Honors, Degrees, Promotions, Invitational Lectures." *Newsletter of the Mycological Society of America* 8 (1957): 28–31.

Bensaude-Vincent, Bernadette. "A Genealogy of the Increasing Gap between Science and the Public." *Public Understanding of Science* 10 (2011): 99–113.

Bensaude-Vincent, Bernadette, and Anne Rasmussen, eds. *Science and Spectacle in the European Enlightenment*. Aldershot: Ashgate, 2008.

Bert, Paul. *La morale des jésuites*. Paris: G. Charpentier, 1883.

Bettencourt, António Viveiros. "As minhas memórias do Professor A. Quintanilha." *Brotéria: Genética* 14 (1993): 43–45.

Biker, Júlio Firmino Júdice, ed. *Supplemento á collecção dos tratados, convenções, contratos e actos publicos celebrados entre a corôa de Portugal e as mais potencias desde 1640*. Lisbon: Imprensa Nacional, 1879.

Boavida, Maria Guida. "Anomalias cromossómicas humanas." *Brotéria: Genética* 4 (1983): 79–80.

Boavida, Maria Guida, C. F. Pinto, R. A. Marques, and R. A. Constant. "Um estudo citogenético em recém-nascidos com síndroma malformativo." *Brotéria: Genética* 4 (1983): 127–45.

Bodmer, Walter. "Recombination and Integration in *Bacillus subtilis*: Involvement of DNA Synthesis." *Journal of Molecular Biology* 14 (1965): 534–57.

Bolufer, Pascual. "Science and Technology in the 20th Century as Seen through the Journal *Ibérica* (1914–2003)." *Contributions to Science* 7 (2011): 185–90.

Bombarda, Miguel. *A consciência e o livre-arbítrio*. Lisbon: Parceria António Maria–Livraria Editora, 1898.

Bombarda, Miguel. *A sciencia e o jesuitismo: Réplica a um padre sábio*. Lisbon: António Maria Pereira, Livraria Editora, 1900.

Bonniot, J. de. "Darwinisme et logique." *Études* 16 (1872): 423–43.

Bousquet, Louis, and H. Roger. *Les métaux colloïdaux en thérapeutique*. Paris: Masson, 1910.

Bowler, Peter J. *The Eclipse of Darwinism: Anti-Darwinian Evolution Theories in the Decades around 1900*. Baltimore: Johns Hopkins University Press, 1983.

Bowler, Peter J., and John Henry. "Evolution." In *Science and Religion: A Historical Introduction*, edited by Gary B. Ferngren, 204–19. Baltimore: Johns Hopkins University Press, 2017.

Braga, Teófilo. *História da Universidade de Coimbra*. Lisbon: Typographia da Academia Real das Ciencias, 1892–1902.

Braga, Teófilo. *As lendas christãs*. Porto: Livraria Internacional de Ernesto Chardron, 1892.

Braga, Teófilo. *Origens poéticas do christianismo*. Porto: Livraria Universal de Magalhães & Moniz, 1880.

Braga, Teófilo. *Soluções positivas da politica portugueza: Obras políticas*. Porto: Livraria Chardron, de Mello & Irmão, 1912.

Brenni, Paolo. "The Evolution of Teaching Instruments and Their Use between 1800 and 1930." *Science and Education* 21 (2012): 191–226.

Brockey, Liam M. *Journey to the East: The Jesuit Mission to China, 1572–1724*. Cambridge, MA: Harvard University Press, 2007.

Brooke, John Hedley. *Science and Religion: Some Historical Perspectives*. Cambridge: Cambridge University Press, 1991.

"A *Brotéria* e as revistas scientificas." *Brotéria* 3 (1904): 366–67.

Burson, Jeffrey D., and Jonathan Wright, eds. *The Jesuit Suppression in Global Context: Causes, Events, and Consequences*. Cambridge: Cambridge University Press, 2015.

Cabral, F. P. "Algumas notas sobre o tremor de terra de 23 de abril." *Brotéria: Vulgarização científica* 8 (1909): 182–90.

Cabral, F. P. "Os terremotos: A proposito da catastrophe de Messina." *Brotéria: Vulgarização científica* 8 (1909): 110–28.

Cabral, Luís Gonzaga. *O eclipse total do Sol no dia 30 de Agosto de 1905: Observações feitas pelas comissões das academias scientificas dos collegios de S. Fiel e Campolide*. Lisbon: La Bécarre, 1905.

Cabral, Luís Gonzaga. *Expulsion of the Jesuits from Portugal: Protest by the Provincial of the Society of Jesus in Portugal*. London: Burns & Oates, 1911.

Caeiro, José. *História da expulsão da Companhia de Jesus da província de Portugal*. 3 Vols. Lisbon: Verbo, 1991–99.

Caeiro, José. *Jesuítas do Brasil e da Índia: Na perseguição do marquês de Pombal (século XVIII)*. Translated by Manuel N. Martins. Baía: Escola Tipográfica Salesiana, 1936.

Câmara, António Sousa da. "A expulsão dos cromossomas pela centrifugação." *Brotéria: Ciências naturais* 11 (1942): 21–28.

Câmara, António Sousa da. "Fragmentos cromosómicos produzidos pelos raios X." *Brotéria: Ciências naturais* 8 (1939): 166–179.

Câmara, António Sousa da, and Sarah Vasconcellos. "Contribuição para o estudo das populações de Drosophila melanogaster em Portugal." *Brotéria: Ciências naturais* 12 (1943): 49–57.

Carayon, Auguste. *Lettres inédites du R.P. Joseph Delvaux sur le rétablissement des jésuites en Portugal: 1829–1834*. Paris: L'Écureux, 1866.

Carbonnelle, I. "Le darwinisme, théorie et objections." *Études* 14 (1869): 472–82.

Carolino, Luís Miguel. "João Delgado S.J. e a 'Quaestio de certitudine mathematicarum' em inícios do século XVII." *Revista brasileira de história da matemática* 6 (2006): 17–49.

Carolino, Luís Miguel, and Henrique Leitão. "Natural Philosophy and Mathematics in Portuguese Universities, 1550–1650." In *Universities and Science in the Early Modern Period*, edited by Mordechai Feingold and Victor Navarro Brotóns, 153–68. Dordrecht: Springer, 2006.

Carolino, Luís Miguel, and Ana Simões. "The Eclipse, the Astronomer and His Audience: Frederico Oom and the Total Solar Eclipse of 28 May 1900 in Portugal." *Annals of Science* 69 (2011): 215–38.

Carvalhais, José. *80 anos na educação (1912–1992): Instituto Nun'Alvres*. Caldas da Saúde: Instituto Nun'Alvres, 1992.

Carvalhais, José. “Padre Alphonse Luisier.” *Brotéria: Ciências naturais* 54 (1958): 3–16.

Carvalhais, José. “Reverendo Padre Alphonse Luisier, S.J.: Homenagem ao cientista e ao mestre.” *Boletim cultural de Santo Tirso* 5 (1957): 223–49.

Carvalho, Mário Santiago de. *Psicologia e etica nel* Cursus Conimbricensis. Rome: Anicia, 2015.

Carvalho, Rómulo de. “As ciências exactas no tempo de Pombal.” In *Como interpretar Pombal? No bicentenário da sua morte,* edited by Manuel Antunes, 215–32. Lisbon: Brotéria, 1982.

Carvalho, Rómulo de. *História da fundação do Colégio Real dos Nobres de Lisboa*. Coimbra: Atlântida-Livraria Editora, 1959.

Carvalho, Rómulo de. *História do ensino em Portugal: Desde a fundação da nacionalidade até ao fim do regime de Salazar-Caetano*. Lisbon: Fundação Calouste Gulbenkian, 1986.

Casalini, Cristiano. *Aristotle in Coimbra. The* Cursus Conimbricensis *and Education at the College of Arts*. New York: Routledge, 2017.

Casalini, Cristiano, and Francesco Mattei. “Acquaviva and the *Ratio studiorum*.” In *The Acquaviva Project: Claudia Acquaviva's Generalate (1581–1615) and the Emergence of Modern Catholicism*, edited by Pierre-Antonio Fabre and Flavio Rurale, 269–90. Chestnut Hill, MA: Institute of Jesuit Sources, 2017.

Casalini, Cristiano, and Claude Pavur, trans. and eds. *Jesuit Pedagogy, 1540–1616: A Reader*. Chestnut Hill, MA: Institute of Jesuit Sources, 2016.

Casimiro, Acácio. “O govêrno de D. João VI e a restauração da Companhia de Jesus (1814–1815).” *Brotéria* 31 (1940): 469–75.

Casimiro, Acácio. *P. António da Costa e Oliveira Pinto*. Porto: Tipografia Pôrto Médico, 1933.

Castellarnau, Ignacio Salla de. “Binomia de los dípteros y ectoparasitos de quirópteros.” *Brotéria: Ciências naturais* 25 (1956): 170–75.

Castellarnau, Ignacio Salla de. “El biólogo Rev. P José Assmuth (S.J.).” *Brotéria: Ciências naturais* 27 (1958): 132–37.

Castellarnau, Ignacio Salla de. “Catálogo de aves anilladas.” *Brotéria: Ciências naturais* 18 (1949): 123–40.

Castellarnau, Ignacio Salla de. “Catálogo de neurópteros de la Índia.” *Brotéria: Ciências naturais* 15 (1946): 114–28.

Castellarnau, Ignacio Salla de. “El embriólogo R.P. Jaime Pujiula.” *Brotéria: Ciências naturais* 28 (1959): 95–104.

Castellarnau, Ignacio Salla de. “La estivación en los peces.” *Brotéria: Ciências naturais* 12 (1943): 85–93.

Castellarnau, Ignacio Salla de. “Un ilustre Botânico: El R.P. Ethelbert Blatter (S.J.).” *Brotéria: Ciências naturais* 4 (1935): 28–34.

Castellarnau, Ignacio Salla de. “In Memoriam. R.P. Longino Navás.” *Brotéria: Ciências naturais* 9 (1940): 131–38.

Castellarnau, Ignacio Salla de. "La invernación en los moluscos y arácnidos." *Brotéria: Ciências naturais* 14 (1945): 72–77.

Castellarnau, Ignacio Salla de. "La invernación en los peces." *Brotéria: Ciências naturais* 12 (1943): 21–28.

Castellarnau, Ignacio Salla de. "El R.P. Henrique Dettmer (S.J.), 1873–1933." *Brotéria: Ciências naturais* 3 (1934): 66–77.

Castellarnau, Ignacio Salla de. "Rev. P. Hermann Schmitz, S.J. (1878–1960)." *Brotéria: Ciências naturais* 30 (1961): 63–66.

Castro, Luís de. "Protesto de homens de sciencias." *Boletim da Associação central da agricultura portuguesa* 1 (1912): 165–167.

Catarino, Fernando. "Carlos das Neves Tavares (1914–1972)." In *Memórias de professores cientistas*, edited by Ana Simões, 102–11. Lisbon: Faculdade de Ciências da Universidade de Lisboa, 2001.

Catarino, Fernando. "Flávio Ferreira Pinto Resende (1907–1967)." In *Memórias de professores cientistas*, edited by Ana Simões, 78–89. Lisbon: Faculdade de Ciências da Universidade de Lisboa, 2001.

Catarino, Fernando. "No centenário de Aurélio Quintanilha." *Brotéria: Genética* 14 (1993): 19–22.

Cenáculo, Manuel do. *Memórias históricas e apendix segundo a disposição da colecção das disposições do superior provincial para a observância e estudos da Congregação da Ordem Terceira de São Francisco*. Lisbon: n.p., 1794.

Chaurasia, Bhagwan Din, and Vijay K. Sharma. "Karyological Studies in *Asphodelus tenuifolius* Cav." *Brotéria: Ciências naturais* 43 (1974): 35–37.

Chaurasia, Bhagwan Din, and Vijay K. Sharma. "Karyological Studies in *Phaseolus mngo* Linn." *Brotéria: Ciências naturais* 43 (1974): 33–34.

Chorão, José Pedro Dias. "A epidemia gripal de 1918 no concelho do Fundão." *Brotéria: Vulgarização científica* 17 (1919): 66–75.

Chorão, José Pedro Dias. "Febre typhoide: Prophylaxia e therapeutica." *Brotéria: Vulgarização científica* 12 (1914): 115–22; 235–43.

Chorão, José Pedro Dias. "Um flagello social: A syphilis." *Brotéria: Vulgarização científica* 13 (1915): 65–74.

Chorão, José Pedro Dias. "Gripe." *Brotéria: Vulgarização científica* 6 (1907): 239–61.

Chorão, José Pedro Dias. "Meningite cerebro-espinal epidémica." *Brotéria: Vulgarização científica* 6 (1907): 50–63.

Chorão, José Pedro Dias. "Os metaes coloidaes electricos na therapeutica." *Brotéria: Vulgarização científica* 11 (1913): 40–50.

Chorão, José Pedro Dias. "A prata dissociada pela corrente electrica, no tratamento da pneumonia." *Brotéria: Vulgarização científica* 8 (1909): 97–99; 251–56.

Chorão, José Pedro Dias. "O tabaquismo e o alcoolismo." *Brotéria: Vulgarização científica* 8 (1909): 5–21; 41–57.

Cicera, Ricard. "La ciencia al alcance de todos." *Ibérica* Spécimen A (1913): 1–5.

Citadino, Eugene. "Botany." In *The Cambridge History of Science*, vol. 6, *Modern Biological and Earth Sciences*, edited by Peter J. Bowler and John V. Pickstone, 225–42. Cambridge: Cambridge University Press, 2009.

Clemente, Manuel. *Igreja e sociedade portuguesa: Do liberalismo à República*. Porto: Assírio & Alvim, 2012.

Clemente, Manuel, and António Matos Ferreira, eds. *Religião e secularização*. Lisbon: Círculo de Leitores, 2002.

Cockerell, Theodore Dru Alison. "The Exiled Naturalists of Portugal." *Science* 34 (1911): 714–15.

The Code of Canon Law. London: Collins, 1983.

Coelho, Francisco Adolfo. *A questão do ensino*. Porto: n.p., 1872; reprinted as Carlos Reis, ed., *As conferências do casino*, 143–77. Lisbon: Alfa, 1990.

Colégio de Campolide. *Á sua celeste padroeira*. Lisbon: n.p., 1906.

Colégio de Campolide. *Prospecto*. Lisbon: Typographia Universal, 1886.

Colégio de Campolide. "Sessão anual de sciências da academia campolidense." *O nosso collegio* 4 (1907–8): 29–31.

Colégio de Campolide. "Sessão solemne de physica." *O nosso collegio* 3 (1906–7): 17–20.

Colégio de São Fiel. *Eclipse do Sol de 28 de Maio de 1900: Observações dos professores do collegio de São Fiel*. Lisbon: La Bécarre, 1900.

Collecção dos negocios de Roma no reinado de El-Rei Dom José I, ministerio do marquez de Pombal. 4 Vols. Lisbon: Imprensa Nacional, 1874–75.

Colombo, Emanuele, and Niccolò Guasti, "The Expulsion and Suppression in Portugal and Spain: An Overview." In *The Jesuit Suppression in Global Context: Causes, Events, and Consequences*, edited by Jeffrey D. Burson and Jonathan Wright, 117–38. Cambridge: Cambridge University Press, 2015.

Cooter, Roger, and Stephen Pumfrey. "Separate Spheres and Public Spaces: Reflection on the History of Science Popularization and Science in Popular Culture." *History of Science* 32 (1994): 269–315.

Corkill, David, and José Carlos Pina Almeida. "Commemoration and Propaganda in Salazar's Portugal: The 'Mundo Português' Exposition of 1940." *Journal of Contemporary History* 44 (2009): 381–99.

Corley, Martin. "The Lepidoptera Collections of Deceased Portuguese Entomologists." *Entomologist's Gazette* 59 (2008): 145–71.

Corley, Martin. "The Lepidoptera Collections of Deceased Portuguese Entomologists: II." *Entomologist's Gazette* 66 (2015): 25–49.

Corn, Joseph J. *The Winged Gospel: America's Romance with Aviation, 1900–1950*. New York: Oxford University Press, 1983.

Correia, Francisco Augusto. *O método missionário dos jesuítas em Moçambique, 1881–1910: Um contributo para a história da missão da Zambézia*. Braga: Livraria Apostolado da Imprensa, 1992.

Côrte-Real, João. "Como conheci o Senhor Professor Serra: Uma experiência de treze anos." *Brotéria: Genética* 13 (1992): 27–31.

Coutinho, Miguel Pereira. "Como conheci o Prof. Aurélio Quintanilha." *Brotéria: Genética* 9 (1988): 19–21.

Coutinho, Miguel Pereira. "Como vejo a figura do Professor Quintanilha." *Brotéria: Genética* 14 (1993): 9–13.

Coutinho, Miguel Pereira. "Homenagem à memória do Prof. António de Sousa da Câmara." *Brotéria: Genética* 2 (1981): 7–9.

Coutinho, Miguel Pereira. "O Professor Antunes Serra, um verdadeiro mestre." *Brotéria: Genética* 13 (1992): 9–12.

Couto, Jorge, ed. *A expulsão dos jesuítas dos domínios portugueses: 250.° aniversário*. Lisbon: Biblioteca Nacional de Portugal, 2009.

Cruz, Manuel Braga da. *O Estado Novo e a Igreja Católica*. Lisbon: Bizâncio, 1998.

Cruz, Miguel Dantas da. "Francisco Xavier de Mendonça Furtado." In *e-Dicionário da terra e do território no império português*, edited by José Vicente Serrão, Márcia Motta, and Susana Münch Miranda. Lisbon: CEHC-IUL, 2015. DOI: 10.15847/cehc.edittip.2015v003.

Cubit, Geoffrey. *The Jesuit Myth: Conspiracy, Theory and Politics in Nineteenth-Century France*. Oxford: Clarendon Press, 1993.

Cunha, Adriano Gonçalves da. "Prof. Doutor Ruy Telles Palhinha." *Boletim da sociedade Broteriana* 32 (1958): vii–xx.

Cunha, J. M. da. "A colheita do vinho nos principais países produtores em 1923: O vinho português em 1923." *Brotéria: Vulgarização científica* 22 (1924): 137–39.

Cunha, J. M. da. "A produção de trigo em Portugal em 1923." *Brotéria: Vulgarização científica* 22 (1924): 135–36.

Cunha, José Anastácio da. *Notícias literárias de Portugal: 1780*. Translated by Joel Serrão. Lisbon: Seara Nova, 1971.

Danner, David B., and Hamilton O. Smith. "Searching a *Haemophilus influenzae* Library for Genes Induced during Competence Development." *Brotéria: Genética* 3 (1982): *19.

Darwin, Charles. *On the Origin of Species by Means of Natural Selection*. London: John Murray, 1859.

De Gabriac, Arthur, "Galilée devant la science, la religion et la littérature." *Études* 12 (1867): 528–47.

Dear, Peter. "Jesuit Mathematical Science and the Reconstitution of Experience in the Early Seventeenth Century." *Studies in History and Philosophy of Science* 18 (1987): 133–75.

Desjardins, E. "M. Mézières et le procès de Galilée." *Études* 21 (1877): 273–81.

Di Gregorio, Mario A. "Zoology." In *The Cambridge History of Science*, vol. 6, *Modern Biological and Earth Sciences*, edited by Peter J. Bowler and John V. Pickstone, 204–24. Cambridge: Cambridge University Press, 2009.

Dinis, Alfredo. "O comentário Conimbricense à *Física* de Aristóteles." *Brotéria* 134 (1992): 398–406.

Dinis, Alfredo. *A Jesuit against Galileo? The Strange Case of Giovanni Battista Riccioli Cosmology*. Braga: Axioma–Publicações da Faculdade de Filosofia, 2017.

Dinis, Alfredo. "Tradição e transição no curso Conimbricense." *Revista portuguesa de filosofia* 47 (1991): 535–60.

Dionel. "Ás leitoras da *Brotéria*." *Brotéria: Vulgarização científica* 16 (1918): 5.

Dionel. "Aviação comercial." *Brotéria: Vulgarização científica* 22 (1922): 29–39.

Dionel. "Os progressos da aviação." *Brotéria: Vulgarização científica* 18 (1920): 180–86.

Divarkar, Parmananda R., trans. *A Pilgrim's Testament: The Memoirs of St. Ignatius of Loyola*. St. Louis, MO: Institute for Jesuit Sources, 1995.

Diz, Alejo. "O tifo exantemático." *Brotéria: Vulgarização científica* 16 (1918): 140–44.

Domingues, Ângela. "Francisco Xavier de Mendonça Furtado." In *Dicionário da história da colonização portuguesa no Brasil*, edited by Maria Beatriz Nizza da Silva, cols. 359–62. Lisbon: Verbo, 1994.

"Donations and Exchanges." *Proceedings of the Linnean Society of New South Wales* 28 (1903): 946–69.

Dooley, Brendan. "The *Storia letteraria d'Italia* and the Rehabilitation of Jesuit Science." In *Jesuit Science and the Republic of Letters*, edited by Mordechai Feingold, 433–73. Cambridge, MA: MIT Press, 2003.

Duarte, Ângela, Manuela Hagenfeldt, Paula Pacheco, Margarida Madureira, Conceição Silva, and João Lavinha. "Fibrose quística em Portugal: Patologia molecular e diagnóstico pré-natal." *Brotéria: Genética* 14 (1993): 87–93.

Duarte, Augusto Simplício. "Notas sobre algumas problemas relacionados com o melhoramento do arroz." *Brotéria: Ciências naturais* 26 (1957): 28–33.

"Duas palavras de introdução." *Brotéria* 1 (1902): v.

Duminuco, Vincent J., ed. *The Jesuit* Ratio studiorum: *400th Anniversary Perspectives*. New York: Fordham University Press, 2000.

Dunkel, M. "Wolfgang Gaede: An Appreciation of His Life on the Occasion of the 50th Anniversary of the Invention of the Diffusion Pump." *Vacuum* 13 (1963): 501–3.

Eckart, Anselmo. *Memórias de um jesuíta prisioneiro de Pombal*. Translated by Joaquim Abranches and Ana Maria Lago da Silva. Braga, São Paulo: Apostolado da Imprensa–Edições Loyola, 1987.

Eddington, Arthur Stanley. *Space, Time and Gravitation: An Outline of the General Relativity Theory*. Cambridge: University Press, 1920.

Estatutos do Collegio Real de Nobres da corte e cidade de Lisboa. Lisbon: Officina de Miguel Rodrigues, 1761.

Études publiées par des pères de la Compagnie de Jésus: Table générale des années 1880 a 1900. Paris: Victor Retaux, Librairie-Éditeur, *c.*1900.

Études religieuses, philosophiques, historiques et littéraires publiée par des pères de la Compagnie de Jésus: Table générale des vingt-cinq premières années (1856–1880). Paris: Victor Retaux & Fils, Éditeurs, *c.*1880.

Études: Tables des années 1911–1924. Paris: Bureau des Études, 1926.

"Exportação do vinho pela barra do Porto em 1913." *Brotéria: Vulgarização científica* 12 (1914): 315.

F.E.A.V. *Manifestação dos crimes e attentados commetidos pelos jesuitas: Em todas as partes do mundo, desde a sua fundação até à sua extinção*. Rio de Janeiro: Typ. de Miranda & Carneiro, 1833.

Fabre, Pierre-Antoine, and Catherine Maires, eds. *Les antijésuites: Discours, figures et lieux de l'antijésuitisme à l'époque modern*. Rennes: Presses Universitaires de Rennes, 2010.

Fabre, Pierre-Antonio, and Flavio Rurale, eds. *The Acquaviva Project: Claudia Acquaviva's Generalate (1581–1615) and the Emergence of Modern Catholicism*. Chestnut Hill, MA: Institute of Jesuit Sources, 2017.

Faria, António Lima de. "Professor Flávio Resende." *Nature* 214 (1967): 435.

Feingold, Mordechai, ed. *Jesuit Science and the Republic of Letters*. Cambridge, MA: MIT Press, 2003.

Feingold, Mordechai. "Jesuits: Savants." In *Jesuit Science and the Republic of Letters*, edited by Mordechai Feingold, 1–45. Cambridge, MA: MIT Press, 2003.

Feingold, Mordechai ed. *The New Science and Jesuit Science: Seventeenth-Century Perspectives*. Dordrecht: Kluwer Academic, 2003.

Feldhay, Rivka. "The Cultural Field of Jesuit Science." In *The Jesuits. Cultures, Sciences and the Arts, 1540–1773*, edited by John W. O'Malley, John, Gauvin Alexander Bailey, Steven J. Harris, and T. Frank Kennedy, 108–30. Toronto: Toronto University Press, 1999.

Fernandes, Abílio. "Lembrando o Prof. Doutor Aurélio Quintanlha." *Brotéria: Genética* 9 (1988): 135–50.

Ferngren, Gary B., ed. *Science and Religion: A Historical Introduction*. Baltimore: Johns Hopkins University Press, 2017.

Ferrão, Pedro. *A educação jesuitica: O Collegio de S. Fiel. Subsidios para a historia contemporanea dos jesuítas*. Lisbon: Guimarães & Ca., 1910.

Ferreira, Jorge. "Teodoro Monteiro." In *Dicionário dos mais ilustres transmontanos e alto durienses*, edited by João Barroso da Fonte, 2:232–33. Guimarães: Editora Cidade Berço, 2011.

Filipe, António Monteiro. "Notas sobre a cultura do cacau em S. Tomé." *Brotéria: Vulgarização científica* 18 (1920): 174–80.

Findlen, Paula. "Scientific Spectacle in Baroque Rome: Athanasius Kircher and the Roman College Museum." In *Jesuit Science and the Republic of Letters*, edited by Mordechai Feingold, 225–48. Cambridge, MA: MIT Press, 2003.

Findlen, Paula, ed. *Athanasius Kircher: The Last Man Who Knew Everything*. London: Routledge, 2004.

Fonseca, Pedro Gouveia da. "Darwin em Portugal (1910–1974): O Darwinismo e a evolução na produção científica de botânicos portugueses." PhD diss., University of Coimbra, 2015.

Foulquier, Joseph. "A borracha brazileira." *Brotéria: Vulgarização científica* 12 (1914): 5–31.

Franco, José Eduardo. "Os catecismos antijesuíticos pombalinos: As obras fundadoras do antijesuitismo do marquês de Pombal." *Revista lusófona de ciência das religiões* 7/8 (2005): 247–68.

Franco, José Eduardo. "Da extinção à restauração dos jesuítas: Construção e reprodução do mito dos jesuítas de matriz pombalina no século XIX." *Brotéria* 179 (2014): 401–28.

Franco, José Eduardo. "Os directores da *Brotéria*: Afonso Luisier." In *Fé, ciência, cultura:* Brotéria; *100 anos*, edited by Hermínio Rico and José Eduardo Franco, 155–57. Lisbon: Gradiva, 2003.

Franco, José Eduardo. "Os directores da *Brotéria*: Carlos Zimmermann." In *Fé, ciência, cultura:* Brotéria; *100 anos*, edited by Hermínio Rico and José Eduardo Franco, 149–50. Lisbon: Gradiva, 2003.

Franco, José Eduardo. "Os directores da *Brotéria*: Joaquim da Silva Tavares." In *Fé, ciência, cultura:* Brotéria; *100 anos*, edited by Hermínio Rico and 143–46. José Eduardo Franco. Lisbon: Gradiva, 2003.

Franco, José Eduardo. "Os directores da *Brotéria*: José Carvalhaes." In *Fé, ciência, cultura:* Brotéria; *100 anos*, edited by Hermínio Rico and José Eduardo Franco, 168–70. Lisbon: Gradiva, 2003.

Franco, José Eduardo. "A evolução editorial da *Brotéria*." In *Fé, ciência, cultura:* Brotéria; *100 anos*, edited by Hermínio Rico and José Eduardo Franco, 103–26. Lisbon: Gradiva, 2003.

Franco, José Eduardo. *O mito dos jesuítas em Portugal, no Brasil e no Oriente (séculos XVI a XX)*. Lisbon: Gradiva, 2006.

Franco, José Eduardo ed. *Obra completa do Padre Manuel Antunes*. 14 Vols. Lisbon: Fundação Calouste Gulbenkian, 2007–12.

Franco, José Eduardo, and Christine Vogel. "Um acontecimento mediático na Europa das Luzes: A propaganda antijesuítica pombalina em Portugal e na Europa." *Brotéria* 169 (2009): 349–506.

Froeschlé-Chopard, Marie-Hélène, and Michel Froeschlé, "'Sciences et arts' dans les *Mémoires de Trévoux* (1701–1762)." *Revue d'histoire moderne et contemporaine* 48 (2001): 30–49.

Fyfe, Aileen, and Bernard Lightman, eds. *Science in the Marketplace. Nineteenth-Century Sites and Experiences* (Chicago: Chicago University Press, 2007).

Gago, Maria do Mar. "The Emergence of Genetics in Portugal: J. A. Serra at the Crossroads of Politics and Biological Communities (1936–1952)." MSc diss., University of Lisbon, 2009.

Gago, Maria do Mar. "Things of Darkness: Genetics, Melanins and the Regime of Salazar (1936–1952)." *Centaurus* 57 (2015): 1–27.

Ganss, George E., ed. *The* Spiritual Exercises *of Saint Ignatius*. St. Louis, MO: Institute of Jesuit Sources, 1992.

Gatto, Romano. "Christoph Clavius's *Ordo servandus in addiscendis disciplinis mathematicis* and the Teaching of Mathematics in Jesuit Colleges at the Beginning of the Modern Era." *Science and Education* 15 (2006): 235–58.

Gatzhammer, Stefan. "Antijesuitismo europeu: Relações político-diplomáticas e culturais entre a Baviera e Portugal (1750–1780)." *Lusitania sacra* 5 (1993): 159–250.

Genescà-Sitjes, Maria. "*Ibérica* Magazine (1913–2004) and the Ebro Observatory." *Contributions to Science* 9 (2013): 159–68.

Ghosh, R. B. "Karyomorphological Studies of Somatic Chromosomes in *Ailanthus excelsa* Roxb. An Ornamental and a Road-Side Plant." *Brotéria: Ciências naturais* 39 (1970): 3–8.

Giménez López, Enrique. "El antijesuitismo en la España de mediados del siglo XVIII." In *Fénix de España: Modernidad y cultura propia en la España del siglo XVIII (1737–1766); Homenaje a Antonio Mestre Sancís*, edited by P. Fernández Albaladejo, 283–326. Madrid: Marcial Pons, 2006.

Giménez López, Enrique. "Portugal y España ante la extinción de los jesuitas." In *Los jesuitas españoles expulsos: Su imagen y su contribución al saber sobre el mundo hispánico en la Europa del siglo XVIII*, edited by Manfred Tietz, 337–57. Madrid-Frankfurt am Main: Iberoamericana-Vervuert, 2010.

"Il giornalismo moderno ed il nostro programma." *La civiltà cattolica* 1 (1850): 5–24.

Gomes, Doriedson Ferreira, Oberdan Caldas, Eduardo Mendes da Silva, Peter Andrew Gell, and David M. Williams. "Father Zimmermann (1871–1950): The First Brazilian Diatomist." *Diatom Research* 27 (2012): 177–88).

Gomes, Inês. "Os gabinetes de história natural dos antigos liceus: Um estudo exploratório a partir dos textos legislativos." In *Actas do Congresso Luso–Brasileiro de História das Ciências*, edited by Carlos Fiolhais, Carlota Simões, and Décio Martins, 1185–1202. Coimbra: Imprensa da Universidade de Coimbra, 2011.

Gomes, Inês. "The Natural History Collection at the Lisbon Military College: Tracing the History of a Teaching Collection." *Journal of the History of Collections* 29 (2017): 409–22. DOI: 10.1093/jhc/fw036.

Gomes, Jesué Pinharanda. *Os Conimbricenses*. Lisbon: Instituto de Cultura e Língua Portuguesa, 1992.

Gomes, João Pereira. *Os professores de filosofia da Universidade de Évora*. Évora: Câmara Municipal de Évora, 1960.

Gomes, João Pereira. "Últimas actividades filosóficas na Universidade de Évora." *Brotéria* 69 (1959): 393–404.

Gomes, Manuel João, ed. *O processo dos Távoras: A expulsão dos jesuítas*. Lisbon: Editora Afrodite, 1974.

Gomez, Francisco Teixidó. "El jesuita Jaime Pujiula Dilmé, científico clave de la investigación biológica barcelonesa." *Llull: Revista de la Sociedad española de historia de las ciencias y de las técnicas* 34, no. 73 (2011): 203–28.

Gonçalves, Nuno da Silva. "Pombal e a restauração da Companhia de Jesus." *Brotéria* 174 (2014): 319–38.

Grainha, Manuel Borges. *História do colégio de Campolide da Companhia de Jesus*. Coimbra: Imprensa da Universidade, 1913.

Grainha, Manuel Borges. *Os jesuitas e as congregações religiosas em Portugal nos ultimos trinta annos*. Porto: Empreza Litteraria e Typographica, 1891.

Grainha, Manuel Borges. *O Portugal jesuita*. Lisbon: Typographia e Sterotypia Moderna, 1893.

Granero, Juan. "El observatorio astronómico, geodinámico y meteorológico de Granada." *Razón y fé* 3 (1902): 222–25.

Granero, Juan. "Observatorio de Granada." *Razón y fé* 3 (1902): 512–20.

Granero, Juan. "Observatorio de Granada: Sección astronómica," *Razón y fé* 5 (1903): 339–47.

Graney, Christopher M. *Setting Aside All Authority: Giovanni Battista Riccioli and the Science against Copernicus in the Age of Galileo*. Notre Dame, IN: University of Notre Dame Press, 2015.

Gray, Irving, and Luís Archer. "The Effect of Anthrax Toxin on Oxidative Metabolism of Lung Tissue." *American Journal of Physiology* 210 (1966): 1313–15.

Gray, Irving, and Luís Archer. "Effect of Anthrax Toxin on Oxidative Metabolism of Rat Lung." *Bacteriological Proceedings* 67 (1966): 52.

Gray, Irving, and Luís Archer. "Metabolic Changes in Nicotinamide Adenine Dinucleotide in Response to Anthrax Toxin." *Journal of Bacteriology* 93 (1967): 36–39.

Grendler, Paul F. "The Culture of the Jesuit Teacher, 1548–1773." *Journal of Jesuit Studies* 3, no. 1 (2016): 17–41.

Grendler, Paul F. *The Jesuits and Italian Universities, 1548–1773*. Washington, DC: Catholic University of America Press, 2017.

Guedes-Pinto, Henrique. "Aurélio Quintanilha: Fragmentos para um esboço." *Brotéria: Genética* 9 (1988): 5–9.

Guedes-Pinto, Henrique. "O Eng.º Tristão de Mello-Sampayo." *Brotéria: Genética* 18 (1997): 5–8.

Guedes-Pinto, Henrique. "O Sr. Professor Miguel Pereira Coutinho: Recordações dispersas." *Brotéria: Genética* 19 (1998): 5–11.

Guerra, Francisco Carvalho. "Tumefacção mitocondrial: Estudo comparativo das mitocôndrias do cérebro e do fígado." *Brotéria: Ciências naturais* 34 (1965): 3–226.

Guerra, Levi, and F. Edmund Hunter. "Sucrose Inhibition of Gramicidin Induced Swelling of Isolated Rat Liver Mitochondria." *Brotéria: Ciências naturais* 34 (1965): 227–46.

Hake, Barry J., and Tom Steele, eds. *Intellectuals, Activists and Reformers: Studies of Cultural, Social and Educational Reform Movements in Europe, 1890–1930*. Leeds: Leeds Studies in Continuing Education, 1998.

Hake, Barry J., Tom Steele, and Alejandro Tiana-Ferrer, eds. *Masters, Missionaries and Militants: Studies of Social Movements and Popular Adult Education 1890–1939*. Leeds: Leeds Studies in Continuing Education, 1996.

Harris, Steven J. "Confession-Building, Long-Distance Networks and the Organization of Jesuit Science." *Early Science and Medicine* 1 (1996): 287–318.

Harris, Steven J. "Jesuit Ideology and Jesuit Science: Scientific Activity in the Society of Jesus, 1540–1773." PhD diss., University of Wisconsin-Madison, 1988.

Harris, Steven J. "Mapping Jesuit Science: The Role of Travel in the Geography of Knowledge." In *The Jesuits: Cultures, Sciences and the Arts, 1540–1773*, edited by John W. O'Malley, Gauvin Alexander Bailey, Steven J. Harris, and T. Frank Kennedy, 212–40. Toronto: Toronto University Press, 1999.

Harris, Steven J. "Networks of Travel, Correspondence, and Exchange." In *The Cambridge History of Science*, vol. 3, *Early Modern Science*, edited by Katharine Park and Lorraine Daston, 341–62. Cambridge: Cambridge University Press, 2006.

Harris, Steven J. "Transposing the Merton Thesis: Apostolic Spirituality and the Establishment of the Jesuit Scientific Tradition." *Science in Context* 3 (1989): 29–65.

Harrison, Peter, ed. *The Cambridge Companion to Science and Religion*. Cambridge: Cambridge University Press, 2010.

Harrison, Peter. *The Territories of Science and Religion*. Chicago: University of Chicago Press, 2015.

Hayes, C. H., and E. M. Sait, "Record of Political Events." *Political Science Quarterly* 25 (1910): 727–60.

Heilbron, John L. *Electricity in the 17th and 18th Centuries: A Study of Early Modern Physics*. Mineola, NY: Dover, 1999.

Hellyer, Marcus. *Catholic Physics: Jesuit Natural Philosophy in Early Modern Germany*. Notre Dame, IN: University of Notre Dame Press, 2005.

Herculano, Alexandre. *Da origem e estabelecimento da Inquisição em Portugal*. Lisbon: Imprensa Nacional, 1854–59; reprinted as: *History of the Origin and Establishment of the Inquisition in Portugal*. Translated by John C. Branner. New York: Ktav Publishing House, 1972.

Herran, Nestor. "'Science to the Glory of God': The Popular Science Magazine *Ibérica* and Its Coverage of Radioactivity." *Science and Education* 21 (2012): 335–53.

Hilgartner, Stephen. "The Dominant View of Popularization: Conceptual Problems, Political Issues." *Social Studies of Science* 20 (1990): 519–39.

História da Universidade em Portugal. 3 Vols. Coimbra and Lisbon: Universidade de Coimbra and Fundação Calouste Gulbenkian, 1997.

Hochadel, Oliver, and Agustí Nieto-Galan, eds. *Barcelona: An Urban History of Science and Modernity, 1888–1929*. London: Routledge, 2016.

Hodge, Jonathan. "Evolution." In *The Cambridge History of Science*, vol. 6, *Modern Biological and Earth Sciences*, edited by Peter J. Bowler and John V. Pickstone, 242–64. Cambridge: Cambridge University Press, 2009.

Hsia, Florence C. *Sojourners in a Strange Land: Jesuits and Their Scientific Missions in Late Imperial China*. Chicago: University of Chicago Press, 2009.

Hsia, Ronnie Po-chia. "Jesuit Foreign Missions: A Historiographical Essay." *Journal of Jesuit Studies* 1, no. 1 (2014): 47–65.

Índices gerais da Brotéria *científica (1902–2002)*. Lisbon: Brotéria, 2002.

Joannis, Joseph de. "Note sur un microlépidoptère nouveau de Portugal." *Bulletin de la Société entomologique de France* 71 (1902): 230–32.

John Paul II. *Evangelium vitae*. Washington, DC: United States Catholic Conference, 1995.

Júnior, Manuel Joaquim de Carvalho. *Nem Deus, nem diabo: Solução da philosophia positiva*. Lisbon: Typographia Elzeviriana, 1884.

Junqueiro, Abílio Guerra. *A velhice do padre eterno*. Porto: Livraria Minerva, 1885.

Junta de Providencia Litteraria. *Compendio historico do estado da Universidade de Coimbra no tempo da invasão dos denominados jesuitas e dos estragos feitos nas sciencias e nos professores, e directores que a regiam pelas maquinações, e publicações dos novos estatutos por eles fabricados*. Lisbon: Regia Officina Typographica, 1771.

Kessner, Thomas. *The Flight of the Century: Charles Lindbergh and the Rise of American Aviation*. New York: Oxford University Press, 2010.

Kieffer, Jean-Jacques. "Synopsis de zoocécidies d'Europe." *Annales de la Société entomologique de France* 70 (1901): 233–580.

King José I of Portugal. *Erros impios e sediciosos, que os religiosos da Companhia de Jesus ensinarão aos reos, que forão justiçados, e pertenderão espalhar nos póvos destes reynos*. Lisbon: Officina de Miguel Rodrigues, impressor do Eminentíssimo Senhor Cardeal Patriarca, 1759.

Kolvenbach, Peter Hans. "Função de uma revista de cultura." *Brotéria* 118 (1984): 483–84.

Lages, António Mendes. *O Sr. Marianno de Carvalho e o collegio de S. Fiel*. Lisbon: Typographia da Cruz do Operário, 1888.

Laird, Charles D., Louise Wang, and Walter Bodmer. "Recombination and DNA Replication in *Bacillus subtilis* Transformation." *Mutation Research* 6 (1968): 205–9.

Lamego, A. "Aeroplanos e suas aplicações." *Brotéria: Vulgarização científica* 22 (1924): 111–16.

Lamego, A. "A aviação ao serviço do comércio." *Brotéria: Vulgarização científica* 20 (1922): 276–80.

Lamego, A. "Os aviões comerciais." *Brotéria: Vulgarização científica* 21 (1923): 186–91.

Lamego, A. "O esforço aeronáutico alemão." *Brotéria: Vulgarização científica* 22 (1924): 216–23.

Lamego, A. "As grandes viagens aéreas." *Brotéria: Vulgarização científica* 22 (1924): 276–79.

Lamego, A. "Heroísmo e aventuras da aviação em 1922." *Brotéria: Vulgarização científica* 21 (1923): 131–34.

Lamego, A. "Os pára-quedas." *Brotéria: Vulgarização científica* 22 (1924): 10–15.

Lamego, A. "Os progressos e serviços da aviação nos diferentes países." *Brotéria: Vulgarização científica* 21 (1923): 163–74.

Lamego, A. "A volta ao mundo em avião." *Brotéria: Vulgarização científica* 22 (1924): 74–77.

Lamego, A. "O voo à vela e suas causas." *Brotéria: Vulgarização científica* 21 (1923): 267–70.

Lamego, A. "O voo à vela ou sem motor." *Brotéria: Vulgarização científica* 20 (1922): 283–89.

Laponce, Jean, "The Jesuits and Science in Eighteenth-Century France: An Analysis of Scientific Writings in the *Journal of Trévoux*." MSc. diss., University of British Columbia, 1990.

Latour, Bruno. *Science in Action: How to Follow Scientists and Engineers through Society*. Cambridge, MA: Harvard University Press, 1987.

Leão, Francisco da Cunha, ed. *Jesuítas na Ásia: Catálogo e guia*. 2 Vols. Macao and Lisbon: Instituto Cultural de Macau and Biblioteca da Ajuda, 1998.

Leitão, Cândido de Mello. "Alimentação da Criança nos primeiros tempos da vida." *Brotéria: Vulgarização científica* 10 (1912): 173–88.

Leitão, Henrique. "Azulejos que testemunham uma tradição de ensino científico." In *Azulejos que ensinam*, edited by António Leal Duarte and Carlota Simões, 16–33. Coimbra: Museu Nacional Machado de Castro and Centro de Matemática da Universidade de Coimbra, 2007.

Leitão, Henrique. *A ciência na "Aula da Esfera" no Colégio de Santo Antão, 1590–1759*. Lisbon: Comissariado Geral das Comemorações do V Centenário do Nascimento de S. Francisco Xavier, 2007.

Leitão, Henrique. "The Contents and Context of Manuel Dias's *Tianwenlüe*." In *History of Mathematical Sciences: Portugal and the East Asia, III: The Jesuits, the Padroado and East Asian Science (1552–1773)*, edited by Luís Saraiva and Catherine Jami, 99–121. Singapore: World Scientific, 2008.

Leitão, Henrique. "Entering Dangerous Ground: Jesuits Teaching Astrology and Chiromancy in Lisbon." In *The Jesuits II: Cultures, Sciences and the Arts, 1540–1773*, edited by John W. O'Malley, Gauvin Alexander Bailey, Steven J. Harris, and T. Frank Kennedy, 371–89. Toronto: Toronto University Press, 2006.

Leitão, Henrique. "Jesuit Mathematical Practice in Portugal, 1540–1759." In *The New Science and Jesuit Science: Seventeenth-Century Perspectives*, edited by Mordechai Feingold, 229–47. Dordrecht: Kluwer Academic, 2003.

Leitão, Henrique. "Longemira: Os primeiros telescópios em Portugal." *Gazeta de física* 33 (2010): 17–21.

Leitão, Henrique. "A Periphery between Two Centers? Portugal in the Scientific Route from Europe to China (16th and 17th Centuries)." In *Travels of Learning: A Geography of Science in Europe*, edited by Ana Simões, Ana Carneiro, and Maria Paula Diogo, 19–46. Dordrecht: Kluwer Academic Publishers, 2003.

Leitão, Henrique ed. *Sphaera mundi: A ciência na "Aula da Esfera"; Manuscritos científicos do Colégio de Santo Antão nas colecções da BNP*. Lisbon: Biblioteca Nacional de Portugal, 2008.

Leitão, Henrique, and José Eduardo Franco, eds. *Jesuítas, ciência e cultura em Portugal: Obra selecta do Pe. João Pereira Gomes S.J.* Lisbon: Esfera do Caos, 2012.

Leitão, Henrique, and Samuel Gessner. "Euclid in Tiles: The Azulejos of the Jesuit College in Coimbra." *Mathematische Semesterberichte* 61 (2014): 1–5.

Leite, António. "Pombal e o ensino secundário." In *Como interpretar Pombal? No bicentenário da sua morte*, edited by Manuel Antunes, 165–81. Lisbon: Brotéria, 1982.

Leite, Serafim. *História da Companhia de Jesus no Brasil*. 10 Vols. Lisbon and Rio de Janeiro: Livraria Portugália and Civilização Brasileira, 1938–50.

Leite, Serafim. "J. S. Tavares." *Brotéria: Fé, ciências, letras* 13 (1931): 273–97.

Lena. "Arte culinária." *Brotéria: Vulgarização científica* 14 (1916): 108–10, 174–75, 227–28, 286–87, 334–35.

Lena. "Arte culinária." *Brotéria: Vulgarização científica* 15 (1917): 31–32, 90–91, 171–72.

Lena. "Arte culinária." *Brotéria: Vulgarização científica* 16 (1918): 1–2.

Lencastre, Hermínia. "Engenharia genética em *Bacillus subtilis* I: Plasmídeos nativos." *Brotéria: Genética* 3 (1982): 15–38.

Lencastre, Hermínia. "Engenharia genética em *Bacillus subtilis* II: Plasmídeos compostos." *Brotéria: Genética* 4 (1983): 81–125.

Lencastre, Hermínia. "Engenharia genética em *Bacillus subtilis* III: Bacteriófagos." *Brotéria: Genética* 6 (1985): 25–55.

Leonardo, António José. "O Instituto de Coimbra e a evolução das ciências físico-químicas em Portugal de 1852 a 1952." PhD diss., Universidade de Coimbra, 2011.

Leonardo, António José, Décio Martins, and Carlos Fiolhais. "A física na Universidade de Coimbra de 1900 a 1960." *Gazeta de física* 34 (2011): 9–15.

Leroy, Michel. *Le mythe jésuite: De Béranger à Michelet*. Paris: Presses Universitaires de France, 1992.

Lima, Américo Pires de. "Educação profissional." *O médico* 1 (1953): 85–91.

Lindberg, David C., and Robert S. Westman, eds. *Reappraisals of the Scientific Revolution*. Cambridge: Cambridge University Press, 1990.

Lister, Arthur, and Gulielma Lister. *A Monograph of the Mycetozoa: A Descriptive Catalogue of the Species in the Herbarium of the British Museum*. 2nd ed. London: Printed by Order of the Trustees of the British Museum, 1911.

Lombardo, Michael F. *Founding Father: John J. Wynne, S.J. and the Inculturation of American Catholicism in the Progressive Era*. Leiden: Brill, 2017.

Lopes, António. *Roteiro histórico dos jesuítas em Lisboa*. Braga: Livraria Apostolado da Imprensa–Editorial Apostolado da Oração, 1985.

Lopes, António, and António Araújo. "Jesuítas." In *Dicionário histórico das ordens, institutos religiosos e outras formas de vida consagrada católica em Portugal*, edited by José Eduardo Franco, 195–206. Lisbon: Gradiva, 2010.

Luisier, Afonso. "Apontamentos sobre a flora da região de Setubal." *Boletim da sociedade Broteriana* 19 (1902): 172–74.

Luisier, Afonso. "Appendice: Listas das plantas colhidas por Tournefort em Setubal e na serra d'Arrabida." *Boletim da sociedade Broteriana* 19 (1902): 272–7.

Luisier, Afonso. "Artrópodes da Madeira, segundo as investigações do Dr. O. Lundblad." *Brotéria: Ciências naturais* 8 (1939): 18–39; 82–95; 101–12; *Brotéria: Ciências naturais* 9 (1940): 184–88; *Brotéria: Ciências naturais* 10 (1941): 61–69; 133–42; 179–84; *Brotéria: Ciências naturais* 11 (1942): 84–93; 137–44; 177–87; *Brotéria: Ciências naturais* 12 (1943): 29–36; 128–34.

Luisier, Afonso. "Catalogo das plantas vasculares de Setubal e da serra d'Arrabida." *Boletim da sociedade Broteriana* 19 (1902): 175–272.

Luisier, Afonso. "Félix de Avelar Brotero." *Brotéria: Ciências naturais* 13 (1944): 145–58.

Luisier, Afonso. "In Memoriam: Le R.P. J. da Silva Tavares, S.J." *Brotéria: Ciências naturais* 1 (1932): 9–34.

Luisier, Afonso. "Les mousses de l'Archipel de Madère et en général des Îles Atlantiques." *Brotéria: Botânica* 23 (1927): 5–53; 129–145; *Brotéria: Botânica* 24 (1930): 18–47; 66–96; 119–40; *Brotéria: Botânica* 25 (1931): 5–20; 123–39; *Brotéria: Ciências naturais* 1 (1932): 164–82; *Brotéria: Ciências naturais* 7 (1938): 78–95; 110–31; *Brotéria: Ciências naturais* 11 (1942): 29–41; *Brotéria: Ciências naturais* 14 (1945): 78–94; 112–27; 156–76.

Luisier, Afonso. "Les mousses de Madère." *Brotéria: Botânica* 15 (1917): 81–98; *Brotéria: Botânica* 16 (1918): 29–71; *Brotéria: Botânica* 17 (1919): 17–142; *Brotéria: Botânica* 18 (1920): 5–22; 78–120; *Brotéria: Botânica* 20 (1922): 76–106.

Luisier, Afonso. "Musci Salmanticenses: Descriptio et distributio specierum hactenus in provincia geographica Salmanticensi cognitarum; Brevi addito conspectu Muscorum totius Peninsulae Ibericae." *Memorias de la Real academia de ciencias exactas físicas y naturales de Madrid* 3 (1924): 1–280.

Luisier, Afonso. "O P. Fernando Theissen." *Brotéria: Botânica* 18 (1920): 73–77.

Luisier, Afonso. "Prólogo." *Brotéria: Ciências naturais* 1 (1932): 7–8.

Luisier, Afonso. "R.P. Cândido Azevedo Mendes (S.J.)." *Brotéria: Ciências naturais* 13 (1944): 44–45.

Luisier, Afonso. "Recherches bryologiques récentes à Madère." *Brotéria: Ciências naturais* 5 (1936): 140–44; *Brotéria: Ciências naturais* 6 (1937): 88–95; *Brotéria: Ciências naturais* 8 (1939): 40–52; *Brotéria: Ciências naturais* 12 (1943): 135–44; *Brotéria: Ciências naturais* 16 (1947): 86–91; *Brotéria: Ciências naturais* 22 (1953): 178–91; *Brotéria: Ciências naturais* 25 (1956): 170–82.

Machado, António. "Contribuição para os Miriápodes de Portugal." *Brotéria: Ciências naturais* 15 (1946): 5–37.

Machado, António. "Miriápodes de Portugal." *Brotéria: Ciências naturais* 21 (1952): 65–170.

Machado, António. "Notas de bryologia portuguesa: 1; Plantas novas para Portugal." *Brotéria: Botânica* 15 (1917): 8–11.

Machado, António. "Notas de bryologia portuguesa: 2; Duas excursões às Serras da Estrela e do Gerês." *Brotéria: Botânica* 15 (1917): 49–63.

Machado, António. "Notas de zoologia." *Brotéria: Ciências naturais* 11 (1942): 97–105.

Machado, António. "Problemas da sistemática (As grandes divisões ou tipos zoológicos)." *Brotéria: Ciências naturais* 14 (1945): 62–71.

Machado, António. "Revisão de algumas hipóteses filogenéticas." *Brotéria: Ciências naturais* 14 (1945): 149–55.

Magone, Rui. "The Textual Tradition of *Tianwenlüe*." In *History of Mathematical Sciences: Portugal and the East Asia, III: The Jesuits, the Padroado and East Asian Science (1552–1773)*, edited by Luís Saraiva and Catherine Jami, 123–38. Singapore: World Scientific, 2008.

Marinho, Júlio. "Higiene escolar." *Brotéria: Vulgarização científica* 13 (1915): 206–9; 247–52.

Marinho, Júlio. "Higiene escolar." *Brotéria: Vulgarização científica* 14 (1916): 47–49.

Marinho, Júlio. "A miopia escolar." *Brotéria: Vulgarização científica* 15 (1917): 217–23.

Marinho, Júlio. "Mobiliário escolar." *Brotéria: Vulgarização científica* 19 (1921): 211–16.

Marinho, Júlio. "Tratamento da lepra." *Brotéria: Vulgarização científica* 22 (1924): 34–36.

Martinez, Ramón. "Observatorio de Granada: Sección meteorológica." *Razón y fé* 4 (1902): 478–90.

Martins, Joaquim Pedro Oliveira. *História da civilização ibérica*. Lisbon: Bertrand, 1879; reprint: Lisbon: Guimarães Editores, 2007.

Martins, Joaquim Pedro Oliveira. *História de Portugal*. Lisbon: Bertrand, 1880; reprint: Lisbon: Guimarães Editores, 2010.

Martins, Manuel Narciso. "O milho grosso em Portugal e seus inimigos." *Brotéria: Vulgarização científica* 7 (1908): 68–83.

Maryks, Robert A., and Jonathan Wright, eds. *Jesuit Survival and Restoration: A Global History, 1773–1900*. Leiden: Brill, 2014.

Masriera, Juliá. "In memoriam: R.P. Ignacio Sala de Castellarnau S.J. (1903–1973)." *Miscellània zoológica* 3–4 (1974): 177–78.

Matheus, J. Alves. "O Canal de Panamá e as outras vias de comunicação entre os dois oceanos." *Brotéria: Vulgarização científica* 11 (1913): 178–80.

Matos, Luís Salgado. *A separação do estado e da igreja*. Lisbon: Dom Quixote, 2011.

Matos, Rolanda Maria Albuquerque de. *José da Silva e Castro: O naturalista e a sua obra, 1842–1928*. Lisbon: Instituto Português de Malacologia, 2009.

Matos, Rolanda Maria Albuquerque de. "Professor José Antunes Serra." *Brotéria: Genética* 12 (1991): 5–44.

Maurício, Domingos. "O casamento de D. Maria I e o incidente diplomático das luminárias em 1760." *Brotéria* 92 (1971): 186–201.

Maurício, Domingos. "O fundador da *Brotéria* na Academia." *Brotéria: Revista contemporânea de cultura* 23 (1936): 449–55.

Maurício, Domingos. "Os jesuítas e o ensino das matemáticas em Portugal." *Brotéria* 20 (1935): 189–205.

Maurício, Domingos. "Obra científico-literária e pedagógica da Universidade de Évora." *Brotéria* 69 (1959): 377–92.

Maurício, Domingos. "Pombal e a Companhia do Grão-Pará e Maranhão." *Brotéria* 80 (1965): 743–50.

Maxwell, Kenneth. *Pombal, Paradox of the Enlightenment*. Cambridge: Cambridge University Press, 1995.

McCoog, Thomas M., ed. *The Mercurian Project: Forming Jesuit Culture 1573–1580*. St. Louis, MO: Institute of Jesuit Sources, 2004.

McCoog, Thomas M., ed. *Promising Hope: Essays on the Suppression and Restoration of the English Province of the Society of Jesus*. Rome: Institutum Historicum Societatis Iesu, 2003.

Mello, F. de. "A propósito de aviação." *Brotéria: Vulgarização científica* 9 (1910): 36–41.

Mendes, Cândido de Azevedo. "Altera nova Mendesia ex Lusitania." *Brotéria: Zoologia* 8 (1909): 65–67.

Mendes, Cândido de Azevedo. *A* Brotéria *no exílio*. Madrid: Imprenta Ibérica, 1913.

Mendes, Cândido de Azevedo. *O collegio de S. Fiel: Resposta ao advogado Sr. José Ramos Preto*. Madrid: López del Horno, 1911.

Mendes, Cândido de Azevedo. "Lepidoptera africana I: Ex Zambezia lusitana." *Brotéria: Zoologia* 10 (1912): 183–91.

Mendes, Cândido de Azevedo. "Lepidoptera africana II: Ex Angola lusitana." *Brotéria: Zoologia* 10 (1912): 191–93.

Mendes, Cândido de Azevedo. "Lepidopteros de S. Fiel (Portugal)." *Brotéria* 1 (1902): 151–71; *Brotéria* 2 (1903): 41–80; *Brotéria* 3 (1904): 223–54; *Brotéria* 4 (1905): 166–77; *Brotéria: Zoologia* 10 (1912): 161–82; *Brotéria: Zoologia* 11 (1913): 15–44.

Mendonça, Marta. "*Brotéria* e a difusão da ciência em Portugal." In *Fé, ciência, cultura:* Brotéria; *100 anos*, edited by Hermínio Rico and José Eduardo Franco, 255–76. Lisbon: Gradiva, 2003.

Meneses, Maria de Fátima. "Museus e ensino: Uma análise histórica sobre os museus pedagógicos e escolares em Portugal (1836–1933)." MSc diss., Universidade Nova de Lisboa, 2003.

Merino, Baltasar. "Viajes de herborización por Galicia." *Razón y fé* 1 (1901): 95–98; *Razón y fé* 2 (1902): 82–89; *Razón y fé* 2 (1902): 367–73; *Razón y fé* 4 (1902): 82–93; *Razón y fé* 5 (1903): 348–60.

Miller, Samuel J. *Portugal and Rome c.1748–1830: An Aspect of the Catholic Enlightenment*. Rome: Gregorian University Press, 1978.

Mir, Gabriel Codina. *Aux sources de la pédagogie des jésuites:* Le Modus Parisiensis. Rome: Institutum Historicum Societatis Iesu, 1968.

Mónica, Maria Filomena. *Fontes Pereira de Melo*. Lisbon: Alêtheia, 2009.

Mónica, Maria Filomena. "Negócios e política: Os tabacos (1800–1890)." *Análise social* 27 (1992): 461–79.

Moniz, António Egas. *A nossa casa*. Lisbon: Paulino Ferreira Filhos Lda, 1950.

Monserrat, Victor. "Longino Navás, His Neuropterological Work and Collection." In *Recent Research in Neuropterology: Proceedings of the 2nd International Symposium on Neuropterology*, edited by Johann Gepp, Host Aspöck, and Herbert Hözel, 173–76. Graz: Gedruckt von Druckhaus Thalerhof, 1986.

Monteiro, Nuno Gonçalo. *D. José: Na sombra de Pombal*. Lisbon: Temas e Debates, 2008.

Monteiro, Nuno Gonçalo. "Pombal's Government: Between Seventeenth-Century Valido and Enlightened Models." In *Enlightened Reform in Southern Europe and Its Atlantic Colonies, c.1750–1830*, edited by Gabriel Paquette, 321–38. Farnham and Burlington: Ashgate, 2009.

Monteiro, Teodoro. "Contribuição para o conhecimento dos lepidópteros de Portugal." *Brotéria: Ciências naturais* 25 (1956): 49–70; *Brotéria: Ciências naturais* 26 (1957): 175–87; *Brotéria: Ciências naturais* 31 (1962): 119–30.

Monteiro, Teodoro. "Pyralideos novos para Portugal." *Brotéria: Ciências naturais* 31 (1962): 181–99.

Moore, James R. *The Post-Darwinian Controversies: A Study of the Protestant Struggle to Come to Terms with Darwin in Great Britain and America, 1870–1900*. New York: Cambridge University Press, 1979.

Moore, James R. "Religion and Science." In *The Cambridge History of Science*, vol. 6, *Modern Biological and Earth Sciences*, edited by Peter J. Bowler and John V. Pickstone, 541–62. Cambridge: Cambridge University Press, 2009.

Morais, Júlio de. *Historiador desconhecido: José Caeiro, grande escritor da época Pombalina*. Braga: Livraria Cruz, 1939.

Mota, Bernardo. "O estatuto da matemática em Portugal nos séculos XVI e XVII." PhD diss., University of Lisbon, 2008.

Mota, Miguel. "A contribuição da *Brotéria* para o desenvolvimento da genética." In *Fé, ciência, cultura:* Brotéria; *100 anos*, edited by Hermínio Rico and José Eduardo Franco, 517–27. Lisbon: Gradiva, 2003.

Mury, Paulo. *História de Gabriel Malagrida da Companhia de Jesus*. Translated by Camilo Castelo-Branco. Lisbon: Livraria Editora de Mattos Moreira, 1875.

Navás, Longino. "Algunos insectos del Museo entomológico de Berlim 'Deutsches Entomologisches Museum.'" *Brotéria: Zoologia* 20 (1922): 87–92.

Navás, Longino. "Algunos insectos del Museo de París." *Brotéria: Zoologia* 21 (1924): 99–114; *Brotéria: Zoologia* 22 (1925): 75–83; *Brotéria: Zoologia* 23 (1926): 85–115; *Brotéria: Zoologia* 24 (1927): 5–36; *Brotéria: Zoologia* 26 (1930): 5–24; 120–44; *Brotéria: Zoologia* 27 (1931): 101–36; *Brotéria: Ciências naturais* 3 (1934): 49–55.

Navás, Longino. "Ascaláfidos (Ins. Neur.) sudamericanos." *Brotéria: Zoologia* 10 (1912): 203–33.

Navás, Longino. "Décadas de insectos nuevos." *Brotéria: Ciências naturais* 1 (1932): 62–85; 109–19; 145–55; *Brotéria: Ciências naturais* 2 (1933): 34–44; 101–10; *Brotéria: Ciências naturais* 3 (1934): 15–24; 133–44; *Brotéria: Ciências naturais* 4 (1935): 97–107; *Brotéria: Ciências naturais* 5 (1936): 161–70.

Navás, Longino. "Neurópteros del Congo (Africa Occidental)." *Brotéria: Zoologia* 8 (1909): 106.

Navás, Longino. "Neurópteros de Egipto." *Brotéria: Zoologia* 8 (1909): 105.

Navás, Longino. "Neurópteros de España y Portugal." *Brotéria* 5 (1906): 145–84; *Brotéria: Zoologia* 6 (1907): 43–100; *Brotéria: Zoologia* 7 (1908): 5–131.

Navás, Longino. "Neurópteros de Zumbo (Africa Oriental, Mozambique) colecionados por el P. Lopes." *Brotéria: Zoologia* 8 (1909): 106.

Navás, Longino. "Neurópteros nuevos de América." *Brotéria: Zoologia* 10 (1912): 194–202.

Navás, Longino. "Neurópteros nuevos de América." *Brotéria: Zoologia* 11 (1913): 45–53.

Navás, Longino. "Neurópteros sudamericanos." *Brotéria: Zoologia* 12 (1914): 45–56; 215–34; *Brotéria: Zoologia* 13 (1915): 5–13; *Brotéria: Zoologia* 14 (1916) 14–35.

Navás, Longino. "Tricópteros nuevos de España." *Brotéria: Zoologia* 14 (1916): 5–11; 139–46; *Brotéria: Zoologia* 15 (1917): 5–17; 63–68.

Nemésio, Vitorino. "Perfil de Aurélio Quintanilha." *Brotéria: Ciências naturais* 44 (1975): 175–88.

Nemésio, Vitorino. "Perfil de Aurélio Quintanilha." *Brotéria: Genética* 14 (1993): 55–65.

Neto, Vítor. "Miguel Bombarda e Manuel Fernandes Santana: Um confronto de ideias." *Revista portuguesa de história* 2 (2003): 303–11.

Neumann, Manuel Sánchez-Navarro. "O recente terremoto de Messina." *Brotéria: Vulgarização científica* 8 (1909): 100–9.

Neumann, Manuel Sánchez-Navarro. "Os tremores de terra estudados sem instrumentos." *Brotéria: Vulgarização científica* 6 (1907): 217–31; 262–80.

Neumann, Manuel Sánchez-Navarro. "Os ultimos descobrimentos em sismologia." *Brotéria: Vulgarização científica* 12 (1914): 36–45; 108–14.

Neves, Maria Luísa. "Recordando o Padre Luisier: Nos 40 anos do seu falecimento." *Brotéria: Genética* 18 (1997): 99–101.

Nieto-Galan, Agustí. "Antonio Gramsci Revisited: Historians of Science, Intellectuals, and the Struggle for Hegemony." *History of Science* 49, no. 4 (2011): 453–78.

Nieto-Galan, Agustí. *Los públicos de la ciencia: Expertos y profanos a través de la historia.* Madrid: Fundación Jorge Juan, Marcial Pons Historia, 2011.

Novo, A. Vieira. "A indústria corticeira." *Brotéria: Vulgarização científica* 21 (1923): 207–12.

Numbers, Ronald L. *The Creationists: From Scientific Creationism to Intelligent Design.* Cambridge, MA: Harvard University Press, 2006.

Numbers, Ronald L. ed. *Galileo Goes to Jail and Other Myths about Science and Religion.* Cambridge, MA: Harvard University Press, 2010.

Ó, Jorge Ramos do. *Ensino liceal (1836–1975).* Lisbon: Ministério da Educação, 2009.

O'Brien, Elinor, and William D. Sullivan. "Electrophoretic Patterns of Proteins Constituents in Tissues of Tumor-Bearing and Non-tumor-bearing Animals." *Brotéria: Ciências naturais* 43 (1974): 3–14.

O'Malley, John W. "The Distinctiveness of the Society of Jesus." *Journal of Jesuit Studies* 3, no. 1 (2016): 1–16.

O'Malley, John W. *The First Jesuits.* Cambridge, MA: Harvard University Press, 1993.

O'Malley, John W., Gauvin Alexander Bailey, Steven J. Harris, and T. Frank Kennedy, eds. *The Jesuits: Cultures, Sciences and the Arts, 1540–1773.* Toronto: Toronto University Press, 1999.

O'Malley, John W., Gauvin Alexander Bailey, Steven J. Harris, and T. Frank Kennedy, eds. *The Jesuits II: Cultures, Sciences and the Arts, 1540–1773.* Toronto: Toronto University Press, 2006.

Olaio, Nuno. "Carlos João Rademaker (1828–1885): Percurso do restaurador da Companhia de Jesus em Portugal." *Lusitania sacra* 12 (2002): 65–119.

Oliveira, Waldir Freitas, and Edilece Souza Couto. *Colégio Antônio Vieira (1911–2011): Vidas e histórias de uma missão jesuíta.* Salvador da Baía: EDUFBA, 2011.

Padberg, John W., S.J., and George E. Ganss, eds. and trans. *The Constitutions of the Society of Jesus and Their Complementary Norms: A Complete English Translation of the Official Latin Texts.* St. Louis, MO: Institute of Jesuit Sources, 1996.

Palhiha, Ruy Telles. "Cartas de Alphonse de Candolle ao conde de Ficalho." *Brotéria: Ciências naturais* 17 (1948): 172–81.

Palhiha, Ruy Telles. "Cartas de G. Schweinfurth para o conde de Ficalho." *Brotéria: Ciências naturais* 18 (1949): 167–72.

Palhiha, Ruy Telles. "As Estufas do Jardim Botânico de Lisboa: Cartas do Dr. Goeze ao conde de Ficalho." *Brotéria: Ciências naturais* 24 (1955): 5–26.

Palhiha, Ruy Telles. "Quatro cartas inéditas de Isaac Newton ao conde de Ficalho." *Brotéria: Ciências naturais* 17 (1948): 69–72.

Park, Katharine, and Lorraine Daston, eds. *The Cambridge History of Science*, vol. 3, *Early Modern Science*. Cambridge: Cambridge University Press, 2006.

Pavone, Sabina. *The Wily Jesuits and the* Monita secreta: *The Forged Secret Instructions of the Jesuits; Myth and Reality*. St. Louis, MO: Institute of Jesuit Sources, 2005.

Pavur, Claude. "The Historiography of Jesuit Pedagogy." In *Jesuit Historiography Online,* edited by Robert A. Maryks. Leiden: Brill, 2016. DOI: 10.1163/2468-7723_jho_COM_194129.

Pavur, Claude trans. *The* Ratio studiorum: *The Official Plan for Jesuit Education*. St. Louis, MO: Institute of Jesuit Sources, 2005.

Pedersen, Olaf. *The Two Books: Historical Notes on Some Interactions between Natural Science and Theology*. Vatican: Vatican Observatory Foundation, 2007.

Pereira, Álvaro Santos. "The Opportunity of a Disaster: The Economic Impact of the 1755 Lisbon Earthquake." *Journal of Economic History* 69 (2009): 466–99.

Pina, Ambrósio. *Carlos Rademaker, 1828–1885: Restaurador dos jesuítas em Portugal no século XIX*. Porto: Livraria Apostolado da Imprensa, 1967.

Pinto, António de Oliveira. "O estudo da radioactividade da materia: Laboratorio de Giff." *Brotéria: Vulgarização científica* 11 (1913): 51–68.

Pinto, António de Oliveira. "O instituto de sciencias naturaes do collegio de Campolide." *O nosso collegio* 5 (1908–9): 99–113.

Pinto, António de Oliveira. "O instituto de sciencias naturaes do collegio de Campolide." *O nosso collegio* 6 (1909–10): 143–49.

Pinto, António de Oliveira. *Primeira contribuição para o estudo da radioactividade das aguas mineraes de Portugal*. Porto: Typographia Occidental, 1910.

Pinto, António de Oliveira. "Première contribution a l'étude de la radioactivité des eaux minérales du Portugal." In *II Congrès international de radiologie et d'électricité*, 3–8. Brussels: Imprimerie Médicale et Scientifique L. Severeyns, 1911.

Pinto, António de Oliveira. "Primeiro congresso internacional de radiologia e ionização." *Brotéria* 5 (1906): 129–34.

Pinto, António de Oliveira. "A radioactividade das aguas medicinaes de fraca mineralisação." *Brotéria: Vulgarização científica* 13 (1915): 244–45.

Pinto, António de Oliveira. "Telegraphia sem fio." *Brotéria: Vulgarização científica* 9 (1910): 181–203.

Pinto, António de Oliveira. "Telegraphia sem fio." *Brotéria: Vulgarização científica* 10 (1912): 14–33; 200–45.

Pinto, Manuel Resende. "Hymeniales de Portugal." *Brotéria: Ciências naturais* 11 (1942): 5–20; 62–80; 106–36; 153–57; *Brotéria: Ciências naturais* 12 (1943): 5–20; 58–75; 116–27; 180–90.

Pinto, Manuel Resende. "Hymeniales de Portugal." *Brotéria: Ciências naturais* 14 (1945): 26–42.

Pinto, Manuel Resende. "Hymeniales de Portugal: Hydnaceae." *Brotéria: Ciências naturais* 13 (1944): 25–36.

Pinto, Manuel Resende. "Hymeniales de Portugal: Polyporaceae." *Brotéria: Ciências naturais* 13 (1944): 30–36; 131–40.

Pinto, Sérgio Ribeiro. *Separação religiosa como modernidade: Decreto-Lei de 20 de Abril de 1911 e modelos alternativos*. Lisbon: Centro de Estudos de História Religiosa–Universidade Católica Portuguesa, 2011.

Preto, José Ramos. *Relatório sobre o extincto collegio de São Fiel da Companhia de Jesus*. Lisbon: Typographia Mauricio, 1911.

Prieto, Andrés I. *Missionary Scientists: Jesuit Science in Spanish South America, 1570–1810*. Nashville, TN: Vanderbilt University Press, 2011.

"A producção vinicola em Portugal em 1913." *Brotéria: Vulgarização científica* 12 (1914): 313–15.

Proença, Maria Cândida. "A reforma de Jaime Moniz. Antecedentes e destino histórico." PhD diss., New University of Lisbon, 1993.

Pujiula, Jaime. "Algunas observaciones citológicas sobre todo en *Hedera helix* y *Solanum tuberosum*, y la primera variante del método tano-argéntico introducida por el. Sr. del Rio Hortega." *Brotéria: Botânica* 17 (1919): 67–96.

Pujiula, Jaime. "Aplicación de algunos métodos de tinción histológica animal al reino vegetal." *Brotéria: Botânica* 25 (1931): 51–64.

Pujiula, Jaime. "La autonomia de la vida." *Brotéria: Ciências naturais* 26 (1957): 194–96.

Pujiula, Jaime. "La biología filosófica." *Brotéria: Ciências naturais* 24 (1955): 191–94.

Pujiula, Jaime. "Contribución anatómico-fisiológica al estudio de la caída de las hojas." *Brotéria: Botânica* 16 (1918): 71–83.

Pujiula, Jaime. "Contribución al conocimiento anatómico-fisiológico de los zarcillos de la zarzaparilla." *Brotéria: Botânica* 19 (1921): 66–72.

Pujiula, Jaime. "Dados citológicos sobre el meristemo radical de *Vicia faba* L." *Brotéria: Botânica* 21 (1924): 90–96.

Pujiula, Jaime. "El hadroma de las hojas del pino." *Brotéria: Ciências naturais* 16 (1947): 173–91.

Pujiula, Jaime. "Observaciones en *Vinca maior*, v. variegata." *Brotéria: Ciências naturais* 22 (1953): 60–65.

"Prólogo." *Brotéria: Vulgarização científica* 6 (1907): 5–9.

Quental, Antero de. *A decadência dos povos peninsulares*. Porto: Typographia Comercial, 1871; reprinted as: Carlos Reis, ed., *As conferências do casino*, 95–128. Lisbon: Alfa, 1990.

Quintanilha, Aurélio. "História da genética em Portugal." *Brotéria: Ciências naturais* 44 (1975): 189–208.

Quintanilha, Aurélio. "In memoriam: Ruy Telles Palhinha." *Boletim da sociedade portuguesa de ciências naturais* 14 (1972): 7–12.

Rabin, Sheila J. "Early Modern Jesuit Science: A Historiographical Essay." *Journal of Jesuit Studies* 1, no. 1 (2014): 88–104.

Rademaker, Carlos. *O discurso fúnebre nas solenes exéquias do Exmo: Sr. D. Carlos Mascarenhas*. Lisbon: Imprensa Nacional, 1861; reprinted in: Marquês de Fronteira, *Memórias do marquês de Fronteira e d'Alorna D. José Trazimundo Mascarenhas Barreto ditadas por ele próprio em 1861*, 5:348–60. Coimbra: Imprensa da Universidade, 1928–32.

Rambo, Arthur Blásio, and Arthur Rabuske. *Pe. João Evangelista Rick, S.J.: Cientista, colonizador, apóstolo social, professor*. São Leopoldo: Unisinos, 2004.

Ramos, Rui, Bernardo Vasconcelos e Sousa, and Gonçalo Monteiro. *História de Portugal*. 8th ed. Lisbon: Esfera dos Livros, 2009.

Rebimbas, Manuel. "Noções de hygiene alimentar." *Brotéria: Vulgarização científica* 6 (1907): 155–69.

Rebimbas, Manuel. "Radioactividade e gazes raros das fontes thermaes." *Brotéria: Vulgarização científica* 7 (1908): 181–84.

"Redacção da *Brotéria*." *Brotéria: Vulgarização científica* 6 (1907): 10–11.

Refóios, Joaquim Augusto de Sousa. *O collegio de São Fiel no Louriçal do Campo e o de Nossa Senhora da Conceição na Covilhã: Apontamentos sobre o jesuitismo no districto de Castello-Branco*. Coimbra: Imprensa da Universidade, 1883.

Reis, Fernando Egídio, ed. *Felicidade, utilidade e instrução: A divulgação científica no Jornal enciclopédico dedicado à Rainha, 1779; 1788–1793; 1806*. Porto: Porto Editora, 2006.

Reis, Fernando Egídio. "Scientific Dissemination in Portuguese Encyclopaedic Periodicals, 1779–1820;" *History of Science* 45 (2007): 83–118.

Relação abbreviada da republica, que os religiosos jesuitas das provincias de Portugal, e Hespanha, estabelecerão nos dominios ultramarinos das duas monarquias, e da guerra, que nelles tem movido, e sustentado contra os exercitos hespanhoes e portuguezes: Formada pelos registos das secretarias dos dous respectivos principaes commissarios, e plenipotenciarios; E outros documentos authenticos. Lisbon, 1757.

Resende, Flávio, and Karl von Poellnitz. "Suculentas africanas: Variedades novas de Haworthia tessellata Haw." *Brotéria: Ciências naturais* 11 (1942): 49–55.

Rick, Johannes. "Fungi gelatinosi et cartilaginosi rio-grandenses." *Brotéria: Botânica* 21 (1924): 131–41.

Rick, Johannes. "Fungos dos arredores de Torres Vedras," *Brotéria* 4 (1905): 159–63.

Rick, Johannes. "Fungos do Rio Grande do Sul." *Brotéria* 3 (1904): 276–93.

Rick, Johannes. "Genus Stereum riograndense." *Brotéria: Ciências naturais* 9 (1940): 42–47; 75–90; 139–48.

Rick, Johannes. "Monographia das Hypoxyleas riograndenses." *Brotéria: Botânica* 25 (1931): 21–50.

Rick, Johannes. "Monographia Pezizinearum Riograndensium." *Brotéria: Botânica* 25 (1931): 77–122.

Rick, Johannes. "Monografia das Rosellinias riograndenses." *Brotéria: Ciências naturais* 1 (1932): 183–92.

Rick, Johannes. "Monografia das Valsíneas do Rio Grande do Sul." *Brotéria: Ciências naturais* 2 (1933): 83–99.

Rick, Johannes. "Pilze aus Rio Grande do Sul (Brasilien)." *Brotéria* 5 (1906): 5–53.

Rick, Johannes. "Polyporeaceae riograndenses." *Brotéria: Ciências naturais* 6 (1937): 153–68.

Rick, Johannes. "Polypori riograndenses." *Brotéria: Ciências naturais* 3 (1934): 180–89; *Brotéria: Ciências naturais* 4 (1935): 17–27; 84–94.

Rick, Johannes. "Polysticti riograndenses." *Brotéria: Ciências naturais* 4 (1935): 121–38; *Brotéria: Ciências naturais* 5 (1936): 171–78; *Brotéria: Ciências naturais* 6 (1937): 81–87.

Rick, Johannes. "Poriae riograndenses." *Brotéria: Ciências naturais* 6 (1937): 128–52.

Robin, Albert. *Les ferments métalliques et leur emploi en thérapeutique*. Paris: J. Rueff Éditeur, 1907.

Robinson, Richard Alan Hodgson. "The Religious Question and the Catholic Revival in Portugal, 1900–30." *Journal of Contemporary History* 12 (1977): 345–62.

Rodrigues, Ana Duarte. *Horticultura para todos*. Lisbon: Biblioteca Nacional de Portugal–Junta de Freguesia de Alvalade, 2017.

Rodrigues, Ana Duarte, and Ana Simões. "Horticulture in Portugal 1850–1900: The Role of Science and Public Utility in Shaping Knowledge." *Annals of Science* 74 (2017): 192–213.

Rodrigues, Francisco. *A formação intellectual do jesuíta: Leis e factos*. Porto: Livraria Magalhães e Moniz, 1917.

Rodrigues, Francisco. *História da Companhia de Jesus na assistência de Portugal*. 7 Vols. Porto: Livraria Apostolado da Imprensa, 1931–50.

Rodrigues, Francisco. *Jesuítas portugueses astrónomos na China, 1583–1805*. Porto: Tipografia Porto Médico, 1925) [reprint Macau: Instituto Cultural de Macau, 1990].

Rodrigues, Francisco. *Jesuitophobia: Resposta serena a uma diatribe*. Porto: Typographia Lusitania, 1917.

Rodrigues, Francisco. "Mathematicos portugueses na China." *Revista de história* 12 (1923): 81–118.

Rodrigues, Teresa Ferreira et al. *História da população portuguesa: Das longas permanências à conquista da modernidade*. Porto: CEPESE–Afrontamento, 2008.

Roehner, Bertrand. "Jesuits and the State: A Comparative Study of Their Expulsions (1590–1990)." *Religion* 27 (1997): 165–82.

Romano, Antonella. *La Contre-Réforme mathématique: Constitution et diffusion d'une culture mathématique jésuite à la Renaissance (1540–1640)*. Rome: École Française de Rome, 1999.

Romano, Antonella. "Teaching Mathematics in Jesuit Schools: Programs, Course Content, and Classroom Practices." In *The Jesuits II: Cultures, Sciences and the Arts, 1540–1773*, edited by John W. O'Malley, Gauvin Alexander Bailey, Steven J. Harris, and T. Frank Kennedy, 355–70. Toronto: Toronto University Press, 2006.

Romeiras, Francisco Malta. "Constituição e percurso das colecções científicas dos jesuítas exilados pela 1ª República: O caso de Carlos Zimmermann S.J. (1871–1950)." *Archivum historicum Societatis Iesu* 168 (2015): 287–327.

Romeiras, Francisco Malta. "The Emergence of Molecular Genetics in Portugal: The Enterprise of Luís Archer S.J." *Archivum historicum Societatis Iesu* (164) (2013): 501–12.

Romeiras, Francisco Malta. "For the Greater Credibility: Jesuit Science and Education in Modern Portugal (1858–2002)." *History of Science* 56 (2018): 97–119. DOI: 10.1177/0073275317738599.

Romeiras, Francisco Malta. "Francisco Saldanha da Gama: The Last Visitor of the Portuguese Assistancy." In *With Eyes and Ears Open: The Role of Visitors in the Society of Jesus*, edited by Thomas M. McCoog, 172–90. Leiden: Brill, 2019.

Romeiras, Francisco Malta. "Jesuit Historiography in Modern Portugal." In *Jesuit Historiography Online*, edited by Robert A. Maryks. Leiden: Brill, 2016. DOI: 10.1163/2468-7723_jho_COM_192570.

Romeiras, Francisco Malta. "The Journal *Brotéria* (1902–2002): Jesuit Science in the 20th Century." *HoST: Journal for the History of Science and Technology* 6 (2012): 100–9.

Romeiras, Francisco Malta. *Ciência, prestígio e devoção: Os jesuítas e a ciência em Portugal (séculos XIX e XX)*. Cascais: Lucerna, 2015.

Romeiras, Francisco Malta. "A última conversa com o P. Luís Archer. Um concílio entre Fé e Ciência?" *Brotéria* 173 (2011): 239–245.

Romeiras, Francisco Malta, and Henrique Leitão. "Jesuítas e ciência em Portugal I: António Oliveira Pinto S.J. e as primeiras experiências com radioactividade em Portugal." *Brotéria* 174 (2012): 9–20.

Romeiras, Francisco Malta, and Henrique Leitão. "Jesuítas e ciência em Portugal II: Carlos Zimmermann S.J. e o ensino da microscopia vegetal." *Brotéria* 174 (2012): 113–25.

Romeiras, Francisco Malta, and Henrique Leitão. "Jesuítas e ciência em Portugal III: As expedições científicas e as observações dos eclipses solares de 1900 e 1905." *Brotéria* 174 (2012): 227–37.

Romeiras, Francisco Malta, and Henrique Leitão, "Jesuítas e ciência em Portugal IV: A revista *Brotéria: Sciencias naturaes* e a sua recepção nacional e internacional." *Brotéria* 174 (2012): 323–33.

Romeiras, Francisco Malta, and Henrique Leitão. "Jesuítas e ciência em Portugal V: Os Colégios de Campolide e de São Fiel e a implantação da República." *Brotéria* 174 (2012): 425–40.

Romeiras, Francisco Malta, and Henrique Leitão eds. *Obra selecta do padre Luís Archer, S.J.* 4 Vols. Lisbon: Fundação Calouste Gulbenkian, 2015–17.

Romeiras, Francisco Malta, and Henrique Leitão. "One Century of Science: the Jesuit Journal *Brotéria* (1902–2002)." In *Exploring Jesuit Distinctiveness: Interdisciplinary Perspectives on Ways of Proceeding within the Society of Jesus*, edited by Robert A. Maryks, 235–58. Leiden: Brill, 2016.

Romeiras, Francisco Malta, and Henrique Leitão. "The Role of Science in the History of Portuguese Anti-Jesuitism." *Journal of Jesuit Studies* 2, no. 1 (2015): 77–99. DOI: 10.1163/22141332-00201004.

Rosa, Giuseppe de. La civiltà cattolica: *150 anni al servizio della chiesa, 1850–1999*. Rome: La Civiltà Cattolica, 1999.

Rozeira, Arnaldo. "Gonçalo Sampaio como sistemata." *Brotéria: Ciências naturais* 15 (1946): 49–55.

Rozeira, Arnaldo. "Plantas novas ou pouco citadas para Trás-os-Montes e Alto-Douro." *Brotéria: Ciências naturais* 12 (1943): 145–71.

Rozeira, Arnaldo. "A secção *Stoechas Gingins*, do género *Lavandula Linn*." *Brotéria: Ciências naturais* 18 (1949): 5–84.

Rudolph, John. *Scientists in the Classroom: The Cold War Reconstruction of American Science Education*. New York: Palgrave, 2002.

Salema, Roberto. "Amido: Estudo ultraestrutural da sua biogénese em plantas superiores." *Brotéria: Ciências naturais* 38 (1969): 1–127.

Salvado, Maria Adelaide Neto. "O colégio de S. Fiel: Centro difusor da ciência." In *Fé, ciência, cultura:* Brotéria; *100 anos*, edited by Hermínio Rico and José Eduardo Franco, 211–31. Lisbon: Gradiva, 2003.

Salvado, Maria Adelaide Neto. *O colégio de S. Fiel: Centro difusor da ciência no interior da Beira*. Castelo Branco: Semedo-Soc. Tipográfica, 2001.

Sampaio, Gonçalo. "*Centaurea luisieri* (sp. n.)." *Brotéria: Botânica* 14 (1916): 104–5.

Sampaio, Joaquim. "As algas marinhas portuguesas e a extracção do iodo." *Brotéria: Ciências naturais* 16 (1947): 165–72.

Sampaio, Joaquim. "Cianófitas da Serra da Estrela." *Brotéria: Ciências naturais* 16 (1947): 105–13.

Sampaio, Joaquim. "Uma localidade nova para o *Ascophyllum nodosum* Le Jol., descoberta pelo Prof. G. Sampaio." *Brotéria: Ciências naturais* 17 (1948): 56–58.

Sampaio, Joaquim. "Sinopse das desmídias conhecidas na flora portuguesa." *Brotéria: Ciências naturais* 12 (1943): 97–115; *Brotéria: Ciências naturais* 13 (1944): 6–24; 88–130; *Brotéria: Ciências naturais* 14 (1945): 128–48.

Sampaio, Joaquim. "Subsídios para a história da botânica em Portugal I: O colector Isaac Newton e o estudo das criptogâmicas celulares portuguesas." *Brotéria: Ciências naturais* 15 (1946): 20–52.

Sampaio, Joaquim. "Subsídios para a história da botânica em Portugal II: O Dr. Romualdo Fragoso, o Dr. Gonçalo Sampaio e a micologia portuguesa." *Brotéria: Ciências naturais* 17 (1948): 115–32; *Brotéria: Ciências naturais* 18 (1949): 86–92; 97–108.

Sampaio, Tristão Mello. "Bactérias fixadores de azoto atmosférico em trigo." *Brotéria: Genética* 1 (1980): 83–85.

Sanches, António Ribeiro. *Cartas sobre a educação da mocidade*. Cologne: n.p., 1760; reprint: Porto: Domingos Barreira, n.d..

Santana, Manuel Fernandes. *Questões de biologia: O materialismo em face da sciencia. A propósito da consciência e livre arbítrio do Sr. Prof. Miguel Bombarda*. 2 Vols. Lisbon: Typographia da Casa Catholica, 1899–1900.

Santo, Joaquim Campo. "Padre Carlos Rademaker." *Novo mensageiro do Coração de Jesus* 23 (1903): 409–10.

Sapega, Ellen W. *Consensus and Debate in Salazar's Portugal: Visual and Literary Negotiations of the National Text, 1933–1948*. University Park, PA: Pennsylvania State University Press, 2008.

Saraiva, Tiago. "Fascist Labscapes: Genetics, Wheat and the Landscapes of Fascism in Italy and Portugal." *Historical Studies in the Natural Sciences* 40 (2010): 457–98.

Saraiva, Tiago. *Fascist Pigs: Technoscientific Organisms and the History of Fascism*. Cambridge, MA: MIT Press, 2016.

Saraiva, Tiago. "Laboratories and Landscapes: The Fascist New State and the Colonization of Portugal and Mozambique." *Journal for the History of Science and Technology* 3 (2009): 7–39.

Sarreira, Raúl Dias. "A propósito de um livro recente: A; Emyieu, *La part des croyants dans les progrès de la science au XIXe siècle*." *Brotéria: Vulgarização científica* 19 (1921): 184–87.

Sarreira, Raúl Dias. "Raios e Pára-raios." *Brotéria: Vulgarização científica* 14 (1916): 215–21.

Sarreira, Raúl Dias. "Os raios X." *Brotéria: Vulgarização científica* 12 (1914): 155–67.

Sarreira, Raúl Dias. "Os raios X." *Brotéria: Vulgarização científica* 15 (1917): 79–90; 118–24; 230–36.

Sarton, George. "Preface to Volume 40: An Appeal for the Republication in Book Form of Father Bosmans' Studies on Belgian Mathematics in the Sixteenth and Seventeenth Centuries." *Isis* 40 (1949): 3–6.

Schlafly, Daniel L. Jr. "General Suppression, Russian Survival, American Success: The 'Russian' Society of Jesus and the Jesuits in the United States." In *The Jesuit Suppression in Global Context: Causes, Events, and Consequences*, edited by Jeffrey D. Burson and Jonathan Wright, 201–15. Cambridge: Cambridge University Press, 2015.

Schul, Hans-Walter. *The Kaiser Wilhelm Institute for Anthropology, Human Heredity and Eugenics, 1927–1935: Crossing Boundaries*. Dordrecht: Springer, 2008.

Schupp, Ambrósio. "A evolução e o homem." *Brotéria: Vulgarização científica* 8 (1909): 81–94; 147–63.

Scott, James Brown. "The Revolution in Portugal: The Recognition of New Governments." *American Journal of International Law* 5 (1911): 178–81.

Seabra, Eurico de. *A egreja, as congregações e a república*. 2nd ed. Lisbon: Clássica, 1914.

Seabra, João. *O estado e a igreja em Portugal no início do século XX: A Lei da Separação de 1911*. Cascais: Princípia, 2009.

Secord, James. "Knowledge in Transit." *Isis* 95 (2004): 654–72.

Sequeiros, Jorge, and Paula Coutinho. "Genetic Aspects of Machado-Joseph Disease." *Brotéria: Genética* 2 (1981): 137–47.

Serra, José Antunes. "Aurélio Quintanilha: Algumas palavras acerca da sua obra como professor e investigador." *Brotéria: Ciências naturais* 44 (1975): 157–74.

Serra, José Antunes. "Constitution of Hair Melanins." *Nature* 157 (1945): 771–72.

Serra, José Antunes. "Cytochemical Demonstration of Masked Lipids." *Science* 128 (1958): 28–29.

Serra, José Antunes. *Modern Genetics*. 3 Vols. London and New York: Academic Press, 1965–68.

Serra, José Antunes. *Moderna genética: Geral e fisiológica*. 2 Vols. Coimbra: Author Edition, 1949.

Serra, José Antunes. "Interpretation of Nucleolar Inclusions." *Nature* 181 (1958): 1544–45.

Serra, José Antunes. "A Position Effect Explanation of the Gene Conversion." *Nature* 183 (1959): 700.

Serra, José Antunes. "A Possible Explanation of the Relations between Hereditary Anemias and Depigmentation in the Mouse." *Nature* 159 (1947): 504–5.

Serra, José Antunes. "Professor Aurélio Quintanilha: Impressões e recordações pessoais de homenagem." *Brotéria: Genética* 9 (1988): 9–17.

Serra, José Antunes, and A. Queiroz Lopes. "Chemical Constitution of Nucleolar Inclusions in Growing Oocyte Cells." *Nature* 155 (1945): 792.

Serra, José Antunes, and Arturo Pereira. "Quantitative Microdetermination of Amino Acids after Paper Chromatography." *Science* 113 (1951): 387–88.

Serra, José Antunes, and Paulo Picciochi, "Demonstration of Canonic Gonial Mitosis and Meiosis in *Parascaris equorum*." *Science* 132 (1960): 1400–1.

Serrão, Joaquim Veríssimo. *História de Portugal*. 18 Vols. Lisbon: Verbo, 1977–2010.

"A servir de prólogo nesta segunda série." *Brotéria: Fé; Sciências; Letras* 1 (1925): 5–6.

Shapin, Steven. *The Scientific Revolution*. Chicago: University of Chicago Press, 1996.

Silva, António da Ferreira. "*Brotéria*." *Revista de chimica pura e applicada* 7 (1911): 377.

Silva, António da Ferreira. "*Brotéria* e os seus naturalistas." *Revista de chimica pura e applicada* (1918): 75–79.

Silva, António da Ferreira. "Director e redatores da *Brotéria*: Um appello aos homens de boa vontade." *Revista de chimica pura e applicada* 6 (1910): 362–63.

Silva, António da Ferreira. "Os redactores da *Brotéria* exilados; as suas collecções scientificas confiscadas e perdidas." *Revista de chimica pura e applicada* 7 (1911): 229–31.

Silva, António da Ferreira. *A suposta salicilagem dos vinhos portugueses no Brasil (1900–1902): Memórias, notas e documentos* (Coimbra: Imprensa da Universidade, 1919).

Silva, António da Ferreira. "Testemunho em favor da *Brotéria*." *Brotéria: Vulgarização científica* 16 (1918): 190–91.

Silva, António da Ferreira. "Os vinhos do Porto." *Brotéria: Vulgarização científica* 13 (1915): 129–39; 181–95.

Silva, César da. *A execução dos Távoras*. Lisbon: João Romano Torres, 1921.

Silva, Isabel Corrêa da. "A Rosa brasileira que incendiou a questão religiosa em Portugal: O Caso Calmon (1899–1901)." *Tempo* 20 (2014): 1–20. DOI: 10.1590/TEM-1980-542X-2014203626.

Silva, Joaquim Candeias da. "Evocação/Memória de alguns médicos notáveis da Beira Interior: Concelho do Fundão (IV)." *Medicina na Beira Interior da Pré-história ao século XX: Cadernos de cultura* 19 (2005): 83–88.

Silva, Joaquim Candeias da. "Evocação/Memória de alguns médicos notáveis da Beira Interior: Concelho do Fundão (V)." *Medicina na Beira Interior da Pré-História ao século XX: Cadernos de cultura* 20 (2006): 70–76.

Silva, José Seabra da. *Deducão chronológica e analítica na qual se manifestão pela sucessiva serie de cada hum dos reynados da monarquia Portuguesa, que decorrerão desde o Governo do Senhor Rey D. João III até o presente, os horrorosos estragos, que a Companhia denominada de Jesus fez em Portugal, e todos os seus domínios por hum plano, e systema por ella inalteravelmente seguido desde que entrou neste Reyno, até que foi delle proscripta, e expulsa pela justa, sabia, e providente Ley de 3 de Setembro de 1759*. 3 Vols. Lisbon: Officina de Miguel Manescal da Costa, 1767–68.

Silva, Lúcio Craveiro da. "A Faculdade de Filosofia de Braga e a sua *Revista portuguesa de filosofia*: Mensagem e evolução." *Revista portuguesa de filosofia* 51 (1995): 281–94.

Silva, Maria de Lourdes Sampaio. "Algumas palavras e recordações de homenagem ao Prof. José Antunes Serra." *Brotéria: Genética* 13 (1992): 19–21.

Simões, Ana, Ana Carneiro, Maria Paula Diogo, Luís Miguel Carolino, and Teresa Salomé Mota. *Uma história da Faculdade de Ciências da Universidade de Lisboa (1911–1974)*. Lisbon: Faculdade de Ciências da Universidade de Lisboa, 2013.

Skuhravá, Marcela, Václav Skuhravý, Javier Blasco-Zumeta, and Juli Pujade Villar. "Gall Midges (Diptera: Cecidomyiidae) of the Iberian Peninsula." *Boletín de la Asociación española de entomología* 20 (1996): 41–61.

Skuhravá, Marcela, Václav Skuhravý, Javier Blasco-Zumeta, and Juli Pujade Villar. "Gall Midges (Diptera: Cecidomyiidae) of the Iberian Península 2: Zoogeographical Analysis of the Gall Midge Fauna." *Boletín de la Asociación española de entomología* 30 (2006): 93–159.

Smolarski, Dennis C. "The Jesuit *Ratio studiorum*, Christopher Clavius, and the Study of Mathematical Sciences in Universities." *Science in Context* 15 (2002): 447–57.

Sommervogel, Carlos. *Table méthodique des* Mémoires de Trévoux *(1701–1775)*. Paris: Imprimerie de E. Donneaud, 1864–65.

Sullivan, William D. "The Spectrophotometric Determination of Malic Dehydrogenase and 'Malic' Enzyme in Normal Populations of *Tetrahymena pyriformis* GL." *Brotéria: Ciências naturais* 33 (1964): 143–58.

Sullivan, William D., and James. V. Boyle. "The Effect of X-Radiation on the Succinic Dehydrogenase Activity on the Dividing Stages of *Tetrahymena pyriformis* GL." *Brotéria: Ciências naturais* 30 (1961): 77–82.

Sullivan, William D., and Adam J. von Knobelsdorff. "The *in vitro* and *in vivo* Effects of Fluoride on Succinic Dehydrogenase Activity." *Brotéria: Ciências naturais* 31 (1962): 3–13.

Tavares, Carlos das Neves. "Líquenes da Serra da Estrela (Contribuição para o seu estudo)." *Brotéria: Ciências naturais* 14 (1945): 14–25, 49–61.

Tavares, Carlos das Neves. "Notas lichénologiques." *Brotéria: Ciências naturais* 16 (1947): 145–57.

Tavares, Carlos das Neves. "Notas liquenológicas." *Brotéria: Ciências naturais* 11 (1942): 42–48.

Tavares, Carlos das Neves. "Notes lichenologiques." *Bulletin de la Société portugaise de sciences naturelles* 13, no. 11 (1939): 53–60; *Bulletin de la Société portugaise de sciences naturelles* 13, no. 28 (1941): 157–67.

Tavares, Carlos das Neves. "Nótulas de Fanerogamia portuguesa." *Brotéria: Ciências naturais* 28 (1959): 69–75.

Tavares, Carlos das Neves. "Vida e actividade científica do Prof. Dr. Flávio Ferreira Pinto Resende." *Portugaliae acta biologica: Serie B; Sistemática, ecologia, biogeografia e paleontologia* 9 (1967–68): 1–63.

Tavares, Joaquim da Silva. "O analphabetismo em Portugal." *Brotéria: Vulgarização científica* 16 (1918): 111–20; 151–64; 229–36.

Tavares, Joaquim da Silva. "A Argentina: Impressões de um excursionista." *Brotéria: Vulgarização científica* 11 (1913): 350–69.

Tavares, Joaquim da Silva. "A *Brotéria* nos primeiros 25 anos da sua existência." In *Brotéria: Número especial no XXV aniversário da sua fundação*, 19–37. Caminha: Brotéria, 1927.

Tavares, Joaquim da Silva. "A *Brotéria* e as revistas scientificas extrangeiras." *Brotéria* 2 (1903): 234–35.

Tavares, Joaquim da Silva. "Capital modelo: Bello Horizonte." *Brotéria: Vulgarização científica* 12 (1914): 65–107.

Tavares, Joaquim da Silva. "A carestia da vida em Portugal." *Brotéria: Vulgarização científica* 20 (1922): 142–49.

Tavares, Joaquim da Silva. "O carvão líquido e a catálise: Vantagens industriais." *Brotéria: Vulgarização científica* 20 (1922): 66–70.

Tavares, Joaquim da Silva. "Cecidologia brasileira." *Brotéria: Zoologia* 16 (1918): 21–28; *Brotéria: Zoologia* 18 (1920): 82–125; *Brotéria: Zoologia* 19 (1921): 76–112; *Brotéria: Zoologia* 20 (1922): 5–48.

Tavares, Joaquim da Silva. "Cécidologie Argentine." *Brotéria: Zoologia* 13 (1915): 88–128.

Tavares, Joaquim da Silva. "O censo da população portuguesa feito em Dezembro de 1920." *Brotéria: Vulgarização científica* 22 (1924): 207–16.

Tavares, Joaquim da Silva. "Uma cidade encantada no Estado do Paraná." *Brotéria: Vulgarização científica* 14 (1916): 201–9.

Tavares, Joaquim da Silva. "A cidade de Parahyba do Norte." *Brotéria: Vulgarização científica* 14 (1916): 81.

Tavares, Joaquim da Silva. "A cidade de Santos." *Brotéria: Vulgarização científica* 10 (1912): 101–38.

Tavares, Joaquim da Silva. "Contributio prima ad cognitionem cecidologiae regionis Braziliae." *Brotéria: Zoologia* 8 (1909): 5–29.

Tavares, Joaquim da Silva. "Contributio prima ad cognitionem cecidologiae regionis Zambeziasis." *Brotéria: Zoologia* 7 (1908): 133–74.

Tavares, Joaquim da Silva. "O Corcovado." *Brotéria: Vulgarização científica* 10 (1912): 189–99.

Tavares, Joaquim da Silva. "A cortiça portuguesa." *Brotéria: Vulgarização científica* 15 (1917): 53–65.

Tavares, Joaquim da Silva. "A criminalidade em Portugal." *Brotéria: Vulgarização científica* 17 (1919): 114–17.

Tavares, Joaquim da Silva. "A cultura da soja." *Brotéria: Vulgarização científica* 22 (1924): 29–34.

Tavares, Joaquim da Silva. *Cynipidae peninsulae Ibericae*. 2 Vols. Lisbon: Brotéria, 1931.

Tavares, Joaquim da Silva. "Os cynípides da Península Ibérica." *Brotéria: Zoologia* 24 (1927): 16–78; *Brotéria: Zoologia* 25 (1928): 11–152; *Brotéria: Zoologia* 26 (1930): 25–53; *Brotéria: Zoologia* 27 (1931): 5–100.

Tavares, Joaquim da Silva. "Eclipse total do sol em 30 de Agosto de 1905." *Brotéria* 5 (1906): 254–60.

Tavares, Joaquim da Silva. "Espécies novas de cynípides e ceccidomyas da Península Ibérica, e descrição de algumas já conhecidas." *Brotéria: Zoologia* 14 (1916): 65–136; *Brotéria: Zoologia* 16 (1918): 130–42; *Brotéria: Zoologia* 17 (1919): 5–101; *Brotéria: Zoologia* 18 (1920): 43–69; *Brotéria: Zoologia* 20 (1922): 97–155; *Brotéria: Zoologia* 21 (1924): 5–48.

Tavares, Joaquim da Silva. "Uma excursão a Minas." *Brotéria: Vulgarização científica* 11 (1913): 126–62.

Tavares, Joaquim da Silva. "Florianópolis, capital do Estado de Santa Catharina." *Brotéria: Vulgarização científica* 14 (1916): 145–46.

Tavares, Joaquim da Silva. "O herbário do Colégio de S. Fiel." *Brotéria: Botânica* 21 (1924): 82–87.

Tavares, Joaquim da Silva. "José Veríssimo de Almeida." *Brotéria: Botânica* 13 (1915): 57–60.

Tavares, Joaquim da Silva. "Maceió, capital do Estado de Alagoas." *Brotéria: Vulgarização científica* 13 (1915): 324.

Tavares, Joaquim da Silva. "Nova contribuição para o conhecimento da cecidologia brazileira." *Brotéria: Zoologia* 22 (1925): 6–55.

Tavares, Joaquim da Silva. "Origem da energia industrial: Gás natural, petróleo e carvão." *Brotéria: Vulgarização científica* 21 (1923): 44–46.

Tavares, Joaquim da Silva. "A pesca do atum em Portugal." *Brotéria: Vulgarização científica* 20 (1922): 15–17.

Tavares, Joaquim da Silva. "A pesca no Rio Minho." *Brotéria: Vulgarização científica* 18 (1920): 262–76.

Tavares, Joaquim da Silva. "A pesca no Rio Minho." *Brotéria: Vulgarização científica* 19 (1921): 21–40.

Tavares, Joaquim da Silva. "A pesca no Rio Minho em 1921." *Brotéria: Vulgarização científica* 20 (1922): 19–23.

Tavares, Joaquim da Silva. "A pesca no Rio Minho em 1922." *Brotéria: Vulgarização científica* 21 (1923): 57–60.

Tavares, Joaquim da Silva. "As pescarias no Rio Minho em 1923." *Brotéria: Vulgarização científica* 22 (1924): 89–95.

Tavares, Joaquim da Silva. "A producção do assucar brasileiro." *Brotéria: Vulgarização científica* 12 (1914): 282–95.

Tavares, Joaquim da Silva. "Producção mundial de lã e particularmente nas colónicas britânicas." *Brotéria: Vulgarização científica* 12 (1914): 305–7.

Tavares, Joaquim da Silva. "A producção vinicola em Portugal em 1915." *Brotéria: Vulgarização científica* 14 (1916): 188–89.

Tavares, Joaquim da Silva. "O salto de Iguaçu (Estado do Paraná)." *Brotéria: Vulgarização científica* 14 (1916): 258–71.

Tavares, Joaquim da Silva. "As sciencias naturais e a religião." *Brotéria: Vulgarização científica* 8 (1909): 33–35.

Tavares, Joaquim da Silva. "A Sociedade Portuguesa de Sciencias Naturais." *Brotéria: Vulgarização científica* 6 (1907): 127–34.

Tavares, Joaquim da Silva. "A substituição da gasolina nos automóveis." *Brotéria: Vulgarização científica* 16 (1918): 73–77.

Tavares, Joaquim da Silva. "A substituição da gasolina nos automóveis pelo álcool." *Brotéria: Vulgarização científica* 16 (1918): 239–40.

Tavares, Joaquim da Silva. "Synopse das zoocecidias portuguesas." *Brotéria* 4 (1905): v–xii; 1–123.

Tavares, Joaquim da Silva. "Victória, capital do Estado do Espírito Santo." *Brotéria: Vulgarização científica* 14 (1916): 16.

Tavares, Joaquim da Silva. "O vinho português em 1916." *Brotéria: Vulgarização científica* 15 (1917): 91–95.

Tavares, Joaquim da Silva. "Zoocecidias da ilha da Madeira." *Brotéria* 2 (1903): 179–86; *Brotéria* 4 (1905): 221–27; *Brotéria: Zoologia* 12 (1914): 193–97.

Tavares, Joaquim da Silva. "As zoocecídias portuguesas: Addenda." *Brotéria* 1 (1902): 5–48.

Tavares, Joaquim da Silva. "As zoocecidias portuguesas: Enumeração das espécies até agora encontradas em Portugal e descrição de dezanove ainda não estudadas." *Anais de ciências naturais* 7 (1900): 17–106.

Theissen, Fernando. "Annotações á mycoflora brazileira I: O género Vizella." *Brotéria: Botânica* 12 (1914): 13–21.

Theissen, Fernando. "Annotações á mycoflora brazileira II: Especies novas." *Brotéria: Botânica* 12 (1914): 22–31.

Theissen, Fernando. "Le genre Asterinella." *Brotéria: Botânica* 10 (1912): 101–24.

Theissen, Fernando. "De hemisphaerialibus notae supplendae." *Brotéria: Botânica* 12 (1914): 73–96.

Theissen, Fernando. "Hipocreaceae riograndenses." *Brotéria: Botânica* 9 (1910): 121–49.

Theissen, Fernando. "Hymenomycetes riograndenses." *Brotéria: Botânica* 10 (1912): 5–28.

Theissen, Fernando. "Marasmii austro-brasilienses." *Brotéria: Botânica* 8 (1909): 53–65.

Theissen, Fernando. "Perisporiales riograndenses." *Brotéria: Botânica* 9 (1910): 5–44.

Tirapicos, Luís. "Ciência e diplomacia na corte de D. João V: A acção de João Baptista Carbone, 1722–1750." PhD diss., University of Lisbon, 2017.

Topham, Jonathan. "Natural Theology and the Sciences." In *The Cambridge Companion to Science and Religion*, edited by Peter Harrison, 59–79. Cambridge: Cambridge University Press, 2010.

Torrend, Camilo. "Contribuições para o estudo dos fungos da região setubalense." *Brotéria* 1 (1902): 93–150; *Brotéria* 2 (1903): 123–48; *Brotéria* 4 (1905): 207–11.

Torrend, Camilo. "Contributions pour l'étude des champignons de l'île de Madère." *Brotéria: Botânica* 8 (1909): 128–44; *Brotéria: Botânica* 10 (1912): 29–49; *Brotéria: Botânica* 11 (1913): 165–81.

Torrend, Camilo. "A cultura do algodão." *Brotéria: Vulgarização científica* 14 (1916): 5–12; 82–89; 137–45.

Torrend, Camilo. "A cultura do cacau." *Brotéria: Vulgarização científica* 16 (1918): 12–19; 82–91; 128–35; 175–83; 274–80.

Torrend, Camilo. "A cultura do cacau." *Brotéria: Vulgarização científica* 17 (1919): 23–28, 41–46; 106–14; 128–38.

Torrend, Camilo. "O eclipse total de 29 de Maio de 1919, no Brazil." *Brotéria: Vulgarização científica* 18 (1920): 40–41.

Torrend, Camilo. "Les myxomycètes: Étude des espèces connues jusqu'ici." *Brotéria: Botânica* 6 (1907): 5–64; *Brotéria: Botânica* 7 (1908): 5–177.

Torrend, Camilo. "Origem da cellula: Geração espontanea." *Brotéria: Vulgarização científica* 15 (1917): 247–53.

Torrend, Camilo. "Origem da cellula: Geração espontanea." *Brotéria: Vulgarização científica* 16 (1918): 103–10.

Torrend, Camilo. "Les Polyporacées du Brésil." *Brotéria: Botânica* 18 (1920): 23–43; 121–42; *Brotéria: Botânica* 20 (1922): 107–12; *Brotéria: Botânica* 22 (1924): 12–42; *Brotéria: Botânica* 24 (1926): 5–19; *Brotéria: Ciências naturais* 31 (1935): 108–20.

Torrend, Camilo. "Primeira contribuição para o estudo da flora mycologica da provincia de Moçambique." *Brotéria* 4 (1905): 212–21.

Trigueiros, António Júlio. "Eusébio da Veiga, o último professor da Aula da Esfera: Reconstituição biográfica de um jesuíta do século XVIII." *Lusitania sacra* 32 (2015): 147–78.

Trigueiros, António Júlio. "A expulsão e exílio dos jesuítas de Évora em 1759." In *Universidade de Évora (1559–2009): 450 anos de modernidade educativa*, edited by Sara Marques Pereira and Francisco Lourenço Vaz, 357–78. Lisbon: Chiado Editora, 2012.

Trigueiros, António Júlio. "'O negócio jesuítico' e o papel da política regalista portuguesa." *Brotéria* 169 (2009): 149–67.

Trigueiros, António Júlio, and Mariagrazia Russo, eds. *I gesuiti dell'Assistenza Lusitana esiliati in Italia (1759–1831)*. Padua: Libraria Editrice Universitá di Padova, 2013.

Udías, Agustín. "Jesuit Contribution to Science 1814–2000: A Historiographical Essay." In *Jesuit Historiography Online*, edited by Robert A. Maryks. Leiden: Brill, 2016. DOI: 10.1163/2468-7723_jho_COM_192552.

Udías, Agustín. *Jesuit Contribution to Science: A History*. Dordrecht: Springer, 2015.

Udías, Agustín. *Searching the Heavens and the Earth: The History of Jesuit Observatories*. Dordrecht: Kluwer Academic Publishers, 2003.

Van Kley, Dale K. "Jansenism and the International Suppression of the Jesuits." In *The Cambridge History of Christianity*, vol. 7, *Enlightenment, Reawakening and*

Revolution 1660–1815, edited by Stewart J. Brown and Timothy Tackett, 302–28. Cambridge: Cambridge University Press, 2006.

Vasconcelos, José Leite de. "Viagem de Pérez Bayer em Portugal, em 1782." *O archeologo portugues* 24 (1920): 108–76.

Veiga, Francisca Branco. "O breve regresso da Companhia de Jesus no reinado de D. Miguel (1829–1834)." *Brotéria* 179 (2014): 387–400.

Verney, Luís António. *Verdadeiro método de estudar, para ser útil à républica, e à igreja: proporcionado ao estilo, e necesidade de Portugal.* Valença: Antonio Balle, 1746; reprint: Porto: Domingos Barreira, 1984.

Vicente, Filipa Lowndes. "Travelling Objects: The Story of Two Natural History Collections in the Nineteenth Century." *Portuguese Studies* 19 (2003): 19–37.

Vieilledent, Paulino. "Perigos do tabaco." *Brotéria: Vulgarização científica* 7 (1908): 151–54.

Novo, A. Vieira "A caminho do Brazil." *Brotéria: Vulgarização científica* 20 (1922): 188–97.

Vilares, Artur. *As congregações religiosas em Portugal, 1901–1926.* Lisbon: Fundação Calouste Gulbenkian–Fundação para a Ciência e Tecnologia, 2003.

Villoslada, Ricardo García. *Storia del Collegio Romano dal suo inizio (1551) alla soppressione della Compagnia di Gesù (1773).* Rome: Pontificia Università Gregoriana, 1954.

Vogel, Christine. *Der Untergang der Gesellschaft Jesu als europäisches medienereignis (1758–1773): Publizistische Debatten im Spannungsfeld von Aufklärung und Gegenaufklärung.* Mainz: Verlag Philipp Von Zabern, 2006; reprinted as: Cristina Osswald, trans. *Guerra aos jesuítas: A propaganda antijesuítica do Marquês de Pombal em Portugal e na Europa.* Lisbon: Temas e Debates–Círculo de Leitroes, 2017.

Von Murr, Christoph Gottlieb. *Geschichte der Jesuiten in Portugal unter der Staatsverwaltung des marquis von Pombal.* Nuremberg: Felsechrischen Buchhandlung, 1787–88; reprinted as: João Bautista Hafkemeyer, ed., *Historia dos jesuitas no ministerio do Marques de Pombal: Extrahida de manuscriptos por Christovão Theophilo de Murr.* Porto Alegre: Typographia do Centro, 1923.

Waddell, Mark A. "A Theater of the Unseen: Athanasius Kircher's *Museum in Rome.*" In World-*Building and the Early Modern Imagination*, edited by Allison B. Kavey, 67–90. New York, Palgrave Macmillan, 2010.

Wagner, Miguel Noronha, and Tristão Mello Sampaio. "Nota acerca da movimentação dos univalentes em híbridos pentaplóides de trigo." *Brotéria: Ciências naturais* 19 (1950): 180–87.

Wallace, William. *Galileo and His Sources: The Heritage of the Collegio Romano in Galileo's Science.* Princeton: Princeton University Press, 1984.

Whitehead, Maurice. "The Jesuit Contribution to Science and Technical Education in Late Nineteenth-Century Liverpool." *Annals of Science* 43 (1986): 353–68.

Worcester, Thomas, ed. *The Cambridge Encyclopedia of the Jesuits.* Cambridge: Cambridge University Press, 2017.

Wright, Jonathan, and Jeffrey D. Burson. "Introduction. Towards a New History of the Eighteenth-Century Jesuit Suppression in Global Context." In *The Jesuit Suppression in Global Context: Causes, Events, and Consequences*, edited by Jeffrey D. Burson and Jonathan Wright, 1–10. Cambridge: Cambridge University Press, 2015.

Year-Book of the Royal Society of London. London: Harrison and Sons, 1904.

Zacharias, Helmut. "Emil Heitz (1892–1965): Chloroplasts, Heterochromatin, and Polytene Chromosomes." *Genetics* 141 (1995): 7–14.

Zimmermann, Carlos. "Beitrag zur Kenntnis der Diatomaceen Flora der Inseln Madeira und Porto Santo." *Brotéria: Botânica* 8 (1909): 114–27; *Brotéria: Botânica* 9 (1910): 103–20.

Zimmermann, Carlos. "Catálogo das diatomaceas portuguezas." *Brotéria* 5 (1906): 245–51; *Brotéria: Botânica* 8 (1909): 89–103; *Brotéria: Botânica* 9 (1910): 95–102; *Brotéria: Botânica* 12 (1914): 115–24.

Zimmermann, Carlos. "Contribuição para o conhecimento das diatomaceas da provincia de Moçambique." *Brotéria: Botânica* 12 (1914): 155–62.

Zimmermann, Carlos. "Contribuições para o estudo das diatomaceas dos Estados Unidos do Brazil." *Brotéria: Botânica* 11 (1913): 149–64; *Brotéria: Botânica* 12 (1914): 5–12; *Brotéria: Botânica* 13 (1915): 37–56; 65–71; 124–46; *Brotéria: Botânica* 14 (1916): 85–103; 130–57; *Brotéria: Botânica* 15 (1917): 30–45; *Brotéria: Botânica* 16 (1918): 8–24; 113–22; *Brotéria: Botânica* 17 (1919): 5–16.

Zimmermann, Carlos. "Microscopia vegetal." *Brotéria* 1 (1902): 49–75; *Brotéria* 2 (1903): 5–40; *Brotéria* 4 (1905): 137–59; *Brotéria* 5 (1906): 229–44.

Zimmermann, Carlos. "Observatorio meteorologico do Collegio de São Fiel." *Brotéria* 1 (1902): 185–188.

Index

www.ingramcontent.com/pod-product-compliance
Lightning Source LLC
Chambersburg PA
CBHW071325230926
51849CB00002B/315
* 9 7 8 9 0 0 4 3 8 2 3 5 0 *